MOLECULAR BIOLOGY
INTELLIGENCE
UNIT

I Domains in Integrins

Donald Gullberg, Ph.D.
Department of Medical Biochemistry and Microbiology
Uppsala University
Uppsala, Sweden

LANDES BIOSCIENCE / EUREKAH.COM
GEORGETOWN, TEXAS
U.S.A.

KLUWER ACADEMIC / PLENUM PUBLISHERS
NEW YORK, NEW YORK
U.S.A.

I Domains in Integrins

Molecular Biology Intelligence Unit

Eurekah.com / Landes Bioscience
Kluwer Academic / Plenum Publishers

Printed in the U.S.A.

Kluwer Academic / Plenum Publishers, 233 Spring Street, New York, New York, U.S.A. 10013
http://www.wkap.nl/

Please address all inquiries to the Publishers:
Eurekah.com / Landes Bioscience, 810 South Church Street
Georgetown, Texas, U.S.A. 78626
Phone: 512/ 863 7762; FAX: 512/ 863 0081
www.Eurekah.com
www.landesbioscience.com

I Domains in Integrins edited by Donald Gullberg, Landes / Kluwer dual imprint / Landes series: Molecular Biology Intelligence Unit

ISBN: 0-306-47836-6

Library of Congress Cataloging-in-Publication Data

I domains in integrins / [edited by] Donald Gullberg.
p. ; cm. -- (Molecular biology intelligence unit)
Includes bibliographical references and index.
ISBN 0-306-47836-6
1. Integrins.
[DNLM: 1. Integrins--physiology. 2. Collagen--physiology. 3. Immune System--physiology. QW 570 I11 2003] I. Gullberg, Donald E. II. Molecular biology intelligence unit (Unnumbered)
QP552.I55I3 2003
572'.68--dc21

2003011220

CONTENTS

EDITOR

Donald Gullberg, Ph.D.
Department of Medical Biochemistry and Microbiology
Uppsala University
Uppsala, Sweden
Chapter 5

CONTRIBUTORS

Antonin R. de Fougerolles
Alnylam Pharmaceuticals
Cambridge, Massachusetts, U.S.A.
Chapter 12

Beate Eckes
University of Cologne
Department of Dermatology
Cologne, Germany
Chapter 11

Susanna Fagerholm
Department of Biosciences
Division of Biochemistry
University of Helsinki
Helsinki, Finland
Chapter 6

Richard W. Farndale
University of Cambridge Department of Biochemistry
Downing Site
Cambridge, United Kingdom
Chapter 9

Carl G. Gahmberg
Department of Biosciences
Division of Biochemistry
University of Helsinki
Helsinki, Finland
Chapter 6

Humphrey Gardner
Biogen Inc.
Cambridge, Massachusetts, U.S.A.
Chapter 2

Jonathan M.G. Higgins
Division of Rheumatology Immunology and Allergy
Brigham and Women's Hospital
Harvard Medical School
Boston, Massachusetts, U.S.A.
Chapter 7

Jyrki Heino
Department of Medical Biochemistry and MediCity Research Laboratory
University of Turku
Turku, Finland
and Cell Biology
University of Jyväskylä
Jyväskylä, Finland
Chapter 10

Mark S. Johnson
Department of Biochemistry and Pharmacy
Åbo Akademi University
Turku, Finland
Chapter 1

Peter J. Kilshaw
The Babraham Institute
Babraham
Cambridge, United Kingdom
Chapter 7

Thomas Krieg
University of Cologne
Department of Dermatology
Cologne, Germany
Chapter 11

Robert C. Liddington
Program on Cell Adhesion
The Burnham Institute
La Jolla, California
Chapter 8

Evy Lundgren-Åkerlund
Cartela AB
Biomedical Center
Lund, Sweden
Chapter 4

Bernhard Nieswandt
Rudolf Virchow Center
for Experimental Biomedicine
University of Wuerzburg
Würzburg, Germany
Chapter 11

Svetlana N. Popova
Upsala Universitet
Department of Medical Biochemistry
and Microbiology
Uppsala, Sweden
Chapter 5

Samuel A. Santoro
Washington University School
of Medicine
Department of Pathology
and Immunology
St. Louis, Missouri, U.S.A.
Chapter 3

Max Schneller
Cartela AB
Biomedical Center
Lund, Sweden
Chapter 4

Carl-Fredrik Tiger
Upsala Universitet
Department of Medical Biochemistry
and Microbiology
Uppsala, Sweden
Chapter 5

Danny Tuckwell
F2G Limited
The Incubator Building
Manchester, United Kingdom
Chapter 1

Mary M. Zutter
Washington University School
of Medicine
Department of Pathology
and Immunology
St. Louis, Missouri, U.S.A.
Chapter 3

PREFACE

Almost two decades ago a group of cell adhesion receptors, previously known by a variety of names, were collectively named integrins.[1] This family, as we know it today, has grown and currently includes 24 members in vertebrates.[2]

As the individual α-chains were sequenced, some integrin α-chains were noted to have an inserted domain present. This domain was in integrins first detected in the integrin α-chains associated with the β2 integrin chain on immune cells, and later in the collagen-binding integrin α1 and α2 integrin α-chains. This domain is similar to the A domain found in a number of matrix and complement proteins. In integrins, the inserted, interactive domain was called the I domain.[3] Unlike the collagen-binding A3 domain in von Willebrand factor,[4] the integrin I domain-mediated ligand binding involves a coordinating metal ion.[5] Nine out of the 18 currently known integrin α chains contain an I domain (see Fig. 1).

Following the crystallization of the Mac-1 integrin α-I domain, it was suggested that a similar structure existed in integrin β chains, which initially was named an I -like domain.[6] From the recent structure of the soluble αvβ3 integrin, it has become clear that this domain in β-chains indeed adopts the typical Rossman fold present in the α-I domains.[7] Although the β-I domain appears to have a longer evolutionary history, and no I domain-less β-chain integrin backbone has been described so far, it is certainly an interactive

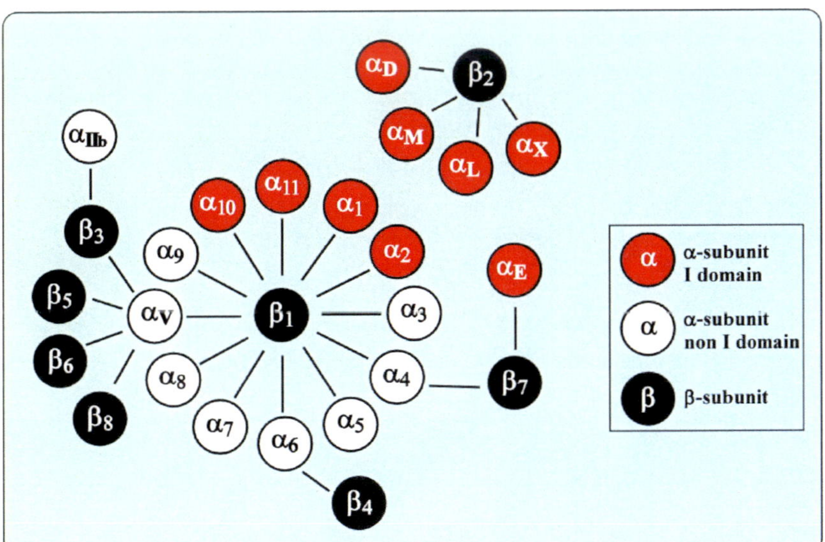

Figure 1. Schematic representation of integrin family. Note the presence of an I domain in 9 out of the 18 integrin α-chains.

region thus, justifying the use of the name I domain also in the β-chain. Throughout this volume we will use the names α-I and β-I domains. With the realization that all integrin heterodimers contain at least one I domain the appearance of a volume on I domains in integrins appears very timely.

In this issue, we will collect the current state-of-the-art regarding I domains in integrins. Some of the experts in the field will systematically go through the integrin α-chains that contain an I domain and evolutionary and structural aspects will be discussed separately.

Donald Gullberg

References

1. Tamkun JW, DeSimone DW, Fonda D et al. Structure of integrin, a glycoprotein involved in the transmembrane linkage between fibronectin and actin. Cell 1986; 46(2):271-82.
2. Gullberg D, Lundgren-Åkerlund E. Collagen-binding I domain integrins-what do they do? Prog Histochem Cytochem 2002; 37:3-54.
3. Larson RS, Corbi AL, Berman L et al. Primary structure of the leukocyte function-associated molecule-1 alpha subunit: an integrin with an embedded domain defining a protein superfamily. J Cell Biol 1989; 108(2):703-12.
4. Nishida N, Sumikawa H, Sakakura N et al. Collagen-binding mode of vWF-A3 domain determined by a transferred cross-saturation experiment. Nat Struct Biol 2003; 10(1):53-58.
5. Lee JO, Rieu P, Arnaout MA, Liddington R. Crystal structure of the A domain from the a subunit of integrin CR3 (CD11b/CD18). Cell 1995; 80(4):631-38.
6. Lee JO, Bankston LA, Arnaout MA et al. Two conformations of the integrin A-domain (I-domain): a pathway for activation? Structure 1995; 3(12):1333-40.
7. Xiong JP, Stehle T, Diefenbach B et al. Crystal structure of the extracellular segment of integrin αVβ3. Science 2001; 294:339-45.

Chapter 1

Evolution of Integrin I-Domains

Mark S. Johnson and Danny Tuckwell

In humans, I domains are ~200 residue long sequences that are "inserted" into the N-terminal domain of all 8 integrin β subunits and 9 of the 18 integrin α subunits. The integrin I-domains are related to the von Willebrand factor A domains (VWFA), found in a wide range of proteins in both prokaryotes and eukaryotes. Analysis of the I-domains and VWFA domains from a range of metazoans shows that the full complement of α and β I-domains found in humans are present in bony fish. Homologues of the αI domains are not present in protostomes although homologues of VWFA domains, α subunits without I domains, and β subunits with I domains are present. Some of the protostome VWFA domain proteins are involved in extracellular matrix function, showing interesting functional parallels with vertebrate I-domain integrins. It seems likely that the present day set of integrin I-domains in vertebrates is a product of the requirements dictated by the large changes in body plan and the evolving immune system initiated in the early deuterostomes and chordates some 600 million years ago.

Introduction

Integrins are cell adhesion receptors which mediate cell-cell and cell-matrix interactions[1] but also function in bi-directional signaling between the cell interior and the extracellular matrix.[2] Integrins are αβ heterodimers: currently, 18 α subunits and 8 β subunits forming more than 24 different heterodimers have been identified in humans.[3] The integrin α subunit consists of a large N-terminal extracellular multi-domain structure, a single transmembrane helix, and a short C-terminal intracellular domain. The extracellular structure[4] consists of the N-terminal β-propeller domain (here β refers to beta strands from which the propeller domain is constructed), followed by the thigh domain and two calf domains (immuglobulin-like folds). Some α subunits, namely α1 (VLA-1, CD49a), α2 (VLA-2, CD49b), α10, α11, αM (Mac-1, CD11b), αL (LFA-1, CD11a), αD, αE (HML-1, CD103) and αX (CR4, CD11c), contain an insertion domain (α I-domain) of approximately 200 amino acids "budding out" from between the second and third repeats of the N-terminal seven bladed propeller domain. While the integrins αM, αL, αD, αE and αX are only expressed in cells of the immune system, the collagen-binding integrins α1 (many mesenchymal cells) and α2 (platelets, fibroblasts and epithelium, among others)[1,5] are more widely distributed. Two other integrins containing I-domains bind collagen, but less is known about them: integrin α10 is found on chondrocytes[6] and α11 was originally isolated from fetal muscle.[7,8]

The integrin β subunit also has a large N-terminal multi-domain structure. A region near the N-terminus was long suspected to be an I-domain (β I-domain),[9,10,11] although others discounted the possibility.[12] It is now clear that I-domains are present in β subunits[4] and they recognize the cell-adhesive "RGD" motif and similar sequences[1] present in numerous proteins

I Domains in Integrins, edited by Donald Gullberg. ©2003 Eurekah.com and Kluwer Academic / Plenum Publishers.

including fibronectin, fibrinogen, von Willebrand factor, laminin, collagens, virus coat proteins, and snake venoms, among many others.[1,13]

The I-domains contain a binding site for ligands referred to as the Metal Ion-Dependent Adhesion Site (MIDAS), where a metal ion, Mg^{2+} or Mn^{2+}, is coordinated by residues, including the ^{140}DXSXS^{144} signature (α_M numbering) plus T^{209} and D^{242}, on three loops of the I-domain. The MIDAS is located near the surface of the I-domain, providing an accessible site for binding the protein ligands recognized by the I-domain; binding displaces a water molecule at the sixth coordination site on the metal and is associated with conformational changes in the I-domain. Not all ligands binding to the α I-domain use a negatively charged residue for coordination to the metal ion (e.g., the "RKKH" cyclic peptide derived from the metalloproteinase of the snake *Bathrops jararaca*[14,15]) and not all ligands bind to the MIDAS site (e.g., binding of echovirus-1 to the α2 I-domain;[16-19] and binding of the cholesterol lowering drug lovastatin to the αL I-domain[20]). Indeed, the structure of echovirus-1 bound α2 I-domain, determined by cryo-electron microscopy followed by fitting of the α2 I-domain structure into receptor density on echovirus-1, shows that the MIDAS site is not in direct contact with the virus, but it is oriented such that collagen can not simultaneously bind to the I-domain (Xing, Huhtala, Pietiäinen, Käpylä, Vuorinen, Heino, Johnson, Hyypiä, and Cheng, submitted).

The structures of four different α I-domains have been published: αM,[9,21,22] αL,[23,24] α2[25,26] and α1[27] (Fig. 1). These domains are classic "Rossmann" folds, found widely throughout the proteome, with five parallel β-strands plus one antiparallel β-strand, all surrounded by six α-helices. Recently, the ectodomain structure of the αvβ3 heterodimer[4] demonstrated that the β3 I-domain shares the same fold as the α I-domains. Moreover, like the αI domains, the β I-domain buds out from a separate domain, in the case of the β I-domain from the B-C loop of an immunoglobulin-like fold. The interaction between the β I-domain and the β-propeller of the α subunit shares structural similarity with the G-protein Gα-Gβ.[4,28]

The α I-domains and β I-domains are homologous to the von Willebrand factor A-domain (VWFA domain). The VWFA domain is found in the blood coagulation protein von Willebrand factor itself as well as in many other proteins such as collagens, complement factors, copines and matrilins.[29] Three-dimensional structures are known for VWFA domains A1 and A3[30-32] and, like the α and β I-domains, the VWFA domains share the same fold. While the VWFA domains do carry the "DXSXS" signature, they are often missing one or both of the remaining two coordinating residues of the MIDAS motif and hence do not bind metal. Indeed, Romijn et al[33] have shown that the collagen binding site on the von Willebrand factor VWF domain A3 is located at the opposite end of the domain (where the N- and C-termini of the domain are found) in comparison with the α and β I-domains. VWFA domain homologues have been identified in invertebrates as well as in both eubacteria and archaea,[34] pointing to the ancient origins of the domain.

Integrins

Vertebrate αI Domains

In humans the α I-domain integrin family comprises integrins expressed by immune system cells: αD, αE, αL, αM and αX; and those I-domain integrins (α1, α2, α10 and α11) that bind proteins of the extracellular matrix such as collagens and laminins. The α3-α9 and αIIb subunits do not have the α I-domain.

Figure 2 presents an alignment of the α I-domains, mainly from human and mouse, together with VWF A3 and sequences from several other species. The alignment of the known three-dimensional structures of α1, α2, αL, αM and von Willebrand factor A3 was used to guide the alignment of the sequences whose structures are presently unknown. The sequence similarity among the collagen-binding α I-domains ranges between 46 % (α2 vs. α10 and

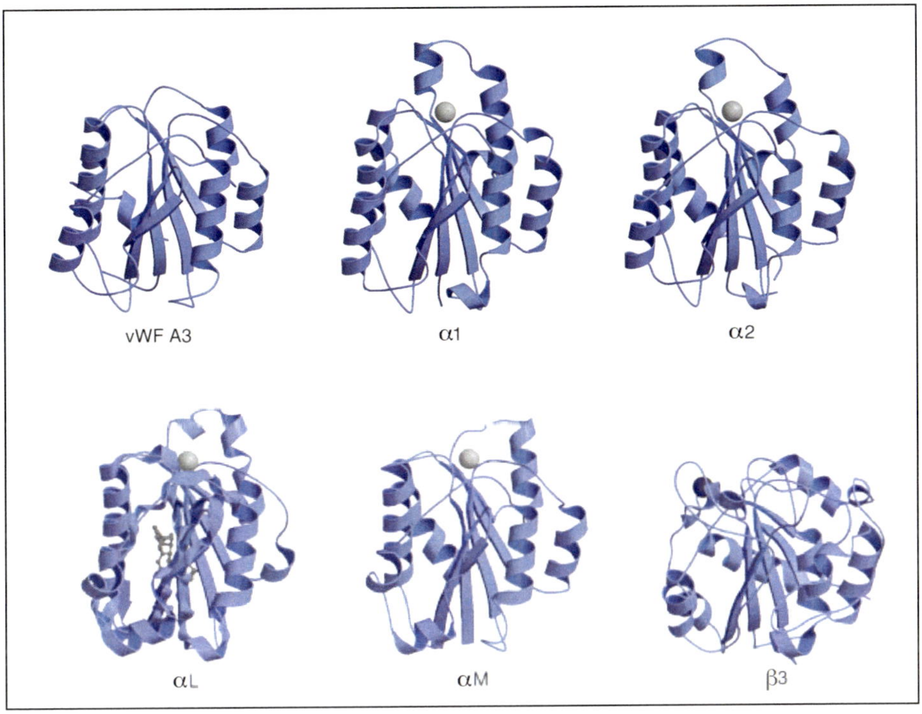

Figure 1. Ribbon diagrams of the structures of von Willebrand factor A3 domain (vWF A3), collagen binding α1 domains (α1, α2), αI domains of immune system cells (αL, αM) and βI domain (β3). The bound cation at the MIDAS is shown as a sphere with the exception of vWF A3 which does not bind a metal ion. The inhibito lovastatin is bound to the αL structure. Both the α1 and α2 I-domains have the αC-helix (top of structures) that is not present in the other domains.

α11) and 59 % (α10 vs. α11) identity, while the α I-domains of the immune system share between 27 % (αL vs. αE) and 61 % (αM vs. αD) identity. The collagen-binding α I-domains and the I-domains of the immune system share between 25 % (α1 and α2 vs. αL) and 33 % (α11 vs. αE) sequence identity. VWFA A3 is 16-24 % identical with the α I-domains. A phylogenetic tree (unrooted) built on the basis of the alignment in Figure 2 for 25 integrin α I-domains is shown in Figure 3. The collagen-binding α I-domains and integrin α I-domains from immune system cells segregate into two separate groups. The αHr1 I-domain (18-23 % identical to the others) and VWF A3 (17-23 % identical to the others) belong to neither group.

Forty-six integrin α subunits were aligned (alignment not shown; those portions of the integrin α subunits corresponding to the I-domain were not included in the alignment), and an unrooted phylogenetic tree constructed (Fig. 4). The nine I-domain -containing integrin α subunits form a combined group that clusters separately from those α subunits that do not contain the I-domain, in agreement with the results of others.[38-41] As is the case for the I-domains themselves (Fig. 3), the αHr1 I-domain does not belong to either the collagen-binding I-domain cluster or to the I-domains of the immune system cells, but is an apparent outlier of both groups, being equally similar (18-23% identical) to all of the other 45 sequences.

The α2 I-domain structure (Fig. 1) described by Emsley et al[25] pinpointed one major difference with the αL and αM I-domain structures: the α2 I-domain has an additional loop in the vicinity of the MIDAS motif whose conformation changes along with adjoining regions in response to collagen binding. Here, we will refer to residues corresponding to ^{284}G-R^{288} in

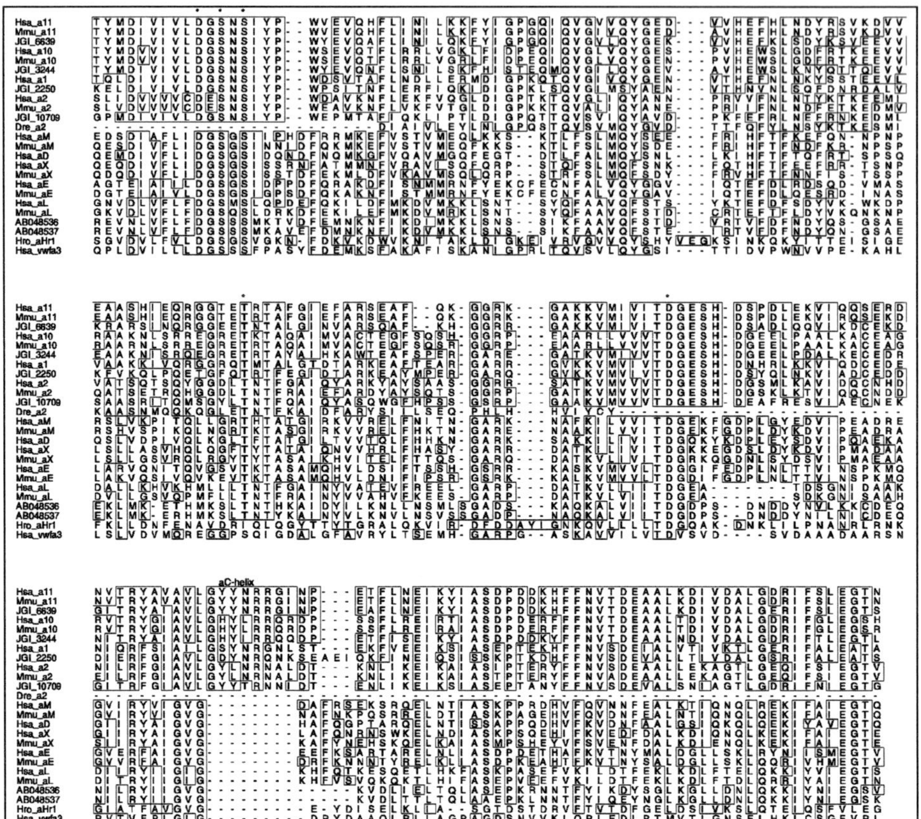

Figure 2. Alignment of the sequences of 24 α I-domains and VWF A3. Hsa = *Homo sapien*; Mmu = *Mus musculus*; Hro = *Halocynthia roretzi* (tunicate). JGI_2250, JGI_6639, and JGI_10709 are fugufish (*Takifugu rubripes*) sequences; AB048536 and AB048537 are from the common carp, *Cyprinus carpio*. The alignment was made using MALIGN[35] but takes into account the alignment of the three-dimensional structures where they are known. The αC-helix present in the collagen binding α I-domain is indicated along with the residues involved in binding metal at the MIDAS site (*).

the α2 I-domain, and present in the structure of the α1 I-domain[27] as the αC-helix. Based on the alignment shown in Figure 2, the αC-helix is diagnostic of collagen-binding αI domains including the α10 and α11 I-domains, but it is not found in I-domains of the immune system or in VWFA domains (Fig. 2).

The immune system of vertebrates is dependent on the complex interaction of many proteins that have, among other functions, the recruitment of both lymphoid (lymphoid cells include B-cells and derived plasma cells, and cytotoxic, helper and suppressor T-cells) and myeloid (myeloid cells include erythrocytes, monocytes (phagocytic), macrophages (phagocytic), the granulocytes: basophils, eosinophils, and neutrophils (phagocytic), and platelets) cells to sites of infection where integrins of these immune system cells mediate cell adhesion. Thus, integrins have roles in targeting phagocytic cells to invading microbes and to sites of injury.

Both lymphoid and myeloid lymphocytes express αLβ2, which recognizes ICAM 1-3. ICAMs, whose ectodomains belong to the immunoglobulin superfamily, are expressed on activated endothelium, functioning to bind lymphocytes at sites of inflammation. The myeloid

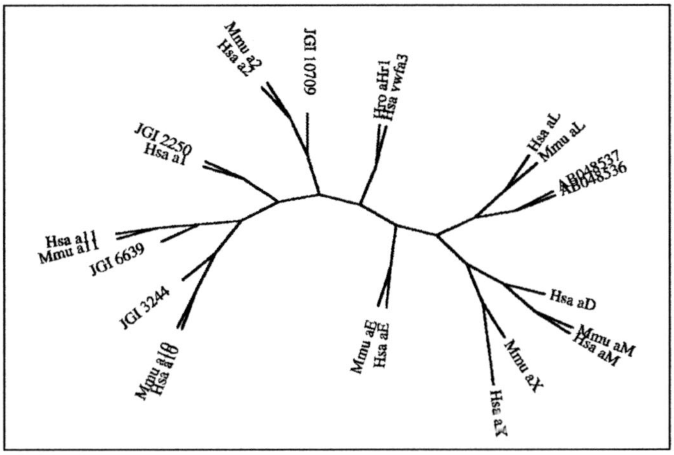

Figure 3. Consensus phylogenetic tree (1000 bootstrapped alignments) of human VWF A3 and the α I-domains from human, mouse, bonyfish (fugufish and carp) and tunicate. The neighbor-joining tree[36] is based on the sequence alignment in Figure 2. Programs of the Phylip package[37] were used except in the calculation of distances from the sequence alignments, where MALFORM[35] was used. All bootstrap values are 90% or greater.

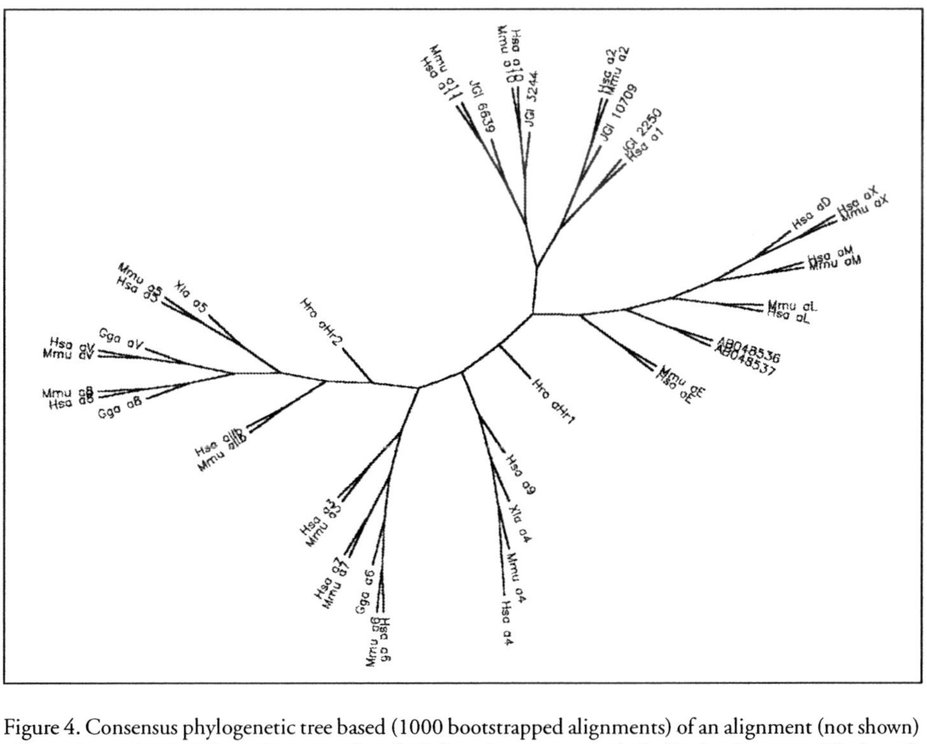

Figure 4. Consensus phylogenetic tree based (1000 bootstrapped alignments) of an alignment (not shown) of integrin α subunits. The regions encoding the I-domains were not included in the alignment. The branch points leading to Hro_αHr2, the α3-α6-α7 cluster, Hro_αHr1 and the α1-α2 cluster have bootstrap values of 40%, 47%, 51% and 68%, respectively; all other nodes have bootstrap values greater than 80%. Gga = *Gallus gallus* (bird); Xla = *Xenopus laevis* (amphibian);

leukocytes: macrophages, monocytes and granulocytes adhere to iC3b-coated cells and to stimulated endothelium using integrin αMβ2. iC3b is the hydrolysis product of complement component C3, functioning as an opsinin to coat cells and targeting them for phagocytosis. In immune adherence, complement proteins, phagocytic lymphocytes and antibodies function to adhere microbes to the endothelium of blood vessels making them easier targets for attack by cells of the immune system. αXβ2 is expressed on myeloid leukocytes, macrophages and monocytes, recognizing fibrinogen and iC3b. αDβ2 is expressed on eosinophils and binds to vascular adhesion protein 1 (VCAM-1; immunoglobulin-like N-terminal domains.[42,43] Eosinophils are active against parasites and (mal)function in allergy and inflammation with a likely role in chronic inflammation. Integrin αE is associated with the β7 subunit, expressed on intestinal intraepithelial T-cells, and adheres to epithelial cells via E-cadherin (epithelial cell cadherin, uvomorulin).[44,45] Cadherins are single chain transmembrane receptors linked to actin filaments through their C-terminal domain and binding other cells in a Ca^{2+} -dependent manner with their N-terminal domains. Cadherins appear to be important for tissue remodelling, especially in embryogenesis and morphogenesis.

The collagen binding integrin α subunits, α1, α2, α10 and α11, mediate interactions between cells and components of the extracellular matrix (ECM), binding to different collagens and laminins. These interactions are important for tissue remodelling in which cells invade the ECM, where the integrins regulate the expression and activation of the metalloproteinases responsible for ECM degradation.[3] The collagen-binding α subunits containing the I-domain associate with the β1 subunit.

The β2-associated α integrins αL, αM, αX and αD each map to chromosome 16 in human (to chromosome 7 in mouse) (Table 1). The clustering of this portion of the phylogenetic tree produced from the alignment of the α-subunits (tree not shown; all branch points have bootstrap values > 99%; I-domain removed) can be summarized as: ((((αX, αD) αM) αL) αE); and separately for the αI-domains (Fig. 3) as: ((((αM, αD) αX) αL) αE). Although the relative positions of αX and αM differ between the two trees, in both trees the αE subunit is the most dissimilar of the set. In line with this, αE associates with the β7 subunit and, like α3 integrin, is found on human chromosome 17 (11 in mouse). Those integrin α subunits that pair with the β1 subunit map to three separate chromosomes in human: both α1 and α2 subunits map to chromosome 5 (13). The human α10 and α11 subunits map to chromosomes 1 and 15, respectively. In mouse, an "α10-like" sequence maps to mouse chromosome 3 and an "α11-like" sequence maps to chromosome 9 (Table 1). The separate trees for the collagen-binding α subunits (not shown) and their αI-domains (Fig. 3) are identical: ((α11, α10) (α1, α2)). Overall, there are clear similarities between the chromosomal disposition of integrin α subunits, including those α subunits without I-domains, within the human and the mouse genomes (Table 1).

While the α1 integrin subunit has been mapped to chromsome 13 in mouse (Table 1), no sequence is available at this time. In mouse, the α10 (LOC213119) and α11 (LOC207653) -like sequences are certainly authentic orthologues of the human α10 and α11 sequences, respectively. In each case the amino acid sequence identity is a high 86 % over the entire length of the α subunit. A fragment of the mouse αD-like sequence includes the I-domain (or part of the I-domain), with high similarity to human αD at the amino terminus of the I-domain, but with very low similarity elsewhere; the mouse αD-like sequence was not included in the alignment in Figure 2. In the case of α9, highly identical sequences were found by searching the human sequence against the mouse genome: α9-like nucleotide sequences (locus NW_00036) with high identity to human α9 (six fragments, 1061 aligned positions, 81 – 94 % identity) were found on chromosome 9. The currently defined amino acid sequences of 15 α subunits in mouse are very similar to their human orthologues over their entire lengths (Table 3, 4) and cluster together (Figs. 3, 4). Taken together, we expect the mouse genome to contain the same complement of α subunits as are found in human: 9 with I-domains and 9 without.

Table 1. Chromosomal locations of human and mouse α integrins

Human[a]					Mouse[b]			
Chromosome	Integrin (+I)	Integrins (+I)	Map Element	Cytogenetic	Chromosome	Integrin (+I)	Integrins (+I)	Map Element
1	α10		ITGA10	1q21	3	α10 like		LOC195504
					3	α10[e] like		LOC213119
2		α6	ITGA6	2q31.1	2		α6	Itga6
2		α4	ITGA4	2q31-q32	2		α4	Itga4
2		αV	ITGAV	2q31-q32	2		αV	Itgav
					2		αV like	LOC228128
					2		αX like	LOC241451
3		α9			9		(α9 like)[d]	
5	α1		ITGA1	5q11.1	13	α1[c]		Itga1
	α2		ITGA2	5q23-31	13	α2		Itga2
10		α8	ITGA8		2		α8[f]	Itga8
12		α5	ITGA5	12q11-q13	15		α5	Itga5
12		α7	ITGA7	12q13	10		α7	Itga7
					10		α7 like	LOC216463
14		α3 like	LOC122789	14q22.1				
15	α11		ITGA11	15q22.3-q23	9	α11[e] like		LOC207653
16	αD		ITGAD	16p13.1-p11	7	αD[f] like		LOC244220
16	αL		ITGAL	16p11.2	7	αL		Itgal
16	αM		ITGAM	16p11.2	7	αM		Itgam
16	αX		ITGAX	16p11.2	7	αX		Itgax
16	αD like		LOC197414	16p11.1	7	αL like		LOC233889
17	αE		ITGAE	17p13	11	αE		Itgae
17		αIIb	ITGA2B	17q21.32	11		α2b	Itg2b
17		α3	ITGA3	17q23.3	11		α3	Itga3
					11	αE like		LOC193032
					11	αE like		LOC216898

[a]Human genome map at www.ncbi.nih.gov; [b]Mouse genome map at www.ncbi.nih. [c]No protein model sequence available at this time.
[d]A blast search with the human α9 sequence identified nucleotides sequences on mouse chromosome 9 having high sequence identity.
[e]Possible orthologues of human sequences used in alignments. [f]Fragment

Table 2. Chromosomal locations of human and mouse β integrins[a,b]

Human				Mouse		
Chromosome	Integrin	Map Element	Cytogenetic	Chromosome	Integrin	Map Element
2	β6	ITGB6	2q24.2	2	β6	Itgb6
3	β5	ITGB5	3q21.3	16	β5	Itgb5
				16	β2 like	Itgb2l
7	β8	ITGB8	7p21.3	12	(β8 like)	XM_127189
10	β1	ITGB1	10p11.2	8	β1	Itgb1
12	β7	ITGB7	12q13.13	15	β7	Itgb7
17	β3	ITGB3	17q21.32	11	β3	Itgb3
17	β4	ITGB4	17q11-qter	11	β4	Itgb4
21	β2	ITGB2	21q22.3	10	β2	Itgb2

[a]Human genome map at www.ncbi.nih.gov; [b]Mouse genome map at www.ncbi.nih.

α I-domains have not been found in the genomes of either *C. elegans* or *D. melanogaster*[46] supporting the notion that the I-domain containing α subunits arose after the divergence of protostomes and deuterostomes more than 600 million years ago (MYA). α I-domains have been previously established in birds (chicken[39,40]) and small fragments of sequences from *X. laevis* support the presence of α2-6 and αIIb in amphibians.[39] So far, α I-domains have not been detected in the earliest deuterostomes, the echinoderms, but a sequence of an I-domain containing an α subunit, designated αHr1, has been sequenced from an invertebrate chordate, the tunicate (ascidian) *Halocynthia roretzi*.[41] The I-domain of αHr1 does not contain the αC-helix signature of the collagen binding I-domains (Fig. 2). In the search for collagen binding I-domains in organisms that have diverged prior to the amniotes (i.e., before reptiles, birds and mammals), we have identified sequences that firmly establish the presence of the I-domain in both amphibians and bony fish.

A fragment of a putative α2 subunit in *X. laevis* (accession code AAA69770), missing the N-terminal 400 amino acids where the α I-domain would be located, is 57 % identical with the human α2 sequence over a span of 380 residues; the alignment contains only one gap of two positions (alignment not shown). For comparison, the percentage amino acid sequence identity (Table 4) between the human α4 and α5 integrin subunits and their counterparts in *X. laevis* are 55 % and 64 %, respectively. Additional searches, using the α1 integrin, as the probe identified four sequence fragments, ~350-450 nucleotides, with high similarity to the probe. These four frog sequences, when back-searched against databanks of nucleotide sequences identified the following high-scoring matches: BI442358 matched α1 integrins from human, rat and chicken; BJ072885: VWFA-domain containing matrilin-4 from human, mouse and chicken; BG234051: human and mouse matrilin-2; and BG579626: VWFA-domain containing collagen 1 from mammals, birds and amphibians. Two additional sequences were identified in another amphibian, *Silurana tropicalis*, but both BQ521443 and BQ521083 score high against matrilin-2 and matrilin-4, respectively. Although a complete set of integrin α subunits has yet to be established in *X. laevis*, given what has been identified so far it is very likely that a full complement of α subunits will be found in amphibians as the genome sequencing efforts progress.

From zebrafish (*Danio rerio*) two ESTs were identified: one sequence of 197 nucleotides (access code: GI 5708002) is similar to the human α2 I-domain, with 37 % sequence identity at the amino acid level, providing support for the presence of collagen-binding αI domains before the amniote-amphibian divergence. Two sequences (AB048536 and AB048537) from the common carp (*Cyrpinus carpio*) are related to the α integrins of the immune system and contain the I-domain.[47] Recent searches of the α1 I-domain against the zebrafish genome identified the following fragments of nucleotide sequences that, when searched against nucleotide and protein sequence databanks matched with high scores I-domain containing sequence fragments and complete sequences: zebrafish BM279735 matches human, rat and chicken α1 integrin; BI326678: mouse αL integrin; AV174735, AW174735 and B888804: human, mouse and frog α5 integrins; AI626570 and AW128160: human α9 integrin; BG884601: αIIb integrin; BI326678: the common carp sequence AB048536; BM812301 and BM155953: human and mouse matrilin-4; AA566274: human matrilin-3; AI958998 and AI958107: mouse collagen α1(VI) and human hypothetical protein XP167306, among others. In another bony fish, the Japanese medaka (*Oryzias latipes*), BJ003847 best matched human collagen α3(VI).

Recently, we made a search of the human genome against the fugufish (*Takifugu rubrides*) genome. This search was made using the Zon Lab Comparative Genomics utility at http://134.174.23.160/compGenomics. From the resultant searches, we identified numerous close matches between human integrins and fugufish sequences containing the N-terminal I-domain; several of these sequences are full-length and together with the carp sequences provide overwhelming evidence that α I-domains exist in bony fishes and very probably in all vertebrates. The tree presented in Figure 3 shows that the fugufish α I-domain sequences aligned in Figure 2 cluster together with human/mouse collagen-binding α I-domains: JGI_2250 (fugufish) with human α1 I-domain; JGI_10709 with human and mouse α2 I-domain; JGI_3244 with the human and mouse α10 I-domain sequences; and JGI_6639 with the human and mouse α11 I-domain sequences. The sequence identity ranges from 58 % (human α2 I-domain versus JGI_10709) to 81% (human α11 I-domain versus JGI_6639) (Table 3). Furthermore, the similarity extends over the length of the α subunit sequences (Table 4), but the overall sequence identity is lower, ranging from 34% (human α1 versus JGI_2250) to 49% (human α11 versus JGI_6639). Moreover, the relationships between the fugufish sequences and human sequences is preserved in a tree constructed from this alignment of 46 α subunits (less the α I-domain) from human, mouse, chicken, frog, fugufish and tunicate (Fig. 4). In our preliminary searches we have not identified any counterparts of the I-domain integrin α subunits expressed on immune system cells except those noted previously in the common carp, nor have we found examples of the α subunits that lack the I-domain (note that our searches were targeted primarily to I-domain –containing sequences). However, the sequence of the fugufish genome is in progress, more extensive investigations need to be made, and based on these results we would be surprised if representatives of most, if not all, of the human integrin α subunit counterparts are not found in the bony fishes.*

The invertebrate chordates and vertebrates diverged about 600 MYA, but the presence of an I-domain containing the α subunit in a urochordate (tunicate) provides strong support for the presence of α subunits containing I-domains in cephalochordates (lancelets), agnatha (hagfish and lamprey) and chondrichyes (sharks and rays). Complement components have been found in sea urchin and tunicates, suggesting that integrins may have a role in targeting

*Note added in proof: Since June 2002 when this chapter was submitted we have identified representatives of nearly all of the α and β subunits present in human/mouse within the teleost fishes (*Danio rerio, Takifugu rubrides, Cyrpinus carpio,* and *Ictalurus punctatus*). In the initial draft genome of the ascidian *Ciona intestinalis* (Dehal P. *et al.* The draft genome of *Ciona Intestinalis*: Insights into chordate and vertebrate evolution. Science 2002; 298:2157-2167) we have tentatively identified at least 8 α subunits (2 containing I-domains) and 5 β subunits. In our preliminary comparisons one putative β subunit of the ascidian clustered as an orthologue of a human/mouse/fish β subunit; however, all other putative α and β subunits from *C. Intestinalis* did not cluster with the human/mouse/fish sequences. (Huhtala, Casciari, de Luisa, Piippo, Pentikäinen, Heino, Johnson, unpublished)

Table 3. Percentage amino acid sequence identity between human α I-domain sequences and sequences from mouse, fugufish, zebrafish and carp

	Mouse	Fugu	Zebrafish[a]	Carp[b]
α11	98	81		
α10	97	67		
α1	-	62		
α2	83	58	37	
αM	79			
αD	-			
αX	69			
αL	74			44, 47
αE	78			

[a]Fragment
[b]The α I-domain region from the carp sequences AB048536 and AB048537 cluster with the human and mouse αL I-domain sequences.

Table 4. Percentage amino acid sequence identity between human α subunit sequences (α I-domain region removed) and sequences from mouse, fugufish and zebrafish

	Mouse	Chicken	Frog	Fugu	Carp*		Mouse	Chicken	Frog
α1	-			34		αIIb	80		
α2	82		57	47		α3	86		
α10	86			46		α4	85		55
α11	86			49		α5	91		64
αD	-					αV	92	83	
αE	69					α6	93	71	
αL	72					α7	87		
αM	74				29, 29	α8	90	77	
αX	71					α9	-		

*The full-length (except for the I-domain) carp sequences AB048536 and AB048537 cluster as a separate branch within the integrin α subunits from immune system cells.

phagocytic hemocytes to C3 coated microbes.[41] Supporting this proposal, Miyazawa et al[41] have shown that antibodies against αHr1 inhibit C3-dependent phagocytosis in *H. roretzi*. Despite this possible role in the immune system of the tunicate, the αHr1 sequence does not cluster separately with either the α subunits of immune system cells or those collagen binding integrins of the ECM, but is an apparent outlier of both groups[41] (Figs. 3, 4). A second tunicate sequence, αHr2, clusters with integrin α subunits (with the αV-α8-α5-αIIb group) as do other invertebrate α subunits that do not contain I domains[41] (Fig. 4).

Vertebrate βI Domains

Eight integrin β subunits are known in human, β1-β8, and most of the counterparts in mouse have been identified (Table 2). Each of the integrin β subunits contains the β I-domain. The β4 subunit in human is much longer, 1822 amino acids, than the other integrin β subunits

Table 5. Percentage amino acid sequence identity between human integrin β subunit sequences and sequences from mouse, chicken, frog and fugufish. Based on the full-length sequences where possible.

	Mouse	Chicken	Frog	Fugu
β1	92	84	71, 70	71, 70
β2	82			44
β3	90			64, 59
β4	-			53
β5	92			57
β6	91			58
β7	86			-
β8	95			58

in human, which are typically about 800 amino acids long. In mouse, we located a sequence (AAH06632) that corresponds to the C-terminal 914 residues of the human β4 subunit and is very likely a fragment of the mouse β4 subunit.

There are similarities in the chromosomal disposition of the genes for the integrin β subunits (Table 2): in human the β3 and β4 subunits, like the αE subunit, are found on chromosome 17; in mouse both are found on chromosome 11. Each of the other β subunits is located on separate chromosomes. The β6 subunit, as well as the integrin α4, α6 and αV subunits, is mapped to chromosome 2 in both human and mouse. The β8 subunit has so far not been mapped in mouse, but a search of the human β8 sequence against the mouse genome identified one strong match locating β8-like sequence fragments (nucleotides 1194323-1193626: 79% identity and 1148282-1148034: 88% identity, at locus NW_000067.1) to chromosome 12 (Table 2).

A preliminary search of the fugufish genome identified nine sequences with high sequence similarity to β subunits (Table 5). JGI_6285 and JGI_7435 are full-length and 70 % identical to the human β1 sequence. JGI_23501 is missing about 120 N-terminal residues but matches the β2 human sequence at a level of 44 % sequence identity. JGI_4993 is a 548 residue fragment that is ~64 % identical with the human and mouse β3 subunit and over 50 % identical with the human and mouse β5 subunit. JGI_5840 is full-length and is 59 % identical with the human β3 sequence. JGI_19019, matches the β4 sequence (53 % identity) over the N-terminal sequence, but little similarity was observed over the C-terminal ~345 residues. The human β5 chain matches the JGI_11952, missing the C-terminal sequence, at a level of 57%. JGI_19899 appears to be full-length and shares 58 % identity with both the human and mouse β6 chain. No match was found for the integrin β7 subunit, but a match to the N-terminus of the human and mouse β8 sequences, 58% identity, was found: JGI_24322 (missing a portion of the C-terminus).

An alignment of the sequences was used to construct a phylogenetic tree using the neighbor-joining method and 1000 bootstrap replicates (Fig. 5). The tree agrees with that presented in Figure 2 of Burke[48] and of Hughes.[40]

Two different functional β1 subunits, β1a and β1b, were cloned long ago in *X. laevis*[49] and do represent integrin β1 subunit isoforms. We have found examples of each of the β integrins in bony fish with the lone exception of the integrin β7 subunit. This clearly establishes the presence of the full complement of integrin β subunits throughout the vertebrates and leaves opens the possibility that they will be found throughout the chordates. As with the *X. laevis* β1 sequences, in fugufish there are pairs of integrin β subunits closely related to the integrin β1 and β3 subunits.

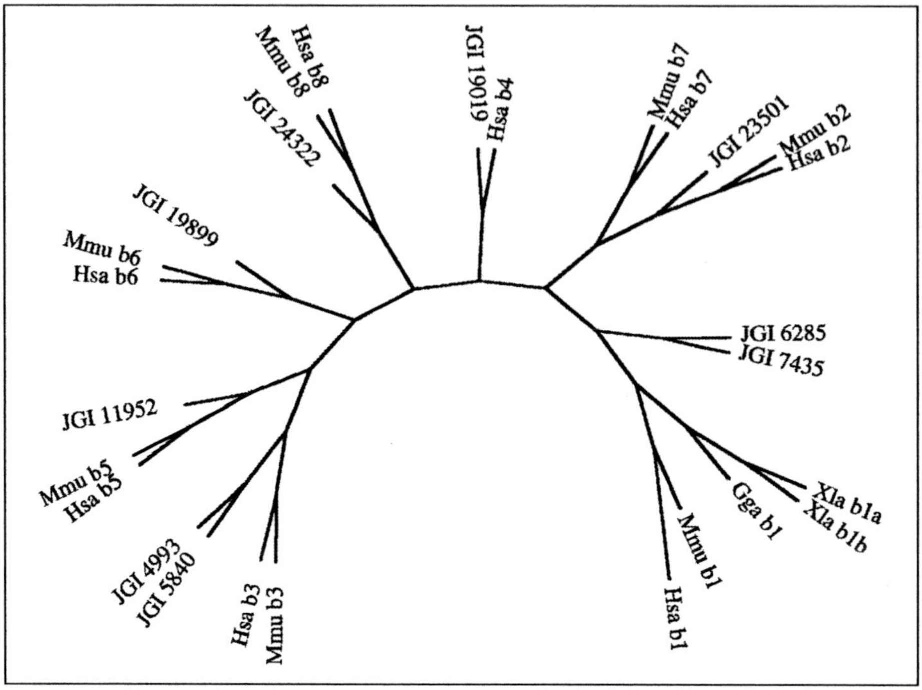

Figure 5. Consensus phylogenetic tree based (1000 bootstrapped alignments) of an alignment (not shown) of integrin β subunits; sequences corresponding to the ~650 N-terminal residues of the human β subunits were compared, matching the lengths of the N-terminal fragments mouse β8, fugufish sequences JGI_1993, JGI_11952, JGI_24322; the sequence of JGI_19019 matched other sequences poorly over the ~345 C-terminal residues. All nodes have bootstrap values over 91 % with the exception of the (JGI_4993, JGI-5880) cluster: 70 %; the β7 – β2 cluster: 59 %; the ((Mmu_b2, Hsa_b2) JGI_23501) cluster: 70 %; the (JGI_6285, JGI_7435) cluster: 48 %; and the ((Xla_b1a, Xla_b1b) Gga_b1) cluster: 69 %. An alignment of the entire sequences produced a similar tree, with the exception of poorer resolution of the truncated sequences.

Integrin β subunits are present throughout the invertebrate phyla,[40,48] with examples found in Porifera (sponges), Cnidaria (coral), Nemotoda (*C. elegans*), various protostomes (molluscs, arthropods), as well as the earliest deuterostomes, the echinoderms. However, none of these invertebrate integrin β subunits appears to be an orthologue of any of the integrin β1-8 subunits found in the vertebrates.[40,48]

Origins of the αI and βI Domains

Given that no orthologs of the individual vertebrate I-domain containing α and β subunits have been observed outside of the vertebrates, it might be tempting to suggest that the complex range of integrins seen in humans and fugufish arose with the divergence of the bony fish or only somewhat earlier. However, much of the evidence for integrins in amphibians and bony fish is the result of focused sequencing efforts on *X. laevis*, *D. rerio* and *T. rubrides*, and representatives of earlier informative divergences are under-represented. We are confident that these domains do not exist in *D. melanogaster* or *C. elegans*, because their genomes have been completely sequenced and no orthologues or even homologues have been found.[46] At present, an I-domain containing integrin α subunit has been cloned in the tunicate *H. roretzi*,[41] but it is not an orthologue of either the collagen-binding α subunits or the α subunits expressed by cells of the immune system. Instead, it appears to be equally similar to both groups but does not have the αC-helix of the collagen binding integrins.

Given the sequence data that are now available we can confidently state that most, if not all of the integrin α and β subunits present in amniotes have orthologues in amphibians as well as bony fish; the bony fish—tetrapod divergence occurred some 450 MYA. The similarity between the orthologues is high. The known bony fish α I-domains are 44-88 % identical to their human orthologues and 29-49 % identical over the integrin α subunits (without the I-domain region). For the integrin β subunit, the overall percentage identity between bony fish and humans range between 44 and 71 % for the identified orthologues. These values suggest that orthologous sequences in earlier diverging groups would be recognizable if their sequences exist. At the moment, the lack of sequence data among the earliest diverging vertebrates and chordates is a likely reason why none have been identified. Nevertheless, we would not be surprised to find them, certainly among the sharks and rays, hagfish and lamprey, and possibly within the invertebrate chordates too.

An apparent total absence of vertebrate orthologues among the invertebrates leads us to suggest that the appearance of the I-domain containing αβ heterodimers present in man today may have arisen as a necessity for the major changes in body plan organization and evolution of the immune system, characteristic of the chordates, which began with the earliest deuterostomes, the echinoderms. In the not too distant future the sequencing of the genomes of selected organisms that diverged early in the evolution of deuterostomes and vertebrates will clearly resolve the origins of the vertebrate integrin α and β I-domains.

Other VWFA Domains in Vertebrates

In addition to the integrin α and β subunits in vertebrates, a variety of different proteins contain VWFA domains[29]. They include von Willebrand factor itself with its three VWFA domains, collagens, the matrilins (cartilage matrix proteins functioning to bind collagen), complement factors, copines (Ca^{2+} -dependent phospholipid binding proteins with a possible role in membrane trafficking), thrombospondins, etc. The tree presented in Figure 6 includes the integrin α and β I-domains, bacterial proteins and representative proteins from protists and arthropods. The VWFA domains for these different sets cluster according to function but it is not reasonable to speculate further on their origins as most of the between cluster-branching is not significant according to the bootstrapping scores, as reported previously by Tuckwell[29].

VWFA Domains in Invertebrates

Caenorhabditis elegans

C. elegans has two integrin α subunits, α_{ina}-1, and α_{pat}-2, and one β subunit, β_{pat}-3.[50-52] These integrins are involved in cell-matrix attachment, e.g., in muscle assembly and neuron migration, paralleling the involvement of integrins in cell adhesion seen in vertebrates. Analysis of the *C. elegans* genome has shown that it contains over 150 collagen genes,[53] and most of these are involved in forming the cuticle, the exoskeleton of the organism. As has been stated above, the *C. elegans* integrin α subunits do not have I-domains, an so it would be predicted that the collagen binding function, which in vertebrates is carried out by α I-domain integrins, would have to be carried out by different molecules in *C. elegans*. The observation that the phenotypes of integrin mutations (muscle and adhesion defects) differ from those of cuticle collagen mutations (morphological defects) supports this hypothesis. Studies from our laboratory and those of others indicate that this is indeed the case: *C. elegans* has similar problems of collagen attachment and ECM assembly as vertebrates, but has solved them in different ways. Significantly, the vertebrate and invertebrate systems have in common the use of VWFA domains.

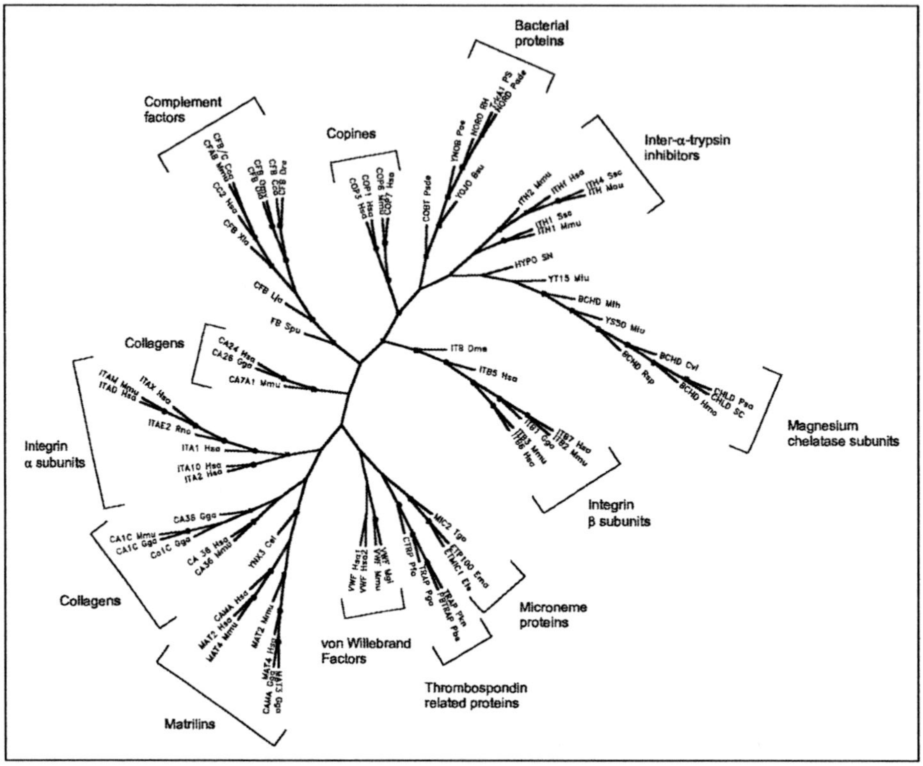

Figure 6. Consensus phylogenetic tree of von Willebrand factor A-domains (1000 bootstrapped alignments) based on the alignment (not shown) of representative sequences, primarily from vertebrates, but including domains from invertebrates, protists and prokaryotes. Black circles indicate the bootstrap values greater than 50% and rectangles values greater than 75%. Bsu, *Bacillus subtilis*; Dre, *Danio rerio*; Cel, *Caenorhabditis elegans*; Cvi, *Chlorobium vibrioforme*; Cca, *Cyprinus carpio*; Dme, *Drosophila melanogaster*; Ema, *Eimeria maxima*; Ete, *Eimeria tenella*; Gga, *Gallus gallus*; Hmo, *Heliobacillus mobilis*; Lja, *Lampetra japonica*; Mgi, *Macropus giganteus*; Mau, *Mesocricetus auratus*; Mth, *Methanobacterium thermoautotrophicum*; Mtu, *Mycobacterium tuberculosis*; Omy, *Oncorhynchus mykiss*; Ola, *Oryzias latipes*; Pade, *Paracoccus denitrificans*; Psa, *Pisum sativum*; Pbe, *Plasmodium berghei*; Pfa, *Plasmodium falciparum*; Pga, *Plasmodium gallinaceum*; Pkn, *Plasmodium knowlesi*; PS, *Pseudomonas sp.*; Pae, *Pseudomonas aeruginosa*; Psde, *Pseudomonas denitrifica*; Rno, *Rattus norvegicus*; RH, *Rhodobacter*; Rsp, *Rhodobacter sphaeroides*; Spu, *Strongylocentrotus purpuratus*; Ssc, *Sus scrofa*; SC, *Synechococcus sp.*; SN, *Synechocystis sp.*; Tgo, *Toxoplasma gondii.*

The short collagens and insoluble proteins (known as cuticulins), which form the *C. elegans* cuticle are secreted by underlying epidermal cells (the hypodermis;[53,54] Fig. 7). The worm progresses through four larval stages (L1-4) before adulthood, and between each stage the cuticle is shed, exposing a new cuticle synthesised during the previous stage. There is an additional stage in the life cycle, the dauer, which is an alternative to the L3, and which is adopted in times of over-crowding or shortage of food. The cuticles from each stage have a complex structure and differ from each other both at the molecular and structural levels.[55,56] These differences include lateral treads, or alae, which are only seen on the L1, dauer and adult cuticles. *C. elegans* also synthesises basement membrane, which has the same basic constituents as human basement membrane, and which surrounds internal organs such as the intestine and gonads.

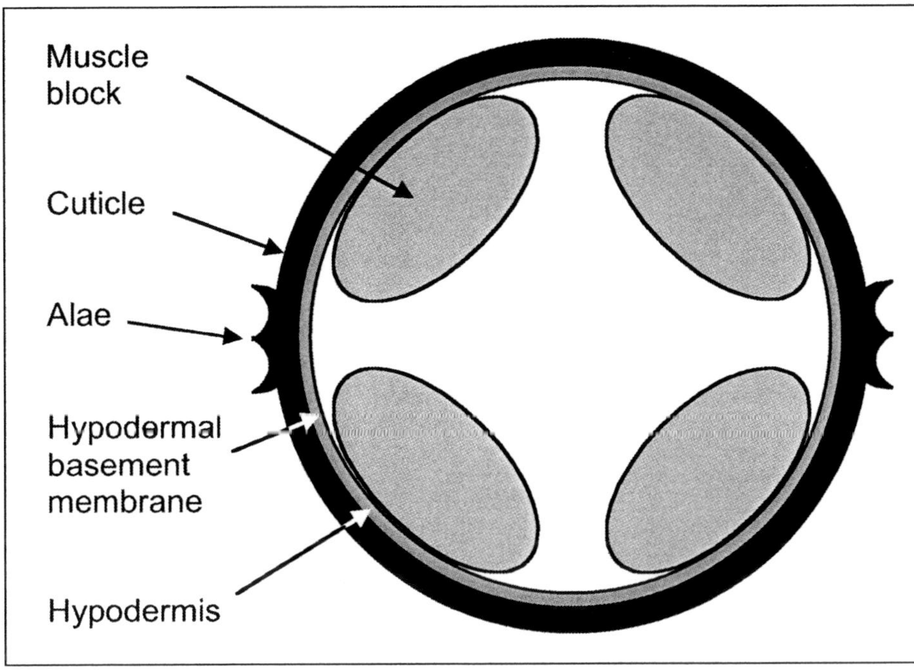

Figure 7. Schematic representation of a cross-section through the nematode *Caenorhabditis elegans*, showing the cuticle, alae, hypodermis, and musculature. Other internal organs are omitted.

The musculature extends in four longditudinal strips, two ventral and two dorsal, which are separated from the hypodermis by the hypodermal basement membrane (Fig. 7). The musculature is linked to the hypodermis, and thereby to the cuticle through hemidesmosomes, and the force generated against the cuticle is important for the movement of the worm. Two homologous VWFA domain-containing proteins, MUA-3 and MUP-4, have been identified which are essential for maintaining the link between the muscles and the cuticle: Null or mutant alleles of the *mua-3* gene show progresive paralysis,[57] and electron microscopy of the mutants shows that the cuticle is separated from the hypodermis. The MUA-3 protein localises to hypodermal hemidesmosomes, where the musculature contacts the hypodermis, and is located on the apical surface of the hypodermis, facing the cuticle. The *mup-4* gene is also required for linking the hypodermis to the cuticle,[58] and probably also co-localises with hemidesmosomes. *mup-4* is crucial at an earlier stage than *mua-3* as the development of *mup-4* mutants is arrested before the worms hatch, although it is expressed in larvae and adults. The two genes are probably responsible for the linkage of the hypodermal hemidesmosomes to the cuticle. These genes seem to be performing an analogous function to the mammalian integrins involved in matrix adhesion, such as $\alpha6\beta4$ in hemidesmosomes, and $\alpha2\beta1$ in matrix attachment.

mua-3 and *mup-4* have a similar modular structure and are clearly the result of a gene duplication. Each comprises a single VWFA domain inserted within a succession of EGF repeats (MUA-3 has more EGF repeats than MUP-4), with a transmembrane region C-terminal to the repeats. The ligands for the VWFA domains of these proteins have yet to be identified, however, with a number of mammalian VWFA domains binding collagens, and with the *C. elegans* cuticle being collagen- rich, it is tempting to speculate that there is an interaction between the VWFA domains and the collagens.

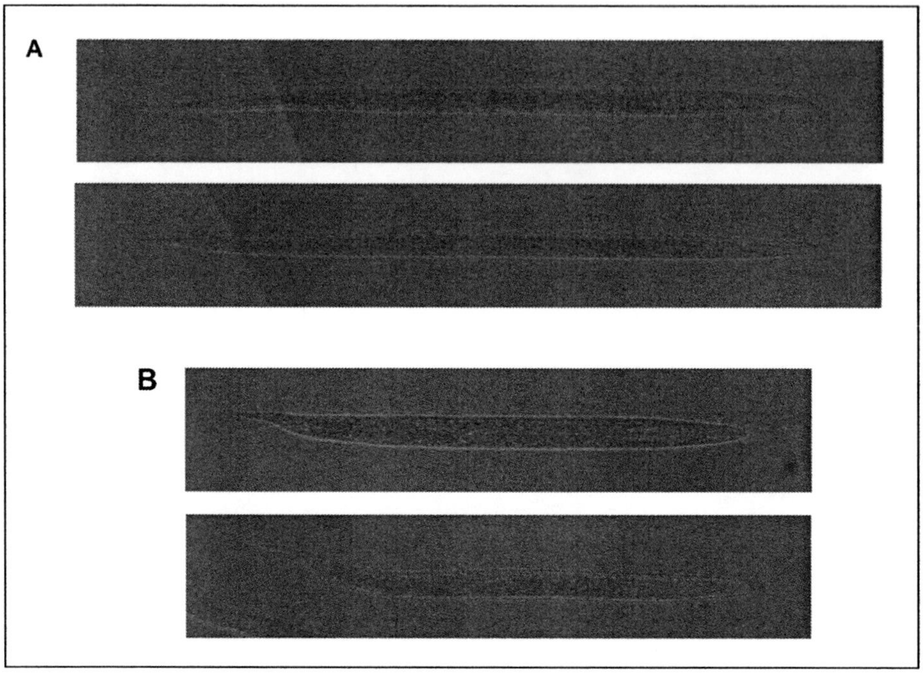

Figure 8. CUT-6 is important for cuticle (ECM) formation in *Caenorhabditis elegans*. Expression of *cut-6* was ablated in worms be RNAi; **A**, wild type N2 *C. elegans*; **B**, worms after RNAi of *cut-6*. Note the "dumpy" morphology.

Our recent work has provided further evidence for the importance of *C. elegans* VWFA domain proteins in ECM function. The VWFA-domain- containing protein CUT-6 is expressed by hypodermal cells and localises to the edges of the alae of dauer larvae (Muriel and Tuckwell, manuscript submitted). When the gene is knocked out by RNA interference, the dauer worms adopt a "dumpy" morphology (Fig. 8). Since this dumpy phenotype has previously been seen in strains with mutations in cuticle collagens,[53,54] our data point to CUT-6 playing an important role in the organisation of the dauer cuticle. As with the MUA-3 and MUP-4 proteins, the binding partner for the VWFA domain has yet to be determined, but again, collagens would seem to be obvious candidates.

Bioinformatics analyses indicate that there are 35 VWFA domain proteins in the genome of the nematode *Caenorhabditis elegans*. Broadly, the genes can be divided into four groups, two $\alpha2\delta$-calcium channel genes, four genes with a $[EGF]_2$-[C-type lectin]-[EGF]-[VWFA] domain structure, 14 genes comprising a VWFA domain followed by a C-type lectin domain, and a further group of 15 genes with disparate modular architectures, including MUA-3, MUP-4 and CUT-6[59] (D. Tuckwell, unpublished observation; http://smart.embl-heidelberg.de/). With the exception of the $\alpha2\delta$-calcium channels, none of these resemble any of the currently known vertebrates VWFA domain proteins. However, it is now clear that, despite the absence of I-domain integrins from the *C. elegans* genome, VWFA domain proteins are exploited by *C. elegans* for matrix adhesion and assembly, indicating an evolutionarily conserved role for these domains. Further molecular characterisation will be required to clarify the extent to which VWFA domain function is conserved between vertebrates and invertebrates, such as determination of the role of the MIDAS and the identification of ligands. We have suggested that collagens could be ligands for some *C. elegans* VWFA domain proteins. Pertinent to this, it is significant that the *Drosophila* genome contains few collagens and no VWFA-domain genes, save the α2δ calcium channels.

Other Metazoans and Parasites

The association of VWFA domains with ECM appears to be very ancient, even if the particular domain arrangements vary between phyla: VWFA domains are clearly involved in the ECM of vertebrates and nematodes, and it now seems likley that they also play a role in molluscs. Proximal thread matrix proteins 1A and B from a tough elastometic tendon of mussels (*Mytilus* spp.; accession numbers AAL17973, AAL17974, AAL83537) each comprise two VWFA domains. While tandem arrangements of VWFA domains are seen in a number of vertebrate and invertebrate proteins, the proximal thread protein does not resemble any other known VWFA domain protein particularly closely. The most conclusive evidence, however, comes from *Hydra*, one of the earliest diverging metazoans. A VWFA domain protein, HT4, has been identified in *Hydra* (accession number AB008935; R. Boot-Handford, personal communication). Interestingly, this incomplete sequence comprises, from the N-terminus, a VWFA domain, a 59 amino acid collagenous region, and a second VWFA domain followed by more collagenous sequence. A second *Hydra* VWFA domain protein has now also been identified containing two VWFA domains in the N-propeptide of a fibrillar collagen (characterised by an extended triple helix and a conserved fibrillar collagen C-propeptide-type domain; R. Boot-Handford, personal communication). *Hydra* does not form an interstitial ECM with fibrillar collagens like that seen in mammals, however, the *Hydra* mesoglia, an ECM between the basement membranes of the epi- and endo-dermises, contains fine fibrils,[60] and the fibrillar collagen-like molecule may localise to this region. These data confirm that VWFA domains had a role in ECM from earliest times in multicellular animal evolution, and also indicate some form of functional link with the collagens. The VWFA domains in integrins are therefore a chordate variation on this ancient theme.

For some time it has been recognised that apicomplexan parasites, some of the simplest unicellular eukaryotes, express VWFA domain proteins.[61] However, the relationship of these proteins to metazoan VWFA domain proteins is not clear. There are at least two VWFA-domain proteins in *Plasmodium* species, the causative agents of malaria; TRAP (thrombospondin-related adhesive protein) is made up of a VWFA domain followed by a thrombospondin type 1 (TSP1) domain, and CTRP (CS-protein-TRAP-related protein) made up of multiple VWFA domains and TSP1 domains. TRAP has been found to be required for the insertion of the parasite sporozoite into mosquito salivary glands and host cells, and mutations in the VWFA domain MIDAS almost completely prevented insertion.[62,63] CTRP also plays a major role in the biology of the parasite, as knockout of the gene affects parasite development and infection.[64,65] Some sort of adhesive role for these proteins seems likely,[63] but it is not known whether they diverged before the establishment of the animal lineage or whether the modules were "stolen" from their hosts. Significantly, TSP1 domains are not otherwise found outside of the animal kingdom, and this would support the more recent evolutionary option. The ligand for the TRAP VWFA domain has yet to be unequivocally demonstrated (compare refs. 63 and 66), and that of CTRP is unknown. Identification of these, presumably cell surface molecules, might enable a relationship with a characterised VWFA domain to be infered. The integrin αL/αM/αX and complement VWFA domains would be ideal candidates because of their roles in the immune system.

Non-Metazoan VWFA Domains

VWFA domains were initially considered to be components of metazoan extracellular molecules, usually in the ECM, and were typically identified on a gene-by-gene basis as a result of biochemical studies. In recent years, however, the advent of powerful and sensitive searching

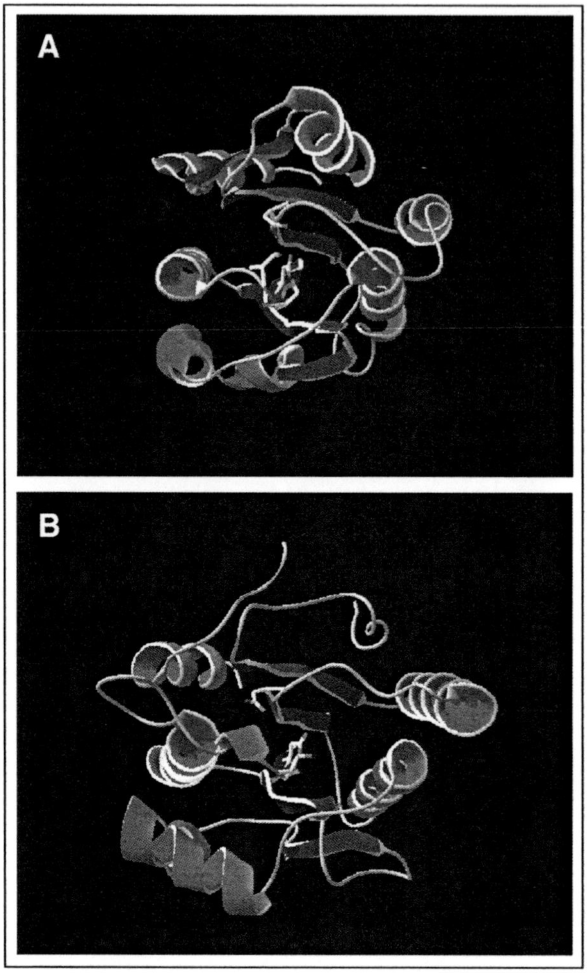

Figure 9. Comparison of structure of VWFA domain and Ku complexes. **A**, Human integrin αM VWFAdomain; **B**, Human Ku70 VWFA domain. MIDAS residues are shown where appropriate. There is an unresolved section in the Ku structure.

tools has enabled novel and distantly related examples of VWFA domains to be identified. The major consequences of this have been, firstly the identification of VWFA domains in intracellular proteins, and secondly their identification in plants, fungi, bacteria and archaebacteria[34] (http://smart.embl-heidelberg.de/). For example, VWFA domains have been identified in subunits of 26S proteasome intracellular proteolytic complex, the chloporhyll synthesis enzyme magnesium chelatase, the *Streptococcus pyrogenes* serum opacity factor, and the Ku DNA repair complex. The Ku complex, for which sequence, structural and functional information are available, provides a good opportunity to consider in detail the implications of the intracellular VWFA domains for the definition and evolution of the domain family.

The Ku complex is involved in the repair of double stranded breaks in eukaryotic chromosomal DNA[67,68] and is made up of two homologous components, Ku70 and Ku80, the N-terminal halves of which resemble VWFA domains. The crystal structure of the Ku complex

```
KU86_HUMAN  GNKAAVVLCMDVGFTMSN.....SQQADFLDALIVSMDVIQHETIGKK-FEKRHIEIFTDLSSRFSK
KU80_DRO    SNKECLIIVLDVRTCAAE.....------EWLNGLQAALKLQSVAATLRVARRRILLLFDFNDFPQD
KU80_TRYP   AFRTSTIFALDVNCSISS.....GVGSSANFIETLTLCVEVFNLKKRRKQFREVLYLLTDAHAEVVR
KU80_ARAB   RNREGLVLVLDVGPAMRS.....TVAGDFLDALIVGMDMLIKMYGNAH-KGKKRMCLITNAACPTKD
KU70_HUMAN  SGRDSLIFLVDASRAMFD.....GHGSDYSLSEVLWVCANLFSDVQVK-MSHKRIMLFTNEDDPHGN
DRO_YF      HGREAILFVVDANLQTAG.....EPDGRGRFDLMIRLCIEMLEKCGKK-LNNAKIAYVTDVSEPHPS
KU70_TRYP   DQFDVVLCLVDFQQRMFQ.....STEAKSHLGDALRAARHLFSQLPSE-VKHRRIFLFTNDVNPHRG
KU70_ARAB   ASKEFVVYLIDASPKMFC.....SDSRENSLYSALWVAQALLRKGSLK-TADKRMFLFTNEDDPFGS
KU_SCHPO1   SDKECTVFVLDLGKDMGT.....TMQGDLISAIVVSFDLMARHCKKNK--WKKKMIVLTAARGIIDF
KU_SCHPO2   IGKYAILFVIEVSPSMLD.....PCSCQVSLSSVLYHCSVIFTTKAEN--FEKRLFLITDNDHPAWD
HDF2_YEAST  MSSESTTFIVDVSPSMMK.....GLQSMIQCLLVVSLDIKQQFQAR---KILKQIVVFTDNLDDLDI
KU_YEAST    DIHEGILFCIELSETMFK.....EKQVRLSVLFTFMLDTFLEEIPGQKQLSNKRVFLFTDIDKPQEA

HR1         SGVDV-LFVLDGSGSVGK.....YTTYTGRALQKVIRDFDDAYIGNK-----QVLLLTDGQ-AKDN
αL hu       GNVDL-VFLFDGSMSLQP.....-LTNTFGAINYVATEVFREELGARP-DATKVLIIITDGE-----
αM hu       EDSDI-AFLIDGSGSIIP.....-RTHTATGIRKVVRELFNITNGARK-NAFKILVVITDGEKFGDP
MIDAS                 *  *  *           *                                *
Structure          beta A                       alpha 4              beta D
```

Figure 10. Alignment of integrin and Ku VWFA domain to show conservation and variation in the region of the MIDAS. Ku proteins; Ku80_ARAB, Arabidopsis Ku80-like, AF283758_1; KU_SCHPO1, S. pombe Ku-like, CAC05245.1 ; S. cerevisiae, Ku-like, HDF2_YEAST ; KU_SCHPO2, Ku-like, S. pombe, CAA22471.1; DRO_YF, Drosophila yolk factor 1 β chain, A53623; KU70_ARAB, Arabidopsis Ku-70-like, AF283759.1; KU70_TRYP, Ku70, Trypanosome, CAC86194.1; KU80_TRYP, Ku80 Trypanosome, CAC86188.1; KU_YEAST, S. cerevisiae, Ku-like, CAA49840.1; KU80_DRO, Ku80, drosophila, AF237722.1. Integrins, HR1, Integrin α subunit VWFA domains from *Halocynthia roretzi*[41]. Structure; integrin αM structure.

shows that the VWFA domains have the same arrangement of sheets and strands as the other VWFA domains for which structures are known (e.g., integrins and von Willebrand factor) indicating that the sequence-based searches identifying the VWFA domains in the Ku components corresponded to an underlying similarity in structure (Fig. 9). Multiple alignments of Ku proteins from a variety of organisms indicate that the residues corresponding to the cation coordinating amino acids of the MIDAS are generally conserved (Fig. 10). Comparison of the structures of Ku in the presence and absence of DNA indicates that the Ku VWFA domains do not under go conformational changes on DNA binding. However, the DNA does not make contacts with the MIDAS face, but interacts with the "bottom" of the domain:[67] Instead, it seems likely that the VWFA domains of Ku are involved in interactions with other proteins.[68] While these observations make a strong case for a resemblance to the "traditional" ECM VWFA domains, not all of the Ku70 and Ku80 subunits match the SMART VWFA domain profile, despite aligning well with each other. The broadening of the VWFA domain consensus to accommodate these proteins might in turn identify yet further examples of the family.

The sequence and structure matches place the Ku subunits within the VWFA domain family, and yet it is clear that these proteins are functionally quite distinct from the ECM VWFA domains such as those of integrins, matrilins and collagens. With the VWFA domain family listed on SMART now containing 1013 proteins, some way of sub-dividing the family would clearly be advantageous. We propose that the VWFA domains could be subdivided into an intracellular and an extracellular group. While this broadly separates the "traditional" ECM and the newer VWFA domains, this working definition also has a sound biochemical basis: The role of cysteine residues differs substantially between the intracellular and extracellular environments as the intracellular environment precludes disulphide bond formation, but in the extracelular environment disulphides form. Placing of cysteines in extracellular proteins is therefore crucial and many of the extracellular VWFA domains contain disulphide bonds, e.g., bridging the N-and C-termini of the domain. However, the Ku subunits and other intracellular VWFA domains contain cysteine residues in which cysteine substitutes for small hydrophobic residues such as valine and leucine. While an archetypal VWFA domain must at some point have transferred from an intracellular to an extracellular function, the disulphide constraints suggest that this was a complex transition and would have only happened infrequently. A working division of the domain family into intracellular and extracellular groups is therefore reasonable. If the movement from an intracellular to an extracellular function occurred more than once, multiple lineages of extracellular VWFA domains could have arisen. While the VWFA domains of integrins, collagen and matrilins all appear similar, a good candidate for a separate origin would be the integrin β subunit I-domains as this would account for the considerable divergence between the VWFA domains of the integrin β subunits and those of other ECM molecules.

Conclusions

The α and β I-domains are clearly part of a large and biologically important family, with relatives, albeit distant ones in bacteria, and archaea. The proliferation of vertebrate integrin αI-domains appears to be associated with key events in vertebrate/chordate evolution, and in the future, sequence information for organisms representing significant branch points in the vertebrate evolutionary tree should help link integrin evolution with physiological and morphological developments. Indeed, the important role that these domains play in ECM and the immune system make them particularly appropriate molecules for evolutionary studies. We have tried to suggest that there may be functional properties conserved across the VWFA domain family. While this can be a subjective process, comparisons of data from *C. elegans* and from the "intracellular" VWFA domains should enable the sequence-based identification of the VWFA domain family to acquire some functional aspects. In conclusion, it looks as if the integrin I-domains and the VWFA domain family in general still have much to tell us.

Acknowledgements

Mirva Piippo and Mikko Huhtala are thanked for there early work on the vertebrate I domains, and for the preparation of Figures 1 and 5. Daniele Casciari and Alessandra de Luise are thanked for their valuable assistance in searching for vertebrate sequences.

References

1. Eble JA, Kühn K, eds. Integrin-ligand interactions. New York: Chapman and Hall, 1997:1-273.
2. Heino J. Cellular signaling by collagen-binding integrins. This volume. 2003.
3. Ivaska J, Heino J. Adhesion receptors and cell invasion: Mechanisms of integrin-guided degradation of extracellular matrix. Cell Mol Life Sci 2000; 57:16-24.
4. Xiong J-P, Stehle T, Diefenbach B et al. Crystal structure of the extracellular segment of integrin $\alpha V\beta 3$. Science 2001; 294:339-345
5. Zutter MM, Santoro SA. Widespread histologic distribution of the alpha 2 beta 1 integrin cell-surface collagen receptor. Am J Pathol 1990; 137:113-120.
6. Camper L, Hellman U, Lundgren-Åkerlund E. Isolation, cloning, and sequence analysis of the integrin subunit alpha10, a beta1-associated collagen binding integrin expressed on chondrocytes. J Biol Chem 1998; 273:20383-20389.
7. Lehnert K, Ni J, Leung E et al. The integrin alpha10 subunit: expression pattern, partial gene structure, and chromosomal localization. Genomics 1999; 60:179-187.
8. Velling T, Kusche-Gullberg M, Serjersen, Gullberg D. cDNA cloning and chromosomal localization of human alpha(11) integrin. A collagen-binding, I domain-containing, beta(1)-associated integrin alpha-chain present in muscle tissues. J Biol Chem 1999; 274:25735-25742.
9. Lee J-O, Rieu P, Arnaout MA et al. Crystal structure of the A domain from the alpha subunit of integrin CR3. Cell 1995a; 80:631-638.
10. Tozer EC, Liddington RC, Sutcliffe MJ et al. Ligand binding to integrin alphaIIbbeta3 is dependent on a MIDAS-like domain in the beta3 subunit. J Biol Chem 1996; 271:21978-21984.
11. Tuckwell D, Humphries MJ. A structure prediction for the ligand-binding region of the integrin beta subunit: evidence for the presence of a von Willebrand factor A domain. FEBS Lett 1997; 400:297-303.
12. Lin EC, Ratnikov BI, Tsai PM et al. Evidence that the integrin beta3 and beta5 subunits contain a metal ion-dependent adhesion site-like motif but lack an I domain. J Biol Chem 1997; 272:14236-14243.
13. Pfaff M. Recognition sites of RGD-dependent ligands. In: Eble JA and Kühn K, eds. Integrin-ligand interactions. New York: Chapman and Hall, 1997:101-121.
14. Ivaska J, Käpylä J, Pentikäinen O et al. A peptide inhibiting the collagen binding function of integrin alpha2 I domain. J Biol Chem 1999; 274:3513-3521.
15. Pentikäinen O, Hoffrén A-M, Ivaska J et al. RKKH peptides from the snake venom metalloproteinase of Bothrops jararaca bind near the MIDAS site of the human integrin $\alpha 2$I -domain. J Biol Chem 1999; 274:31493-31505.
16. Bergelson JM, St. John NF, Kawaguchi S et al. The I domain is essential for echovirus 1 interaction with VLA-2. Cell Adhes Commun 1994; 2:455-464.
17. Kamata T, Puzon W, Takada Y. Identification of putative ligand binding sites within I domain of integrin alpha 2 beta 1 (VLA-2, CD49b/CD29). J Biol Chem 1994; 269:9659-9663.
18. King SL, Kamata T, Cunningham JA et al. Echovirus 1 interaction with the human very late antigen-2 (integrin alpha2beta1) I domain. Identification of two independent virus contact sites distinct from the metal ion-dependent adhesion site. J Biol Chem 1997; 272:28518.
19. Dickeson KS, Mathis NL, Rahman M et al. Determinants of ligand binding specificity of the $\alpha 1\beta 1$ and $\alpha 2\beta 1$ integrins. J Biol Chem 1999; 274:32182-32191.
20. Kallen J, Welzenbach K, Ramage P et al. Structural basis for LFA-1 inhibition upon lovastatin binding to the CD11a I-domain. Mol Biol 1999; 292:1-9.

21. Lee J-O, Bankston LA, Arnaout MA et al. Two conformations of the integrin A-domain (I-domain): a pathway for activation? Structure 1995b; 3:1333-1340.
22. Baldwin ET, Sarver RW, Bryant GL Jr et al. Cation binding to the integrin CD11b I domain and activation model assessment. Structure 1998; 6:923-935.
23. Qu A, Leahy DJ. Crystal structure of the I-domain from the CD11a/CD18 (LFA-1, alpha L beta 2) integrin. Proc Natl Acad Sci USA 1995; 92:10277-10281.
24. Qu A, Leahy DJ. The role of the divalent cation in the structure of the I domain from the CD11a/CD18 integrin. Structure 1996; 4:931-941.
25. Emsley J, King SL, Bergelson JM et al. Crystal structure of the I domain from integrin alpha2beta1. J Biol Chem 1997; 272:28512-28517.
26. Emsley J, Knight CG, Farndale RW et al. Structural basis for collagen recognition by integrin $\alpha 2\beta 1$. Cell 2000; 101:47-56.
27. Salminen T, Nymalm Y, Kankare J et al. Large-scale purification and preliminary x-ray analysis of the human integrin a1I domain. Acta Crystallogr 1999; D55:1365-1367.
28. McDowall A, Leitinger B, Stanley P et al. The I domain of integrin leukocyte function-associated antigen-1 is involved in a conformational change leading to high affinity binding to ligand intercellular adhesion molecule 1 (ICAM-1) J Biol Chem 1998; 273:27396-27403.
29. Tuckwell D. Evolution of von Willebrand factor A (VWA) domains. Biochem Soc Trans 1999; 27:835-840.
30. Bienkowska J, Cruz M, Atiemo A et al. The von willebrand factor A3 domain does not contain a metal ion-dependent adhesion site motif. J Biol Chem 1997; 272:25162-25167.
31. Huizinga EG, Martijn van der Plas R, Kroon J et al. Crystal structure of the A3 domain of human von Willebrand factor: implications for collagen binding. Structure 1997; 5:1147-1156.
32. Celikel R, Varughese KI, Madhusudan et al. Crystal structure of the von Willebrand factor A1 domain in complex with the function blocking NMC-4 Fab. Nat Struct Biol 1998; 5:189-194.
33. Romijn RAP, Bouma B, Wuyster W et al. Identification of the collagen-binding site of the von Willebrand factor A3-domain. J Biol Chem 2001; 276:9985-9991.
34. Ponting CP, Aravind L, Schultz J et al. Eukaryotic signalling domain homologues in archaea and bacteria. Ancient ancestry and horizontal gene transfer. J Mol Biol 1999; 289:729-745.
35. Johnson MS, Overington JP. A structural basis for sequence comparisons. J Mol Biol 1993; 233:716-738
36. Saitou N, Nei M. The neighbor-joining method: A new method for reconstructing phylogenetic trees. Mol Biol Evol 1987; 4:406-425.
37. Felsenstein J. Confidence limits on phylogenies: An approach using the bootstrap. Evolution 1985; 39:783-791.
38. Hughes AL. Coevolution of the vertebrate α- and β- chain genes. Mol Biol Evol 1992; 9:216-234.
39. Whittaker CA, DeSimone DW. Integrin alpha subunit mRNAs are differentially expressed in early Xenopus embryos. Development 1993; 117:1239-1249.
40. Hughes AL. Evolution of the integrin α and β protein families. J Mol Evol 2001; 52:63-72.
41. Miyazawa S, Azumi K, Nonaka M. Cloning and characterization of integrin alpha subunits from the solitary ascidian, Halocynthia roretzi. J Immuno 2001; 166:1710-1715.
42. Grayson MH, Van der Vieren M, Sterbinsky SA et al. Alpha d beta2 integrin is expressed on human eosinophils and functions as an alternative ligand for vascular cell adhesion molecule 1 (VCAM-1). J Exp Med 1998; 188(11):2187-91.
43. Van der Vieren M, Crowe DT, Hoekstra D et al. The leukocyte integrin alpha D beta 2 binds VCAM-1: evidence for a binding interface between I domain and VCAM-1. J Immunol 1999; 163:1984-1990.
44. Cepek KL, Parker CM, Madara JL et al. Integrin alpha E beta 7 mediates adhesion of T lymphocytes to epithelial cells. J Immunol 1993; 150:3459-3470.
45. Taraszka KS, Higgins JM, Tan K et al. Molecular basis for leukocyte integrin alpha(E)beta(7) adhesion to epithelial (E)-cadherin. J Exp Med 2000; 191(9):1555-1567.
46. Hynes RO, Zhao Q. The evolution of cell adhesion. J Cell Biol 2000; 150:F89-F95.
47. Fujiki K, Bayne CJ, Shin DH et al. Molecular cloning of carp (Cyprinus carpio) C-type lectin and pentraxin by use of suppression subtractive hybridisation. Fish Shellfish Immunol. 2001; 11:275-279.

48. Burke RD. Invertebrate integrins: Structure, function and evolution. Int Rev Cytol 1999; 191:257-284.
49. DeSimone DW, Hynes RO. Xenopus laevis integrins. Structure and evolutionary divergence of the b subunits. J Biol Chem 1988; 163:5333-5340.
50. Gettner SN, Kenyon C, Reichardt LF. Characterisation of βpat-3 heterodimers, a family ofessential integrin receptors in C. elegans. J Cell Biol 1995; 129, 1127-1141.
51. Baum PD, Garriga G. Neuronal migrations and axon fasciculation are disrupted in ina-1 integrin mutants. Neuron 1997; 19:51-62.
52. Brown NH. Cell-cell adhesion via the ECM: integrin genetics in fly and worm. Matrix Biol 2000:19; 191-201.
53. Johnstone IL. Cuticle collagen genes. Expression in Caenorhabditis elegans. Trends Genet 2000; 16:21-27.
54. Kramer J. Extracellular matrix. In: Riddle DL, Blumenthal T, Meyer BJ, eds. C. elegans II. Cold Spring Harbor Laboratory Press, 1997:471-500.
55. Cox GN, Staprans S, Edgar RS. The cuticle of Caenorhabditis elegans. II. Stage-specific changes in ultrastructure and protein composition during postembryonic development. Dev Biol 1981; 86:456-470.
56. Johnstone IL, Barry JD. Temporal reiteration of a precise gene expression pattern during nematode development. EMBO J 1996; 15:3633-3639.
57. Bercher M, Wahl J, Vogel B et al. mua-3, a gene required for mechanical tissue integrity in Caenorhabditis elegans, encodes a novel transmembrane protein of epithelial attachment complexes. J Cell Biol 2001; 154:415-426.
58. Hong L, Elbl T, Ward T et al. MUP-4 is a noel transmembrane protein with functions in epithelial cell adhesion in Caenorhabditis elegans. J Cell Biol 2001; 154:403-414.
59. Hutter H, Vogel B, Plenefisch JD et al. Conservation and novelty in the evolution of the evolution of cell adhesion and extracellular matrix genes. Science 2000; 287:989-994.
60. Sarras MP Jr, Deutzmann R. Hydra and Niccolo Paganini (1782-1840)—two peas in a pod? The molecular basis of extracellular matrix structure in the invertebrate. Hydra Bioessays 2001; 8:716-24.
61. Bork P, Rhode K. More von Willebrand factor type A domains? Biochem J 1991; 279:908-910.
62. Mota MM, Thathy V, Nussenzweig RS et al. Gene targeting in the rodent malaria parasite Plasmodium yoelii. Mol Biochem Parasitol 2001; 113(2):271-8.
63. Matuschewski K, Nunes AC, Nussenzweig V et al. Plasmodium sporozoite invasion into insect and mammalian cells is directed by the same dual binding system. EMBO J 2002; 21:1597-606.
64. Yuda M, Sakaida H, Chinzei Y. Targeted disruption of the plasmodium berghei CTRP gene reveals its essential role in malaria infection of the vector mosquito. J Exp Med 1999; 190:1711-1716.
65. Templeton TJ, Kaslow DC, Fidock DA. Developmental arrest of the human malaria parasite Plasmodium falciparum within the mosquito midgut via CTRP gene disruption. Mol Microbiol 2000; 36:1-9.
66. McCormick CJ, Tuckwell DS, Crisanti A et al. Identification of heparin as a ligand for the A-domain of Plasmodium falciparum thrombospondin-related adhesion protein. Mol Biochem Parasitol 1999; 100:111-124.
67. Walker JR, Corpina RA, Goldberg J. Structure of the Ku heterodimer bound to DNA and its implications for double-strand break repair. Nature 2001; 412:607-614.
68. Doherty AJ, Jackson SP. DNA repair: How Ku makes ends meet. Curr Biol 2001; 11:R920-R924.

Chapter 2

Integrin $\alpha 1\beta 1$

Humphrey Gardner

Introduction

The integrin $\alpha 1$ subunit was first discovered by Hemler et al as the α component of the Very Late Antigen I (VLA1) expressed on a subset of T cells in the joints of patients with rheumatoid arthritis,[1] as well as in a subset of lymphocytes after long term in vitro culture.[2] $\alpha 1$ is the largest of the α subunits, with an apparent mw of 190 kDa nonreduced and 210 kDa reduced. $\alpha 1$'s larger size compared to $\alpha 2$ is most likely due to a higher degree of glycosylation (see also Gullberg et al chapter 6). At the C terminus, the intracellular portion of $\alpha 1$ is the shortest of the α subunits, at 13 residues.

Functionally, $\alpha 1$ is one of four collagen binding I domain containing $\beta 1$ partners, along with $\alpha 2$, $\alpha 10$ and $\alpha 11$. The tissue distributions of these four, along with their differing relative affinities for different collagen types, suggest both the differences and potential overlaps in their physiological roles. None of the four are known to partner with any β subunit other than $\beta 1$. The $\alpha 1$ I domain is likely to show, like $\alpha 2$, $\alpha 10$ and $\alpha 11$, affinity modulation of ligand binding activity in the same way as has been described for αL,[4,5] but experimental evidence for this has not yet been published.

Tissue Distribution

$\alpha 1\beta 1$, like $\alpha 11\beta 1$, is predominantly present in mesenchyme. In the adult $\alpha 1$ is most abundant in vascular and visceral smooth muscle. This smooth muscle expression has been shown, in the chicken, to be due to a unique combination of transcription factors, gata6, SRF, and Nkx3.2.[6] The latter is not found in mammals, but similar factors such as Bapx1 and its family members may play the same role. The regulation of $\alpha 1$ baseline expression in other tissues has not been extensively explored. Other sites of $\alpha 1$ expression include fibroblasts[7,8] and, particularly, specialized fibroblast related cells such as hepatic stellate (Ito) cells,[9] pericytes[8] and mesangial cells;[2,10] bone marrow mesenchymal stem cells;[11,12] chondrocytes,[13] in concert with integrin $\alpha 10$[14] and $\alpha 2$;[15] neural cells including undifferentiated schwann cells[16] and neurons;[17] and many white blood cells.[18,19] Microvascular endothelium shows abundant $\alpha 1$ expression,[20] which is upregulated during angiogenesis. Surprisingly, immunoelectron microscopy shows the presence of abundant $\alpha 1\beta 1$ on the luminal, as well as abluminal, endothelial surface (Raivich G, pers comm), where no canonical $\alpha 1$ ligand would be expected to be. $\alpha 1$ is generally absent from normal epithelia, other than the endoderm derived hepatocytes,[21-23] and endometrial glands,[24] where it is cyclically expressed.

I Domains in Integrins, edited by Donald Gullberg.

Expression During Development

During development, there is abundant and dynamic expression of α1 in embryonic tissues. It is first seen at the leading edge of invading trophoblast shortly after implantation,[25] and antibody blockade of α1 inhibits trophoblast invasion in vitro.[26] During early to mid embryogenesis α1 is expressed transiently by neurons of the CNS[17], by maturing skeletal and cardiac muscle,[27] in the skin,[28] throughout the developing kidney,[29] and in neural crest cells as they mature to dorsal root ganglia.[17]

Expression in Tumors

Dysregulation of α1 has been noted in tumors. Some studies of melanoma have shown a correlation of worse clinical behavior with the presence of α1,[30,31] and others with the absence of α1.[32] Leiomyosarcomas often show loss of α1 and gain of α2.[33] Bronchoalveolar carcinomas sometimes show gain of α1.[34] Lastly, dermatotrophic T cell lymphomas show expression of α1,[35] probably consistent with their derivation from the T cell lineage. In short, there is no consistent relationship between α1 expression and tumor behavior, in contrast to, for example, the well characterized and functionally significant α6β4 to α6β1 switch seen in some epithelial malignancies.

Integrin α1β1 Regulation by Cytokines

Most studies of regulation of α1 expression in the adult relate to expression during lymphocyte ontogeny, and in fibroblasts in response to a variety of cytokines. Like the α2, α3, α4, and α5 integrin chains, α1β1 is upregulated in fibroblast lineages by TGFβ,[36] as well as interleukin 1β,[37] TNF α, and interferon gamma.[38] The only cytokine which appears to cause differential regulation of α1 is platelet derived growth factor, which causes downregulation of α1 and upregulation of α5 in fibroblasts[38] and mesangial cells.[39]

Integrin α1β1 Ligands

The best known ligands for α1 are the collagens, investigated mostly in fibroblast studies, and laminin-1, investigated primarily in studies of neural cells. Other α1 ligands include matrilin-1, expressed in cartilage, galectins 1 and 3, and the NC1 domain of collagen IV(1), which will be discussed in the context of endothelial regulation.

Collagen

α1β1 and α2β1 integrins have collagen binding preferences that are at first glance discordant with their tissue distributions. α1β1, predominantly expressed on connective tissue, has a higher affinity for collagen type IV than for type I; whereas α2β1, predominantly on epithelial cells, favors collagen I, which epithelial cells do not normally see, over the collagen IV abundant in epithelial basement membranes. α1 and α2 (and probably α10 and α11) bind the triple helical domains of the collagens with highest affinity, and biochemical, cell biological and crystallographic studies show that this binding is contributed to by more than one chain of the triple helix.[40,41] As such, the binding is dependent on the chains being in register, and would thus be exquisitely sensitive to melting. As collagen melting has recently been shown to occur at or below physiological temperatures in a very dynamic fashion,[42] it is likely that α1 ligand binding, and hence signalling, can be affected by events distal to the receptor along the collagen fibril. This might be especially important in tissue remodelling.

In a groundbreaking series of analyses of collagen I binding by the α1 I domain Rich et al and Xu et al,[43,44] using crystallography, surface plasmon resonance, and rotary shadowing, showed that the α1 I domain can bind the collagen triple helix at multiple different sites, with the relative affinities being divisible into several classes. Among these there are approximately three of the highest binding affinity, with Kds of ~0.25uM, and about 13 in the next affinity class, with Kds of ~14uM. The highest binding class regions are adjacent to or overlap with the sites occupied by α2 I domain, and these can be competed off both α1 and α2 I domains by triple helical model molecules containing the core sequence GLOGER or GFOGER, the latter of which was also independently identified by Knight et al as an inhibitor of α1 and α2 binding to collagen I.[45] This core peptide is not effective in blocking α1 binding to collagen IV, but is effective in blocking α2. Thus the collagen IV binding site for α1 is unique and of higher affinity,[46] and has been shown by elegant biochemical work to require asp 461 in the α1 chain of collagen IV and arg 461 in the α2 chain.[40] The binding of integrin α1, α2 and α10 I domains to other collagens has also been explored.[47,48] The relationship between the probable affinity modulation of α1, the multiplicity of sites on the collagen fibril along which α1 can bind, and the dynamic instability of the triple helix, suggest a highly dynamic interaction between integrin and collagen. For example, one could see how fibroblast motility along collagen I might be contributed to by detachment and reattachment of the integrin along the fibril. Another possibility is that collagen fibril assembly and extrusion from the fibroblast might be aided by α1β1 protruding from the plasmamembrane surrounding the fibril, and supporting it like the gantries supporting a rocket. Indeed, the α1 null mouse has narrower and less well formed collagen fibrils than the wild type animal (Gardner, unpublished). Other aspects of integrin α1β1-collagen interactions are discussed in Chapter 10.

Laminin

Laminin-1/-2 binding by integrin α1β1 is seen in fibroblasts, and is particularly evident on neural cells, for which the pheochromocytoma line PC12 is used as a prototype.[49,50] These cells show integrin α1β1-dependent adhesion to domain VI of the laminin α1/α2 chains at sites located at the opposite end of the laminin molecules compared with the binding regions of the epithelial laminin receptor α3β1 and the hemidesmosome integrin α6β4.[51] This seems reasonable in the context of an epithelial basement membrane, where epithelial cells would bind at one end of the molecule and mesenchymal cells at the other (although binding to laminin 5 by α1 is not seen). In vitro, α1 has been.[52] found to be important for neurite outgrowth on laminin[53] and neural crest cell attachment to collagen.[54] Neural crest cell attachment to laminin can be inhibited by antisense oligonucleotides to α1 mRNA.[55] The α1 null mouse, however, appears neurologically and neuroanatomically normal except for an ill explored sensitivity to ketamine/xylazine anaesthesia (Davidson J, unpublished observations) which may have a neurological basis. Whether α2, possibly coexpressed on neurons, provides an adequate alternative for neurite outgrowth, will be seen in the α1/α2 double null animal.

Matrilin and Galectin

Matrilin-1 is found in cartilage, and appears to cause increased chondrocyte adhesion to collagen II, via its association with α1β1.[56] Galectins 1 and 3, secreted by vascular smooth muscle also appear to bind integrin α1, the latter in a lactose dependent manner.[57,58] These glycoproteins, in contrast to matrilin-1, appear to inhibit cell attachment to other matrix components.

Coreceptors

Many non-I domain containing integrins have been shown to associate in the membrane with other receptors, the best examples being the tetraspanins[59] and Integrin Associated Protein.[60] These may modulate integrin behavior and binding to ligands. α1 has not been shown to associate with such proteins, but this is an area meriting further exploration. On the other hand, α1 is one of a subset of integrins (including α5β1, αvβ3, and α6β4) which associate in the membrane with caveolin and stimulate the Erk pathway via Fyn and Shc.[61,62]

The α1 Null Mouse

Aspects of the α1 null mouse will be discussed in subsequent sections. However, a brief overview is provided here to provide perspective on the known and suspected roles of α1. Integrin α1 null mice are viable and fertile, and embryogenesis proceeds normally despite the broad and dynamic expression in trophoblast and developing nervous system. Superficially, adult animals are remarkably normal with a mild decrease in weight, normal smooth muscle function, normal rates of wound healing, normal liver function, normal behaviour, and no blatant immunodeficiency in a laboratory environment.[63] With these data in mind, the role of α1 in various organ systems will be discussed.

Integrin α1β1 Signalling and the Cell Cycle

The potential role of α1 as a cell cycle regulator was suggested by studies showing that α1β1 was a member of a small group of integrins which could activate the adaptor protein Shc, resulting ultimately in MAP kinase activation.[61] Several observations from the α1 null mouse confirmed this, including a reduction in fibroblast proliferation rate in embryonic skin and dermal fibroblast number in the adult, as well as the observations that embryonic fibroblasts from the α1 null failed to activate Shc in response to adhesion to collagen, and that they failed to grow on collagen in conditions of limiting serum, whereas growth on the α5 ligand fibronectin or the αv ligand fibrinogen, was normal.[64] As α2 and probably α11 are present on these cells, this suggests that α1 is unique among collagen binding integrins in mesenchyme in being able to stimulate proliferation. More recently, it has been observed that fracture calluses are smaller in α1 null mice, concomitant with a deficiency in bone marrow derived mesenchymal stem cell proliferation.[65] Interestingly, the number and proliferation of mesenchymal stem cell derived hypertrophic chondrocytes in this model is normal - suggesting a specific and transient dependence on α1 for proliferation in the mesenchymal stem cell differentiation pathway. A role for α1 has also been described in osteoblast differentiation.[66] Similar phases of α1 dependence for proliferation appear to be present at some stages of lymphocyte ontogeny.[67] Overall, the subtlety of the proliferative deficit in the α1 null mouse is accounted for by the vast number of overlapping proliferative pathways, ligands, and integrins present in the organism.

In contrast to these studies ascribing a pro-proliferative role for α1 signalling, it has been shown that overexpression of α1 in mesangial cells leads to activation of p27Kip and cell cycle arrest.[68] This may be due to some kind of integrin crosstalk, where the excess integrin α1 activates a pathway normally associated with another integrin.[69] A potential corollary of this is that monomer and polymer collagen have different effects on mesangial cell growth; on the latter substrate growth is inhibited, α1 is excluded from focal contacts, and ERK1/2 phosphorylation is diminished.[70] The significance of this in vivo is not yet clear, but one possibility is that as collagen fibrillogenesis proceeds the higher affinity α2 displaces α1, and that the α1, no longer in focal contacts, ceases to provide a proliferative signal, permitting the fibroblast to quiescence as the matrix matures.

The α1 cytoplasmic domain is very short, and other than Shc activation, no signalling functions have been uniquely ascribed to α1. However, the α1 tail is required for α1β1 migration into focal adhesions,[71] and has a role in binding cytoskeletelal components,[72] pp125FAK, and phospholipase C gamma.[73] As these features are not unique to the α1 tail, deficiency in FAK activation or in focal adhesion formation has not been noted in cells from the α1 null animal.

Integrin α1β1, Fibroblasts, and Collagen and Collagenase Regulation

Many studies have shown that α1β1 is a negative feedback regulator of collagen synthesis by fibroblasts. These were initiated by Langholtz et al, who showed that an activating antibody to α1 accentuated the normal downregulation of collagen synthesis seen when fibroblasts are suspended in collagen gels.[74] It was also noted that α1 levels appeared to be reduced on scleroderma fibroblasts, in conjunction with their upregulation of collagen synthesis.[75] Data from the α1 null mouse lent strong support to this role: in vivo the mice show a 20% increase in the rate of collagen incorporation into the skin, and fibroblasts from these animals are deficient in downregulating synthesis in response to gel suspension.[76] Encouraged by this data, we subsequently examined keloids to determine whether loss of α1 could account for the increased collagen expression in these lesions.[8] A high proportion of lesional fibroblasts expressed α1 (in contrast to scleroderma lesions), although the levels expressed were somewhat lower than seen in chronic wounds with low collagen production. Thus, absence of α1 could not account for the excess collagen production in keloids, but there may be a relative deficiency compared to normal wounds, which could not be examined in the study for technical reasons.

The mechanism for downregulation of collagen synthesis mediated by α1 has been at least partially dissected. α1β1 stimulation by ligand activates the MAP kinases Erk1 and 2 via Fyn and Shc,[61,77] and Erk1/2 activation reduces collagen synthesis.[78] Reciprocally, we have found that the Erk1/2 inhibitor PD98059 causes upregulation of fibroblast collagen synthesis (Pozzi and Gardner, unpublished). This is the reverse of the effect of α2β1 stimulation, which activates p38 and causes induction of collagen synthesis.[79,80] Thus, α1b1 and α2β1 are opponents in their effect on collagen synthesis, the former inhibitory and the latter activatory.

The regulation of metalloproteases has similar themes but appears more complex. Firstly, the structure and function of the mouse and human collagenases is not congruent: MMP1, which is the major fibroblast collagenase in humans, is upregulated by Erk1/2 activation, but the mouse MMP1 structural equivalents, McolA and McolB, are not seen in skin fibroblasts, and have a restricted expression in the placenta and uterus.[81] On the other hand human MMP13, found in chronic ulcers,[82] is downregulated by Erk1/2 activation as well as being upregulated by p38 activation,[80] and MMP13 is the major fibroblast collagenase in mouse.[83] Although functionally equivalent to human MMP1, mouse MMP13 appears to be regulated like human MMP13, as it is markedly upregulated in α1 null mice where there is loss of Erk1/2 signalling but normal α2-p38 signalling. The α1 null animal shows an increase in expression of several other MMPs, including 7, 9 and 2 in endothelial cells, and 9 and 2 in fibroblasts.[76,84] For want of other evidence, this may be attributed to reduced Erk1/2 activation. However, whereas α1 stimulation is always inhibitory to collagen synthesis, it is sometimes activatory to MMP synthesis. In some systems integrin α1β1 activation by laminin-1[85] or by collagen IV (Pozzi and Gardner, unpublished) causes an increase in MMP synthesis. This may be dependent on a preexisting pro-MMP signal from α2 activation, and is presumably unrelated to α1 mediated Erk1/2 activation (as the latter is inhibitory to at least some MMPs), but the pathway has not been elucidated.

In many studies of fibroblast-collagen interaction, the complex process of collagen gel contraction is addressed. In dermal fibroblasts integrin α2β1 is seen to be the dominant player in this process,[74] which can be uncoupled from MMP synthesis,[86] and is dependent on a functional cytoskeleton. However, in studies of specialised cardiac fibroblasts,[87] smooth muscle cells,[88] stellate cells[9] and mesangial cells,[89] α1 blockade has been shown to prevent gel contraction, as has integrin αvβ3 blocklade in other cell types.[90] It is possible that whereas α2 is structurally more suited to gel contraction (having a higher affinity for collagen I), α1 expression may be required for maintenance of the contractile myofibroblastoid phenotype. It is striking that α1 expression is upregulated in vivo in all activated contractile myofibroblastoid cells including myofibroblasts in wound repair, mesangial cells, pericytes, myoepithelial cells,[91] and hepatic stellate cells.

In summary, there may be several roles for α1 and its interplay with α2 in the fibroblast during dermal wound healing and other episodes of mesenchymal repair. α1 upregulation in fibroblasts contributes to collagen stimulated cell proliferation, and probably to the myofibroblast transition. α2 is the major contributor to the synthetic phenotype, where it contributes the major part of collagen matrix contraction and activates collagen synthesis, as well as activating MMP synthesis for matrix remodelling. α1 fine tunes the MMP response, possibly providing

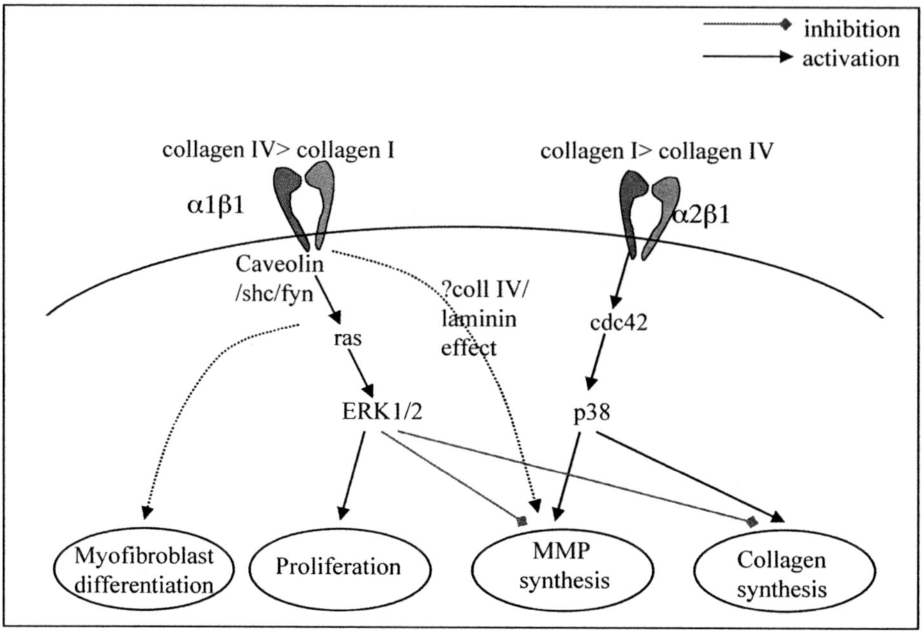

Figure 1. Interplay between integrins α1β1 and α2β1 in fibroblasts.

general inhibition of MMP release, but allowing for specific activation near the epidermal boundary where there is a greater abundance of the α1 high affinity ligand, collagen IV. α1 also provides feedback inhibition against excessive collagen synthesis. Consistent with these suggestions, the α1 null shows excessive collagen and collagenase synthesis at overlapping phases of wound healing,[92] and collagen fibrils are densely aggregated and irregular in the dermis of the α1 null, while being individually smaller (Gardner, unpublished observations). The role of integrin α1 in the fibroblast is summarized in Fig. 1.

Integrin α1β1 and Angiogenesis

Immunohistochemical analysis of murine and human tissue shows that α1 is present on at least some normal microvascular endothelium. α1 has also been shown to be upregulated on endothelia in MS lesions.[93] New tumor microvessels appear always to express α1, while a smaller proportion of them, predominantly the slightly larger ones, also express α2.[84] Senger et al have shown that vascular endothelial growth factor (VEGF/VPF) can induce α1 on endothelial cells, and that, not surprisingly, as the only collagen receptors expressed, α1 and α2 are required for endothelial haptotaxis through collagen. They also observe that antibodies to α1 and α2 reduce angiogenesis in response to subcutaneously implanted gels of fibrin or collagen containing VEGF/VPF, or to tumor xenografts.[94,95] However, tumor matrix contains a great variety of alternative integrin ligands. As we have learned from fibroblasts, α1β1 can activate an Erk1/2 proliferation pathway mediated by Shc. α2β1 can also positively regulate the progression through the cell cycle in epithelial cells by non overlapping mechanisms.[96] Thus, α1 and α2 blockade in vivo may cause a simple reduction in endothelial proliferation. Indeed, Senger's work suggests that antibody treatment affects vessel diameter more severely than vessel number, consistent with a proliferative defect rather than a migration deficit in new sprouts. With this in mind, there is no deficency in normal vasculo- and angio-genesis in α1 null mice. Analysis of the null mice, however, revealed another, subtler role for α1 in angiogenesis.

Detailed analysis of endothelial cells and tumor vasculature in α1 null animals[84] led to independent verification of the significance of plasminogen fragments, the angiostatins,[97] in endothelial growth regulation. Pulmonary microvascular endothelial cells from α1 null mice grew poorly compared to wild type, regardless of the substratum on which they were grown. This growth deficiency could be completely rescued by frequent media change even if the cells were grown on collagen. Furthermore, media conditioned by α1 null endothelial cells was inhibitory to the growth of wild type cells. This suggested that α1 null endothelial cells generated a humoral growth inhibitory factor. The growth deficiency in α1 null endothelial cells was also corrected by antibodies to angiostatin, or growth in media containing serum from plasminogen null mice (from which no angiostatin could be generated) instead of fetal calf serum. Lastly, the growth deficiency could be rescued by MMP9 blockade. Analysis of conditioned medium from α1 null endothelial cells as well as plasma from wounded (but not unwounded) or tumor bearing α1 null mice showed an increase in MMP9 and angiostatin compared to wild type animals. These findings in endothelial cells were consistent with the increased MMP expression seen in α1 null fibroblasts, due to loss of α1-Erk1/2 inhibitory signalling with normal α2-p38 activatory signalling. The increased MMP9 released by the α1 null cells would then cleave plasminogen[98] to yield the endothelial inhibitor, angiostatin.

These ex vivo observations were paralleled in vivo, where α1 null mice, with higher plasma MMP9 and angiostatin levels, were less able to vascularize subcutaneous tumors than wild type, but their deficit could be reversed by oral treatment of the animals with the MMP9 inhibitor doxycycline, and consequent reduction of their angiostatin levels.[99] MMP9 levels in the vasculature were also shown to correlate inversely with tumor vascularization even in wild type mice. These studies showed that the interplay between α1 and α2 integrins has significant consequences in the vascular system. Thus, during vascular remodelling, upregulation of endothelial α1 and α2 occurs, and the balance between them regulates MMP release, and ultimately vessel number.

While plasminogen is an MMP9 target, and we showed that its cleavage product angiostatin was entirely responsible for endothelial growth inhibition in vitro, it is possible that other MMP targets might be of importance in this feedback system in vivo. These include the collagens themselves. In this regard the finding that a collagen NC1 domain is a ligand for α1 may be of significance. The NC1 domain of collagen IV α3, also known as tumstatin, causes endothelial cytostasis and blocks angiogenesis by binding to integrin αvβ3, and a similar mechanism appears to exist for α1 binding to the collagen IV(1) NC1 domain (arresten).[100] This might be an explanation for the presence of α1 on the luminal surface of endothelium, where it could act as a detector of collagen fragments released during remodelling, and provide negative feedback to angiogenesis. This suggestion begs a question: how might ligand binding to α1 inhibit endothelial growth when we know that α1 binding to collagen in fibroblasts causes activation of Erk1/2 via Shc? There is evidence that overexpression of α1 in mesangial cells is growth inhibitory,[68] and therefore signalling from α1 can be growth inhibitory. The recent work of Springer's group (eg[4]) strongly suggests that conformational changes induced by ligands in integrins may be of sufficient variety that different ligands induce different signals through the same integrin. Collagen and laminin binding to α1 appear to have different outcomes in at least one experimental system.[85] Thus, arresten may provide a growth inhibitory signal, whereas intact collagen IV provides a stimulatory one. The absence of both signals in the α1 null could explain why normal angiogenesis is unaltered in α1 null mice. The relationship of α1 to angiogenesis is summarized in Fig. 2.

The Mystery of Smooth Muscle

α1 is extremely abundant on smooth muscle, both visceral and vascular[101], and, in vivo, smooth muscle expresses no other collagen binding integrin (explanted smooth muscle rapidly

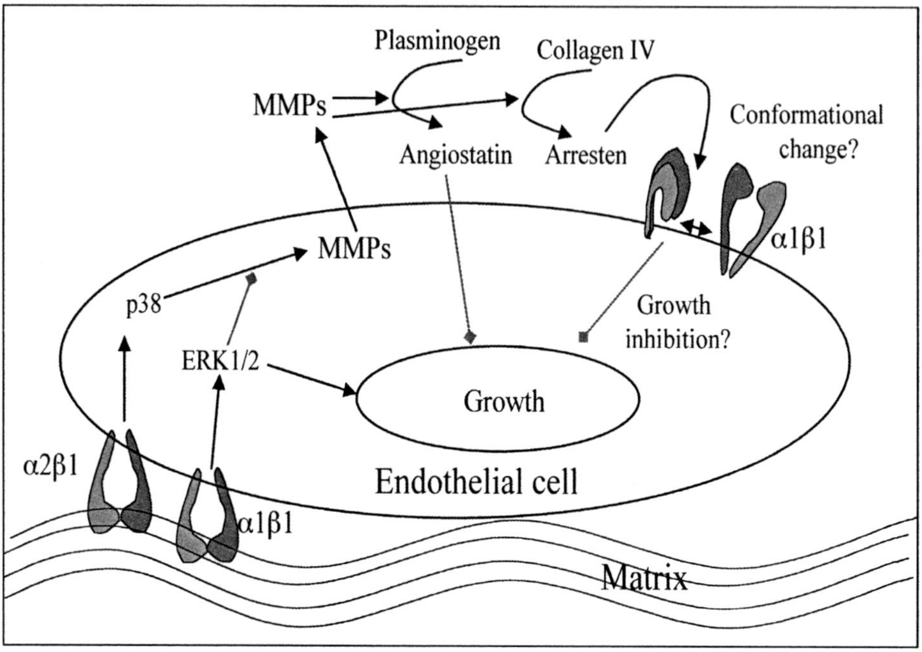

Figure 2. Integrin α1 and angiogenesis.

upregulates α2, complicating studies[102]). Furthermore, smooth muscle basal lamina has abundant collagen IV. There is no upregulation of α2 or α10 in the α1 null smooth muscle in vivo, as assessed by immunostaining.[63] Yet in the α1 null mouse digestion and parturition is entirely normal, and EM studies reveal no alterations in smooth muscle structure. Ongoing studies have detected some subtle changes in the tone and responsiveness of the mesenteric vascular bed, but this seems to be a very subtle phenotype for the lack of such an abundant structural component. Integrin α8, a fibronectin receptor, is also abundant in smooth muscle, but the double knockout α1/α8 animal also had histologically normal smooth muscle (Gardner and Brandenburger, unpublished). Further collagen binding integrin double knockouts may reveal the answer to this mystery.

Integrin α1β1 and the Immune System

The extensive role of α1 in the immune system is discussed in detail in Chapter 12. It is therefore only very briefly summarized here.

Integrin α1 was first discovered as a very late antigen on cultured T cells, and being the largest of the α subunits, was named Very Late Antigen 1 (VLA1), a name which persists in immunological studies. Hemler et al subsequently showed that VLA1 was present on a large proportion of T cells in the joints of rheumatoid arthritics, but was almost absent from the circulation, giving a first clue to a role for α1 in tissue migration and T cell activation. More detailed study of the immune system revealed that α1 is also expressed on a subset of NK-T cells as well as populations of activated monocytes and NK cells.

α1-deficiency induced by knockout or antibody blockade has dramatic consequences in the immune system. The α1-deficient mice show no overt immunodeficiency, but they show resistance to many different disease models which involve monocyte function or peripheral T cell memory. These include a resistance to anti collagen II antibody induced and mycobacterium

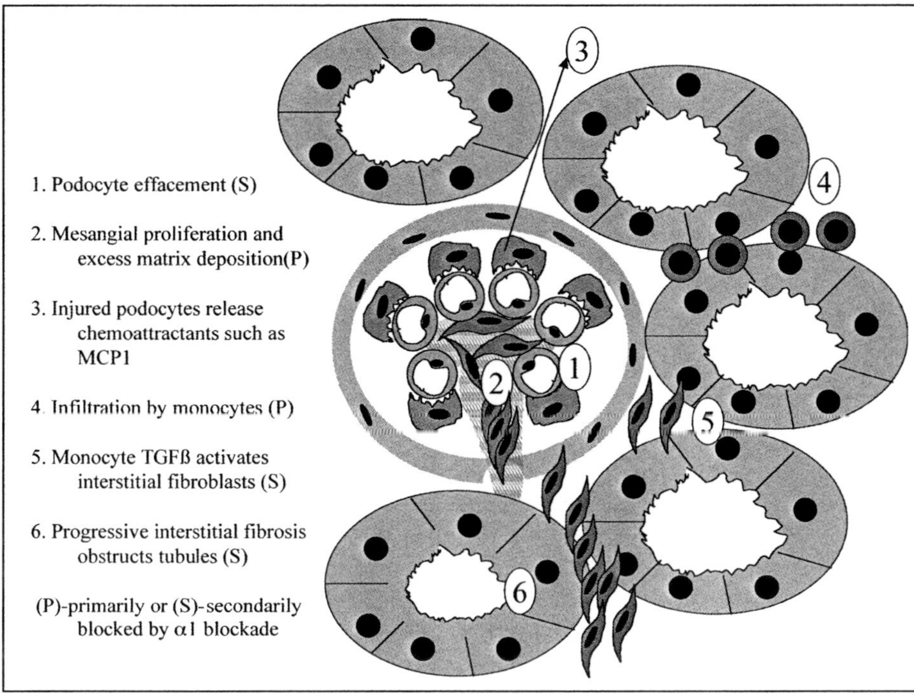

Figure 3. Integrin α1 deficiency rescues Alport's syndrome.

induced arthritis,[19,103] colitis, DTH, contact hypersensitivity, and LCMV induced encephalopathy. Inflamed tissues in these models, as well as the normal gut mucosal epithelium,[67] show reduced infiltration by T cells and monocytes. Furthermore, cultured splenocytes from α1 null animals show reduced proliferation in response to collagen, and fail to express integrin α2 upon long term culture. Taken together, the results suggest that α1 is needed both for lymphocyte migration into the collagen rich periphery, and for the proliferation of activated T cells in those locations, or their long term survival as mediators of peripheral T cell memory.[104]

Integrin α1β1 and the Kidney

Expression of α1 by glomerular mesangial cells[2,105] as well as the developing kidney[29] led to a great deal of interest in the role of this integrin in the kidney. Surprisingly, α1 null mice showed no functional or anatomic renal abnormality. In order to challenge renal function in the α1 null, we crossed it with the collagen IV α3 chain null (COL4A3/Alports) mouse.[106] Reduced glomerular basement membrane stiffness in the COL4A3 null leads to a progressive glomerulonephritis with mesangial expansion and secondary tubulointerstitial fibrosis. To our amazement, the double null animal lived twice as long as the COL4A3 null, due to a delay in the progression of renal failure. Subsequent analysis has revealed two independent protective effects of α1 deficiency in the kidney.

Firstly, in the normal progression of murine Alports, there is a marked influx of monocytes into the interstitium in response to glomerular epithelial damage. α1 null monocytes are defective in migrating into the renal interstitium, and are therefore reduced in number in the double null kidney. This reduces delivery of TGFβ to the kidney, delaying the onset and progression of interstitial fibrosis.[107] Reduction of monocyte migration is a major beneficial effect of $\alpha 1$ blockade or deficiency in a variety of inflammatory diseases, as discussed above. Secondly, mesangium appears to be dependent on $\alpha 1$ for proliferation, contraction, and matrix accumulation. In the $\alpha 1$/COL4A3 double null, the mesangial expansion is greatly reduced.[106] The importance of $\alpha 1$ in mesangial expansion has recently been explored in another glomerulonephritis model, anti-Thy-1 GN in the rat.[89] Here, direct injection of anti-$\alpha 1$ in the renal artery caused a marked reduction in mesangial proliferation and matrix accumulation, an important in vivo validation of a series of studies of the role of $\alpha 1$ in mesangial cells.[39,68,108] The renal effects of $\alpha 1$ blockade are shown in (Fig. 3). Some preliminary data suggests that $\alpha 1$ blockade is potentially valuable in cirrhosis through similar mechanisms.

Summary and Prospects

Integrin $\alpha 1$ has major roles as a modulator of mesenchymal proliferation and differentiation, matrix turnover, and immune function. Its roles in the immune system make it a clear target for therapy. In its biochemical properties, $\alpha 1$ appears to have a unique role in binding basement membrane collagens, the significance of which in vivo is not yet clear. Like the other collagen binding I domain containing integrins, $\alpha 2$, $\alpha 10$ and $\alpha 11$, its absence is not associated with major structural deficits in the mouse. Crosses to generate multiple knockouts of these integrins are eagerly awaited, to discern the functional overlaps between them.

References

1. Hemler ME, Glass D, Coblyn JS et al. Very late activation antigens on rheumatoid synovial fluid T lymphocytes. Association with stages of T cell activation. J Clin Invest 1986; 78(3):696-702.
2. Hemler ME, Sanchez-Madrid F, Flotte TJ et al. Glycoproteins of 210,000 and 130,000 m.w. on activated T cells: cell distribution and antigenic relation to components on resting cells and T cell lines. J Immunol 1984; 132(6):3011-8.
3. Beglova N, Blacklow SC, Takagi J et al. Cysteine-rich module structure reveals a fulcrum for integrin rearrangement upon activation. Nat Struct Biol 2002; 9(4):282-7.
4. Shimaoka M, Takagi J, Springer TA. Conformational regulation of integrin structure and function. Annu Rev Biophys Biomol Struct 2002; 31:485-516.
5. Ma Q, Shimaoka M, Lu C et al. Activation-induced conformational changes in the I domain region of lymphocyte function-associated antigen 1. J Biol Chem 2002; 277(12):10638-41.
6. Nishida W NM, Mori S, Takahashi M et al. A triad of serum response factor and the GATA and NK families governs the transcription of smooth and cardiac muscle genes. J Biol Chem 2002; 277(9):7308-17.
7. Soligo D SR, Luksch R, Manara G et al. Expression of integrins in human bone marrow. Br J Haematol 1990; 76(3):323-32.
8. Szulgit G RR, Wandel A, Tenenhaus M et al. Alterations in fibroblast alpha1beta1 integrin collagen receptor expression in keloids and hypertrophic scars. J Invest Dermatol 2002; 118(3):409-15.
9. Racine-Samson L, Rockey DC, Bissell DM. The role of alpha1beta1 integrin in wound contraction. A quantitative analysis of liver myofibroblasts in vivo and in primary culture. J Biol Chem 1997; 272(49):30911-7.
10. Mendrick DL KD, duMont SS, Sandstrom DJ. Glomerular epithelial and mesangial cells differentially modulate the binding specificities of VLA-1 and VLA-2. Lab Invest 1995; 72(3):367-75.
11. Gronthos S SP, Graves SE, Robey PG. Integrin-mediated interactions between human bone marrow stromal precursor cells and the extracellular matrix. Bone 2001; 28(2):174-81.
12. Deschaseaux F CP. Human marrow stromal precursors are alpha 1 integrin subunit-positive. J Cell Physiol 2000; 184(3):319-25.
13. Loeser RF, Carlson CS, McGee MP. Expression of beta 1 integrins by cultured articular chondrocytes and in osteoarthritic cartilage. Exp Cell Res 1995; 217(2):248-57.
14. Camper L, Hellman U, Lundgren-Akerlund E. Isolation, cloning, and sequence analysis of the integrin subunit alpha10, a beta1-associated collagen binding integrin expressed on chondrocytes. J Biol Chem 1998; 273(32):20383-9.

15. Wu JE, Santoro SA. Complex patterns of expression suggest extensive roles for the alpha 2 beta 1 integrin in murine development. Dev Dyn 1994; 199(4):292-314.
16. Stewart HJ TD, Jessen KR, Mirsky R. Expression and regulation of alpha1beta1 integrin in Schwann cells. J Neurobiol 1997; 33(7):914-28.
17. Duband JL, Belkin AM, Syfrig J et al. Expression of alpha 1 integrin, a laminin-collagen receptor, during myogenesis and neurogenesis in the avian embryo. Development 1992; 116(3):585-600.
18. Hemler ME, Jacobson JG, Brenner MB et al. VLA-1: a T cell surface antigen which defines a novel late stage of human T cell activation. Eur J Immunol 1985; 15(5):502-8.
19. de Fougerolles AR SA, Nickerson-Nutter CL, Chi-Rosso G et al. Regulation of inflammation by collagen-binding integrins alpha1beta1 and alpha2beta1 in models of hypersensitivity and arthritis. J Clin Invest 2000; 105(6):721-9.
20. Defilippi P, van Hinsbergh V, Bertolotto A et al. Differential distribution and modulation of expression of alpha 1/beta 1 integrin on human endothelial cells. J Cell Biol 1991; 114(4):855-63.
21. Gullberg D TD, Borg TK, Terracio L et al. Different beta 1-integrin collagen receptors on rat hepatocytes and cardiac fibroblasts. Exp Cell Res 1990; 190(2):254-64.
22. Loster K VS, Heidrich C, Hofmann W et al. Cell-collagen adhesion is inhibited by monoclonal antibody 33.4 against the rat alpha 1-integrin subunit. Exp Cell Res 1994; 212(1):155-60.
23. Stamatoglou SC BS, Johansson S, Forsberg E et al. Affinity of integrin alpha 1 beta 1 from liver sinusoidal membranes for type IV collagen. 1 FEBS Lett 1991; 288(1-2):241-3.
24. Bhat KP PJ. Resveratrol exhibits cytostatic and antiestrogenic properties with human endometrial adenocarcinoma (Ishikawa) cells. Cancer Res 2001; 61(16):6137-44.
25. Sutherland AE, Calarco PG, Damsky CH. Developmental regulation of integrin expression at the time of implantation in the mouse embryo. Development 1993; 119(4):1175-86.
26. Damsky CH, Librach C, Lim KH et al. Integrin switching regulates normal trophoblast invasion. Development 1994; 120(12):3657-66.
27. Terracio L, Rubin K, Gullberg D et al. Expression of collagen binding integrins during cardiac development and hypertrophy. Circ Res 1991; 68(3):734-44.
28. Hertle MD, Adams JC, Watt FM. Integrin expression during human epidermal development in vivo and in vitro. Development 1991; 112(1):193-206.
29. Korhonen M, Ylanne J, Laitinen L et al. The alpha 1-alpha 6 subunits of integrins are characteristically expressed in distinct segments of developing and adult human nephron. J Cell Biol 1990; 111(3):1245-54.
30. Schadendorf D GC, Haney U, Ostmeier H et al. Tumour progression and metastatic behaviour in vivo correlates with integrin expression on melanocytic tumours. J Pathol 1993; 170(4):429-34.
31. Schadendorf D FI, Makki A, Alijagic S et al. Metastatic potential of human melanoma cells in nude mice—characterisation of phenotype, cytokine secretion and tumour-associated antigens. Br J Cancer 1996; 74(2):194-9.
32. Gilhar A UY, Kalish RS, Berkutski T et al. Favorable melanoma prognosis associated with the expression of the tumour necrosis factor receptor and the alpha1beta1 integrin: a preliminary report. Melanoma Res 1997; 7(6):486-95.
33. Mechtersheimer G, Barth T, Quentmeier A et al. Differential expression of beta 1 integrins in nonneoplastic smooth and striated muscle cells and in tumors derived from these cells. Am J Pathol 1994; 144(6):1172-82.
34. Koukoulis GK WW, Virtanen I, Gould VE. Immunolocalization of integrins in the normal lung and in pulmonary carcinomas. Erratum in: Hum Pathol 1997 Dec; 28(12):1442. Hum Pathol 1997; 28(9):1018-25.
35. Sterry W, Mielke V, Konter U et al. Role of beta 1-integrins in epidermotropism of malignant T cells. Am J Pathol 1992; 141(4):855-60.
36. Heino J, Ignotz RA, Hemler ME et al. Regulation of cell adhesion receptors by transforming growth factor- beta. Concomitant regulation of integrins that share a common beta 1 subunit. J Biol Chem 1989; 264(1):380-8.
37. Pirila L, Heino J. Altered integrin expression in rheumatoid synovial lining type B cells: in vitro cytokine regulation of alpha 1 beta 1, alpha 6 beta 1, and alpha v beta 5 integrins. J Rheumatol 1996; 23(10):1691-8.

38. Gailit J XJ, Bueller H, Clark RA. Platelet-derived growth factor and inflammatory cytokines have differential effects on the expression of integrins alpha 1 beta 1 and alpha 5 beta 1 by human dermal fibroblasts in vitro. J Cell Physiol 1996; 169(2):281-9.
39. Kagami S KS, Loster K, Reutter W et al. Alpha1beta1 integrin-mediated collagen matrix remodeling by rat mesangial cells is differentially regulated by transforming growth factor-beta and platelet-derived growth factor-BB. J Am Soc Nephrol 1999; 10(4):779-89.
40. Eble JA, Golbik R, Mann K et al. The alpha 1 beta 1 integrin recognition site of the basement membrane collagen molecule [alpha 1(IV)]2 alpha 2(IV). Embo J 1993; 12(12):4795-802.
41. Emsley J, King SL, Bergelson JM et al. Crystal structure of the I domain from integrin alpha2beta1. J Biol Chem 1997; 272(45):28512-7.
42. Leikina E, Mertts MV, Kuznetsova N et al. Type I collagen is thermally unstable at body temperature. Proc Natl Acad Sci USA 2002; 99(3):1314-8.
43. Rich RL DC, Owens RT, Carson M et al. Trench-shaped binding sites promote multiple classes of interactions between collagen and the adherence receptors, alpha(1)beta(1) integrin and Staphylococcus aureus cna MSCRAMM. J Biol Chem 1999; 274(35):24906-13.
44. Xu Y GS, Rich RL, Owens RT et al. Multiple binding sites in collagen type I for the integrins alpha1beta1 and alpha2beta1. J Biol Chem 2000; 275(50):38981-9.
45. Knight CG ML, Peachey AR, Tuckwell DS et al. The collagen-binding A-domains of integrins alpha(1)beta(1) and alpha(2)beta(1) recognize the same specific amino acid sequence, GFOGER, in native (triple-helical) collagens. J Biol Chem 2000; 275(1):35-40.
46. Calderwood DA, Tuckwell DS, Eble J et al. The integrin alpha1 A-domain is a ligand binding site for collagens and laminin. J Biol Chem 1997; 272(19):12311-7.
47. Nykvist P TH, Ivaska J, Kapyla J et al. Distinct recognition of collagen subtypes by alpha(1)beta(1) and alpha(2)beta(1) integrins. Alpha(1)beta(1) mediates cell adhesion to type XIII collagen. J Biol Chem 2000; 275(11):8255-61.
48. Tulla M PO, Viitasalo T, Kapyla J et al. Selective binding of collagen subtypes by integrin alpha 1I, alpha 2I, and alpha 10I domains. J Biol Chem 2001; 276(51):48206-12.
49. Tawil NJ HM, Blacher R, Esch F et al. Alpha 1 beta 1 integrin heterodimer functions as a dual laminin/collagen receptor in neural cells. Biochemistry 1990; 29(27):6540-4.
50. Zhang Z, Tarone G, Turner DC. Expression of integrin alpha 1 beta 1 is regulated by nerve growth factor and dexamethasone in PC12 cells. Functional consequences for adhesion and neurite outgrowth. J Biol Chem 1993; 268(8):5557-65.
51. Colognato-Pyke H, O'Rear JJ, Yamada Y et al. Mapping of network-forming, heparin-binding, and alpha 1 beta 1 integrin-recognition sites within the alpha-chain short arm of laminin-1. J Biol Chem 1995; 270(16):9398-406.
52. Briesewitz R, Epstein MR, Marcantonio EE. Expression of native and truncated forms of the human integrin alpha 1 subunit. J Biol Chem 1993; 268(4):2989-96.
53. Tomaselli KJ, Doherty P, Emmett CJ et al. Expression of beta 1 integrins in sensory neurons of the dorsal root ganglion and their functions in neurite outgrowth on two laminin isoforms. J Neurosci 1993; 13(11):4880-8.
54. Perris R, Syfrig J, Paulsson M et al. Molecular mechanisms of neural crest cell attachment and migration on types I and IV collagen. J Cell Sci 1993; 106(Pt 4):1357-68.
55. Lallier T, Bronner-Fraser M. Inhibition of neural crest cell attachment by integrin antisense oligonucleotides. Science 1993; 259(5095):692-5.
56. Makihira S YW, Ohno S, Kawamoto T et al. Enhancement of cell adhesion and spreading by a cartilage-specific noncollagenous protein, cartilage matrix protein (CMP/Matrilin-1), via integrin alpha1beta1. J Biol Chem 1999; 274(16):11417-23.
57. Moiseeva EP SE, Baron JH, de Bono DP. Galectin 1 modulates attachment, spreading and migration of cultured vascular smooth muscle cells via interactions with cellular receptors and components of extracellular matrix. J Vasc Res 1999; 36(1):47-58.
58. Ochieng J L-BM, Warfield P. Regulation of cellular adhesion to extracellular matrix proteins by galectin-3. Biochem Biophys Res Commun 1998; 246(3):788-91.
59. Berditchevski F, Bazzoni G, Hemler ME. Specific association of CD63 with the VLA-3 and VLA-6 integrins. J Biol Chem 1995; 270(30):17784-90.

60. Brown E, Hooper L, Ho T et al. Integrin-associated protein: a 50-kD plasma membrane antigen physically and functionally associated with integrins. J Cell Biol 1990; 111(6 Pt 1):2785-94.
61. Wary KK, Mainiero F, Isakoff SJ et al. The adaptor protein Shc couples a class of integrins to the control of cell cycle progression. Cell 1996; 87(4):733-43.
62. Wary KK, Mariotti A, Zurzolo C et al. A requirement for caveolin-1 and associated kinase Fyn in integrin signaling and anchorage-dependent cell growth. Cell 1998; 94(5):625-34.
63. Gardner H, Kreidberg J, Koteliansky V et al. Deletion of integrin alpha 1 by homologous recombination permits normal murine development but gives rise to a specific deficit in cell adhesion. Dev Biol 1996; 175(2):301-13.
64. Pozzi A, Wary KK, Giancotti FG et al. Integrin alpha1beta1 mediates a unique collagen-dependent proliferation pathway in vivo. J Cell Biol 1998; 142(2):587-94.
65. Ekholm E, Hankenson KD, Uusitalo H et al. Diminished callus size and cartilage synthesis in alpha 1 beta 1 integrin-deficient mice during bone fracture healing. Am J Pathol 2002; 160(5):1779-85.
66. Jikko A HS, Chen D, Mendrick DL et al. Collagen integrin receptors regulate early osteoblast differentiation induced by BMP-2. J Bone Miner Res 1999; 14(7):1075-83.
67. Meharra EJ SM, Hassett D, Parker C et al. Reduced gut intraepithelial lymphocytes in VLA1 null mice. Cell Immunol 2000; 201(1):1-5.
68. Kagami S KS, Urushihara M, Loster K et al. Overexpression of alpha1beta1 integrin directly affects rat mesangial cell behavior. Kidney Int 2000; 58(3):1088-97.
69. Schwartz MA, Ginsberg MH. Networks and crosstalk: integrin signalling spreads. Nat Cell Biol 2002; 4(4):E65-8.
70. Schocklmann HO LS, Kralewski M, Hartner A et al. Distinct structural forms of type I collagen modulate cell cycle regulatory proteins in mesangial cells. Kidney Int 2000; 58(3):1108-20.
71. Briesewitz R, Kern A, Marcantonio EE. Ligand-dependent and -independent integrin focal contact localization: the role of the alpha chain cytoplasmic domain. Mol Biol Cell 1993; 4(6):593-604.
72. Loster K VD, Hofmann W, Reutter W et al. alpha1 Integrin cytoplasmic domain is involved in focal adhesion formation via association with intracellular proteins. Biochem J 2001; 356(Pt 1):233-40.
73. Vossmeyer D HW, Loster K, Reutter W et al. Phospholipase Cgamma binds alpha1beta1 integrin and modulates alpha1beta1 integrin-specific adhesion. J Biol Chem 2002; 277(7):4636-43.
74. Langholz O, Rockel D, Mauch C et al. Collagen and collagenase gene expression in three-dimensional collagen lattices are differentially regulated by alpha 1 beta 1 and alpha 2 beta 1 integrins. J Cell Biol 1995; 131(6 Pt 2):1903-15.
75. Ivarsson M, McWhirter A, Black CM et al. Impaired regulation of collagen pro-alpha 1(I) mRNA and change in pattern of collagen-binding integrins on scleroderma fibroblasts. J Invest Dermatol 1993; 101(2):216-21.
76. Gardner H BA, Pozzi A, Laato M et al. Absence of integrin alpha1beta1 in the mouse causes loss of feedback regulation of collagen synthesis in normal and wounded dermis. J Cell Sci 1999; 112 (Pt 3):263-72.
77. Pozzi A WK, Giancotti FG, Gardner HA. Integrin alpha1beta1 mediates a unique collagen-dependent proliferation pathway in vivo. J Cell Biol 1998; 142(2):587-94.
78. Reunanen N, Foschi M, Han J et al. Activation of extracellular signal-regulated kinase 1/2 inhibits type I collagen expression by human skin fibroblasts. J Biol Chem 2000; 275(44):34634-9.
79. Riikonen T, Westermarck J, Koivisto L et al. Integrin alpha 2 beta 1 is a positive regulator of collagenase (MMP-1) and collagen alpha 1(I) gene expression. J Biol Chem 1995; 270(22):13548-52.
80. Ravanti L, Heino J, Lopez-Otin C et al. Induction of collagenase-3 (MMP-13) expression in human skin fibroblasts by three-dimensional collagen is mediated by p38 mitogen- activated protein kinase [In Process Citation]. J Biol Chem 1999; 274(4):2446-55.
81. Balbin M, Fueyo A, Knauper V et al. Identification and enzymatic characterization of two diverging murine counterparts of human interstitial collagenase (MMP-1) expressed at sites of embryo implantation. J Biol Chem 2001; 276(13):10253-62.
82. Vaalamo M, Mattila L, Johansson N et al. Distinct populations of stromal cells express collagenase-3 (MMP-13) and collagenase-1 (MMP-1) in chronic ulcers but not in normally healing wounds. J Invest Dermatol 1997; 109(1):96-101.

83. Liu X, Wu H, Byrne M et al. A targeted mutation at the known collagenase cleavage site in mouse type I collagen impairs tissue remodeling. J Cell Biol 1995; 130(1):227-37.
84. Pozzi A MP, Miles LA, Wagner S et al. Elevated matrix metalloprotease and angiostatin levels in integrin alpha 1 knockout mice cause reduced tumor vascularization. Proc Natl Acad Sci USA 2000; 97(5):2202-7.
85. Lochter A NM, Werb Z, Bissell MJ. alpha1 and alpha2 integrins mediate invasive activity of mouse mammary carcinoma cells through regulation of stromelysin-1 expression. Mol Biol Cell 1999; 10(2):271-82.
86. Broberg A, Heino J. Integrin alpha2beta1-dependent contraction of floating collagen gels and induction of collagenase are inhibited by tyrosine kinase inhibitors. Exp Cell Res 1996; 228(1):29-35.
87. Carver W, Molano I, Reaves TA et al. Role of the alpha 1 beta 1 integrin complex in collagen gel contraction in vitro by fibroblasts. J Cell Physiol 1995; 165(2):425-37.
88. Gotwals PJ, Chi-Rosso G, Lindner V et al. The alpha1beta1 integrin is expressed during neointima formation in rat arteries and mediates collagen matrix reorganization. J Clin Invest 1996; 97(11):2469-77.
89. Kagami S, Urushihara M, Kondo S et al. Effects of anti-alpha1 integrin subunit antibody on anti-Thy-1 glomerulonephritis. Lab Invest 2002; 82(9):1219-27.
90. Cooke ME, Sakai T, Mosher DF. Contraction of collagen matrices mediated by alpha2beta1A and alpha(v)beta3 integrins. J Cell Sci 2000; 113(Pt 13):2375-83.
91. Koukoulis GK, Virtanen I, Korhonen M et al. Immunohistochemical localization of integrins in the normal, hyperplastic, and neoplastic breast. Correlations with their functions as receptors and cell adhesion molecules. Am J Pathol 1991; 139(4):787-99.
92. Gardner H, Broberg A, Pozzi A et al. Absence of integrin alpha1beta1 in the mouse causes loss of feedback regulation of collagen synthesis in normal and wounded dermis. J Cell Sci 1999; 112(Pt 3):263-72.
93. Sobel RA HJ, Maeda A, Chen M. Endothelial cell integrin laminin receptor expression in multiple sclerosis lesions. Am J Pathol 1998; 153(2):405-15.
94. Senger DR, Perruzzi CA, Streit M et al. The alpha(1)beta(1) and alpha(2)beta(1) integrins provide critical support for vascular endothelial growth factor signaling, endothelial cell migration, and tumor angiogenesis. Am J Pathol 2002; 160(1):195-204.
95. Senger DR CK, Benes JE, Perruzzi CA et al. Angiogenesis promoted by vascular endothelial growth factor: regulation through alpha1beta1 and alpha2beta1 integrins. Proc Natl Acad Sci USA 1997; 94(25):13612-7.
96. Klekotka PA, Santoro SA, Ho A et al. Mammary epithelial cell-cycle progression via the alpha(2)beta(1) integrin: unique and synergistic roles of the alpha(2) cytoplasmic domain. Am J Pathol 2001; 159(3):983-92.
97. O'Reilly MS, Holmgren L, Shing Y et al. Angiostatin: a novel angiogenesis inhibitor that mediates the suppression of metastases by a Lewis lung carcinoma. Cell 1994; 79(2):315-28.
98. Patterson BC, Sang QA. Angiostatin-converting enzyme activities of human matrilysin (MMP-7) and gelatinase B/type IV collagenase (MMP-9). J Biol Chem 1997; 272(46):28823-5.
99. Pozzi A LW, Gardner HA. Low plasma levels of matrix metalloproteinase 9 permit increased tumor angiogenesis. Oncogene 2002; 21(2):272-81.
100. Colorado PC, Torre A, Kamphaus G et al. Anti-angiogenic cues from vascular basement membrane collagen. Cancer Res 2000; 60(9):2520-6.
101. Belkin VM, Belkin AM, Koteliansky VE. Human smooth muscle VLA-1 integrin: purification, substrate specificity, localization in aorta, and expression during development. J Cell Biol 1990; 111(5 Pt 1):2159-70.
102. Skinner MP, Raines EW, Ross R. Dynamic expression of alpha 1 beta 1 and alpha 2 beta 1 integrin receptors by human vascular smooth muscle cells. Alpha 2 beta 1 integrin is required for chemotaxis across type I collagen-coated membranes. Am J Pathol 1994; 145(5):1070-81.
103. Ianaro A CC, Calignano A, Koteliansky V et al. Anti-very late antigen-1 monoclonal antibody modulates the development of secondary lesion and T-cell response in experimental arthritis. Lab Invest 2000; 80(1):73-80.
104. Dustin ML, de Fougerolles AR. Reprogramming T cells: the role of extracellular matrix in coordination of T cell activation and migration. Curr Opin Immunol 2001; 13(3):286-90.

105. Cosio FG SD, Nahman NS Jr. Cellular receptors for matrix proteins in normal human kidney and human mesangial cells. Kidney Int 1990; 38(5):886-95.
106. Cosgrove D RK, Meehan D, Miller C et al. Integrin alpha1beta1 and transforming growth factor-beta1 play distinct roles in alport glomerular pathogenesis and serve as dual targets for metabolic therapy. Am J Pathol 2000; 157(5):1649-59.
107. Sampson NS RS, Enke DA, Cosgrove D et al. Global gene expression analysis reveals a role for the alpha 1 integrin in renal pathogenesis. J Biol Chem 2001; 276(36):34182-8.
108. Kagami S UM, Kondo S, Loster K et al. Requirement for tyrosine kinase-ERK1/2 signaling in alpha 1 beta 1 integrin-mediated collagen matrix remodeling by rat mesangial cells. Exp Cell Res 2001; 268(2):274-83.

Chapter 3

Function of α2β1 Integrin

Mary M. Zutter and Samuel A. Santoro

The α2β1 integrin serves as an extracellular matrix receptor for collagen and /or laminin, as well as several other nonmatrix ligands.[1-3] Interactions between the α2β1 integrin and the extracellular matrix have been implicated in a number of biologic and pathobiologic processes.[4] The important role that the α2β1 integrin serves as a collagen receptor on platelets for platelet function and hemostasis is well established.[4-7] In some cell types such as endothelial cells and epithelial cells, a more activated form of the receptor may serve as both a collagen and laminin receptor.[3,8] Extensive in vitro studies on pulmonary, gastrointestinal and breast epithelia suggest that signals downstream of α2β1 integrin ligation contribute to branching morphogenesis.[9-12] Experiments using explant cultures and three-dimensional collagen gels have implicated the α2β1 integrin in keratinocyte migration and fibroblast contraction during wound healing.[13-15] The role of the α2β1 integrin in these multiple complex biologic processes will be discussed in this chapter.

Collagen Receptors

The α2β1 integrin is one of four integrins including the α1β1, α2β1, α10β1 and α11β1 integrins that serve as receptors for native collagens.[16-26] Work from our laboratory has focused primarily on the α2β1 integrin. While the α1β1 and α2β1 integrins have been the most extensively characterized, the α10β1 and α11β1 integrins were only recently identified.

All four alpha integrin subunits form covalently associated heterodimers only with the β1 subunit. The α1, α2, α10, and α11 subunits have many similarities in structure and ligand binding, including Mg^{2+} dependence for binding, and enhancement of integrin function by Mn^{2+}.[16,17,21] In addition, the four α subunits contain a highly conserved (inserted) I domain that, in large part, mediates ligand binding.

The α subunit I domain is an autonomously folding domain of approximately 220 amino acids.[27] The I domain is the critical determinant for ligand recognition and binding for the α1β1, α2β1 and α10β1 integrins. A similar function is expected for the α11 subunit I domain. The intergrin I domains are highly homologous to the A domain found in Von Willebrand factor, in cartilage matrix protein, in some collagen subtypes and in components of the complement system.

Although cells expressing either the α1β1, α2β1, α10β1 or α11β1 integrin adhere to several different collagens, they exhibit different relative preferences depending on the identity of the substrate. For instance, cells expressing the α1β1 integrin preferentially adhere to collagen IV, VI, and XIII, whereas cells expressing the α2β1 integrin preferentially bind to collagen I-III.[28] The binding of purified recombinant α integrin I domain to collagen type I or IV reflects the same relative affinity for this ligand as does the parent integrin.[29] Therefore, the differences in I domain structure probably account for the differential recognition of distinct collagen subtypes.

I Domains in Integrins, edited by Donald Gullberg. ©2003 Eurekah.com and Kluwer Academic / Plenum Publishers.

In addition to differences in collagen recognition, there are major differences in the expression pattern of the different collagen receptors. The $\alpha1\beta1$ integrin is primarily expressed in vivo by cells of mesenchymal origin including smooth muscle cells and neuronal cells.[30,31] In contrast, the $\alpha2\beta1$ integrin is primarily expressed in vivo by epithelial cells and platelet/megakaryocytes.[32] Both the $\alpha1\beta1$ and $\alpha2\beta1$ integrins are both expressed by activated T-cells, fibroblasts and some endothelial cells. Neither the $\alpha10\beta1$ or the $\alpha11\beta1$ integrin is expressed by epithelial cells or cells of hematopoietic origin.

In cell types where both the $\alpha1\beta1$ and $\alpha2\beta1$ integrins are expressed the $\alpha1\beta1$ and $\alpha2\beta1$ integrins appear to play distinct roles. Studies by several groups including our own have shown that the $\alpha1\beta1$ and $\alpha2\beta1$ integrins mediate different functions and are not simply redundant receptors.[11,33-36] When cultured in three-dimensional collagen gels, primary fibroblasts that express both the $\alpha1\beta1$ and $\alpha2\beta1$ integrins down-regulate collagen biosynthesis and up-regulate metalloproteinase expression. Ligation of the $\alpha1\beta1$ integrin results in down-regulation of collagen gene expression, whereas ligation of the $\alpha2\beta1$ integrin up-regulates matrix metalloproteinase-1 (MMP-1) gene expression.[33,37,38] As mentioned earlier, recent studies carried out in fibroblasts and endothelial cells demonstrate that the $\alpha1\beta1$ integrin promotes cell cycle progression and prevents apoptosis but the $\alpha2\beta1$ integrin fails to promote cell survival or cell cycle progression.[39,40] Reexpression of the $\alpha2\beta1$ integrin in a tumorigenic mammary epithelial cell line restored the ability of the cells to form glandular structures and diminished their malignant potential despite the fact that the cells expressed high levels of endogenous $\alpha1\beta1$ integrin.[11] In addition, the $\alpha2\beta1$ integrin supports proliferation of mammary epithelial cells when plated on monolayer type I collagen or embedded in three-dimensional collagen gels.[34]

The majority of earlier work defined the role and function of the $\alpha2\beta1$ integrin by studies of human platelets and in vitro models. These in vitro studies implicated the $\alpha2\beta1$ integrin in a wide range of biologic and pathobiologic functions including platelet adhesion required for hemostasis and thrombosis, epithelial differentiation and branching morphogenesis, tumor biology, wound healing, angiogenesis, and inflammation and immunity. A rigorous analysis of $\alpha2\beta1$ function in vivo was hindered by lack of a mouse model. Recent characterization of the $\alpha2$ subunit-deficient mouse has shed new insight on the role of the $\alpha2\beta1$ integrin. Our discussion of $\alpha2\beta1$ integrin function will begin with studies carried out in vitro. Additional information gained from the $\alpha2$ integrin-subunit deficient mouse will then be discussed.

Platelet Function

Many studies have suggested that the $\alpha2\beta1$ integrin is a critical mediator of platelet adhesion to collagen within the vessel wall after vascular injury and that the interactions of platelets with subendothelial collagen after vascular injury is required for proper hemostasis. Collagen not only serves as an adhesive substrate, but in its fibrillar form, also activates platelets leading to secretion of granule contents and platelet aggregation.[4] The critical role that the $\alpha2\beta1$ integrin plays in platelet adhesion to collagen substrates was initially established by three complementary sets of experiments and observations.

One avenue of investigation that delineated the role of the $\alpha2\beta1$ integrin as a platelet collagen receptor came from patient studies. Indeed, the initial suggestion that the $\alpha2\beta1$ integrin might serve as a platelet surface collagen receptor came from the description by Nieuwenhuis of a patient with a bleeding disorder.[41,42] The patient's platelets showed no response to collagen, but exhibited normal responsiveness to other platelet agonists. Electrophoretic analysis of the platelets revealed the deficiency of glycoprotein 1a ($\alpha2$ subunit). In addition, other patients with either reduced levels of platelet expression of the $\alpha2\beta1$ integrin or the presence of autoantibodies to the integrin have been described to exhibit impaired platelet activation by collagen but not by other agonists.

Second, purified human platelets adhere to collagens I-VIII via an α2β1 integrin-dependent mechanism that requires Mg^{++}.[1,16] Adhesion is inhibited by monoclonal inhibitory antibodies directed against the α2β1 integrin. Effective inhibition of platelet deposition onto collagen substrates under both static and flow conditions using whole blood demonstrates the importance of the α2β1 integrin for platelet deposition onto collagen substrates under physiologic conditions.[17]

Finally, purified α2β1 integrin incorporated into liposomes adheres to collagen substrate in a Mg^{2+}-dependent manner and exhibits the same collagen-type specificity as intact platelets.[17] The α2β1 integrin containing liposomes do not adhere to any other of several extracellular matrix proteins confirming the specificity of collagen adhesion via the α2β1 integrin platelet receptor.

Platelet adhesion to collagen is complex and the α2β1 integrin is not the only collagen receptor on platelets. The results of a series of in vitro and in vivo studies are consistent with a two-step, two-site model of collagen-induced platelet activation.[43,44] In this model the α2β1 integrin supports the initial rapid platelet-collagen interaction that mediates platelet adhesion to the vessel wall under conditions of flow. This allows a second low-affinity signal transducing co-receptor to bind collagen and mediate collagen-induced platelet activation and aggregation. Studies from several investigators suggest that glycoprotein VI (GPVI) serves the role as the signal transducing co-receptor.[45-50] In this model, the α2β1 integrin mediates strong adhesion but does not contribute to platelet activation.

The signaling pathways downstream of GPVI that lead to platelet activation have been extensively studied using GPVI-mimetics such as collagen related peptide (CRP) and snake venoms like convulxin that bind GPVI but not the α2β1 integrin (reviewed in ref. 43). GPVI, a member of the immunoglobulin superfamily, noncovalently but constitutively associates with the F_c receptor γ ($F_cR\gamma$) chain forming a multimeric signaling complex. Ligation of GPVI with collagen, collagen related peptide or crosslinking by antibody signals activation. The $F_cR\gamma$ contains an immunoreceptor type activation motif (ITAM), that is phosphorylated by a Src family kinase, either Fyn or Lyn, following platelet adhesion to collagen. Phosphorylation of the ITAM of the F_cR stimulates recruitment and activation of Syk. Syk phosphorlyation initiates a signaling cascade involving LAT-Gads-SLP-76 [Src homology 2 (SH2) domain-containing leukocyte protein of 76 kDa], downstream activation of PI3K and phospholipase Cγ2 ($PLC\gamma_2$) and ultimate granule release, activation of the $\alpha_{IIb}\beta_{IIIa}$ integrin, and platelet aggregation.

A discreet subset of signals is initiated by α2β1 integrin binding to collagen. Studies using intact collagen, snake venom, GPVI-deficient platelets, and/or inhibitory monoclonal antibodies suggest that the α2β1 integrin plays a role independent of GPVI in mediating signals downstream of platelet adhesion to collagen. Collagen-induced phosphorylation of $PLC\gamma_2$ and Syk is inhibited by antibodies that block α2β1 integrin adhesion to collagen[51] and jararhagin, a snake venom that selectively cleaves the β1 integrin subunit of the α2β1 integrin.[52] In contrast, phosphorylation of $PLC\gamma_2$ and Syk is maintained in platelets from F_cR γ-deficient mice and GPVI-deficient patients.[53,54] In other studies collagen-induced phosphorylation of c-Src is also maintained in GPVI-deficient human platelets.[55] Recent studies using the snake venom rhodocytin demonstrate that Src and Lyn constitutively associate with the α2β1 integrin and that Src kinase activity increases following rhodocytin stimulation of platelets. Src activation, presumably via the α2β1 integrin is accompanied by activation of the p130 Crk-associated substrate (Cas).[56] Additional signaling cascades are initiated via the α2β1 integrin following platelet adhesion to collagen under flow conditions. Platelet adhesion to collagen under flow conditions causes rapid tyrosine phosphorylation of pp125 focal adhesion kinase.[57] Other studies suggesting that the α2β1 integrin mediates signals distinct from GPVI come from comparing signals initiated by collagen versus GPVI-mimetics. Platelet adhesion to intact collagen stimulates a different response than platelet adhesion to GPVI-mimetics.[58,59] Although the role that signals downstream of $\alpha 2\beta 1$ integrin ligation play in platelet aggregation remains controversial, the signals downstream of $\alpha 2\beta 1$ integrin-ligation have been well elucidated in other cell types such as epithelial cells and fibroblasts.[36,60-62]

The important role that the $\alpha 2\beta 1$ integrin plays in mediating adhesion to type I collagen and thrombus formation is supported by epidemiologic data that links levels of $\alpha 2\beta 1$ integrin expression with pathologic thrombosis. Kunicki et al demonstrated that the number of $\alpha 2\beta 1$ integrin molecules per platelet correlates with the ability of platelets to adhere to Type I collagen.[63] In fact, the normal population expresses variable levels of the $\alpha 2\beta 1$ integrin on the platelet surface and these differences are associated with three alleles of the $\alpha 2$ integrin subunit gene.[64,65] The three $\alpha 2$ gene alleles have been defined by 8 nucleotide polymorphisms in the coding region of the $\alpha 2$ gene at nucleotide 807(C or T) and 873(G or A). Individuals carrying the 807T/873A allele express high levels of platelet $\alpha 2\beta 1$ integrin, whereas individuals carrying the 807C/873G allele exhibit low levels of $\alpha 2\beta 1$ integrin expression. Clinical and epidemiologic studies now demonstrate a direct clinical significance of these allelic differences in levels of $\alpha 2\beta 1$ integrin expression. High levels of $\alpha 2\beta 1$ integrin expression are associated with nonfatal myocardial infarction in individuals less than a mean age of 62 years, with an increased risk of developing diabetic retinopathy in patients with Type II diabetes mellitus, and with an increased risk of stroke.[66,67] To understand the phenomena, the epidemiologic studies must now be supported with studies using animal models.

Epithelial Differentiation and Branching Morphogenesis

The $\alpha 2\beta 1$ integrin is expressed at high levels on numerous epithelial cells including keratinizing and nonkeratinizing stratified squamous epithelium, ciliated columnar epithelium of the respiratory tract, the epithelial cells of the gastrointestinal tract and urinary tract, and the glandular epithelium of the breast.[32] Expression is pronounced at sites of terminal duct branching in the mammary gland and lung. A recurrent finding in our early studies was the association of the increased expression of the $\alpha 2\beta 1$ integrin with the orderly regulated proliferation of epithelial cells.

These observations prompted an investigation of $\alpha 2\beta 1$-integrin expression in malignancy using breast cancer as a model. This initial consideration revealed that the $\alpha 2\beta 1$-integrin was highly expressed in the epithelium of ducts and ductules of normal breast tissue.[32,68] Normal or close to normal levels of expression were observed in benign lesions such as fibroadenomas or papillomas. In contrast, markedly reduced or undetectable $\alpha 2\beta 1$-integrin expression was seen in poorly-differentiated carcinomas. In fact, $\alpha 2\beta 1$-integrin expression was diminished or lost in a manner that correlated with a loss of epithelial differentiation and tumor progression in mammary carcinoma as well as other adenocarcinomas, including those of the colon, prostate, lung, pancreas, and skin.[69-73] Review of the existing literature revealed that decreased expression of $\alpha 2\beta 1$ is the most common change of integrin expression in epithelial malignancies.[74]

The dramatic and consistent alterations in expression of the $\alpha 2\beta 1$ in epithelial malignancy that correlated with the differentiated epithelial phenotype suggested that $\alpha 2\beta 1$-integrin expression may contribute to maintenance of the differentiated epithelial phenotype and that altered $\alpha 2\beta 1$-expression may contribute to the malignant behavior of breast cancer cells. "Gain of function" and "loss of function" models, described by our group, as well as the laboratories of our colleagues defined the importance of $\alpha 2\beta 1$-integrin expression to the differentiated phenotype of mammary and other epithelial cells.[10,11,75]

Re-expression of the human $\alpha 2$ integrin cDNA in a poorly-differentiated murine mammary carcinoma cell line, that expressed no detectable $\alpha 2$ integrin subunit protein or mRNA, resulted in dramatic phenotypic alterations from the fibroblastoid, spindle-shaped, noncontact inhibited cell to an epithelioid, polygonal-shaped and contact-inhibited cell in culture. Although adhesion to collagen (mediated by the $\alpha 1\beta 1$-integrin) of the parental and control cells was comparable to the adhesion of the α2-tranfectants, only the α2-transfectants formed organized structures including alveolar-like and elongated multilayered duct-like structures in three-dimensional floating collagen gels. Re-expression of the α2β1 integrin by a poorly-differentiated breast carcinoma cell that expressed high levels of endogenous α1β1 not only restored epithelial morphogenetic capacity but also profoundly suppressed growth in soft agar and tumor development in vivo.[11] In two separate "loss of function" models, anti-sense mRNA experiments reduced α2β1-integrin expression by the well-differentiated estrogen-responsive breast cancer cell line T47D and the Maden-Darby kidney cell line MDCK. Loss of α2β1-integrin expression resulted in loss of the ability to undergo branching morphogenesis. These results further established a critical role for the α2β1 integrin in branching morphogenesis in three-dimensional collagen gels.[10,75]

The critical observation that α2β1-integrin expression is required for maintenance of the differentiated epithelial phenotype and glandular differentiation in vitro has been confirmed by studies from a number of investigators. Using a primary human nonmalignant, but immortalized, mammary epithelial cell line, Berdichsky et al and D'Souza et al demonstrated that branching morphogenesis can be blocked by inhibitory monoclonal antibodies directed against the α2 integrin subunit or by altered α2β1-integrin expression mediated by the expression of the *c-erbB2* proto-oncogene, respectively.[9,76] Studies from other investigators support a similar role for the α2β1 integrin in the morphologic differentiation of colonic epithelial cells.

The observation that reexpression of the α2β1 integrin in a tumorigenic cell line restored the ability of the cells to form glandular structures and diminished their malignant potential despite the fact that the cells expressed high levels of endogenous α1β1 integrin further supported the idea that the α1β1 and α2β1 integrins mediate different functions and are not simply redundant receptors.[11,33-36] In addition, the ability of the α2, but not the α1-integrin subunit to mediate branching morphogenesis suggested that the signals downstream of the α2β1 integrin were involved in the complex biology of branching and epithelial differentiation.

The unique contributions that the α1β1 and the α2β1 integrins play in cell phenotype and branching morphogenesis were evaluated using primary mammary epithelial cells that endogenously expressed the integrin or an immortalized, but nontransformed, mammary epithelial cell model lacking endogenous α1β1 and α2β1 collagen receptors but expressing either the full-length human α2 integrin subunit cDNA (X2C2), a chimeric cDNA consisting of the α2 extracellular and transmembrane domains fused to the α1 cytoplasmic domain (X2C1), or a cDNA encoding the α2 subunit with a cytoplasmic domain truncation after the conserved GFFKR motif (X2C0).[34,60] The endogenous murine α2β1 integrin on the primary mammary epithelial cells or the transfected human α2β1 integrin on the cell lines mediated effective adhesion, spreading, formation of focal adhesion complexes and proliferation on type I collagen matrices. The cytoplasmic domain of the α2 integrin subunit uniquely supported S-phase entry in the presence of insulin alone. The α2, but not the α1, cytoplasmic domain mediated activation of the cyclin E/cdk2 complex and entry into S-phase in the absence of growth factors other than insulin. These results suggest that the α2 integrin cytoplasmic domain and the insulin receptor synergize to regulate cell cycle progression.[60]

Branching morphogenesis requires not only cell proliferation but also cell migration. The α2, but not the α1, integrin cytoplasmic domain mediated signals, via the p38 MAP kinase pathway, that lead to a migratory phenotype.[61,62] Expression of the constitutively active small G protein Racl augmented p38 MAP kinase phosphorylation and migration of the X2C2 transfectants in the absence and presence of EGF. In addition, expression of constitutively active Racl by the X2C1 transfectants restored activation of the p38 MAP kinase pathway and a migratory phenotype. Thus, the activation of the p38 MAP kinase pathway that occurs upon adhesion to collagen is uniquely dependent on the α2-cytoplasmic domain, but not the α1-cytoplasmic domain, and is required for cell migration of mammary epithelial cells. The role of the $\alpha 2$-cytoplasmic domain in activation of the p38 MAP kinase pathway was also established in fibroblasts. Fibroblasts grown in three-dimensional collagen gels require the $\alpha 2$-cytoplasmic domain for p38 MAP kinase activation that leads to $\alpha 2\beta 1$ integrin-mediated up-regulation of collagen gene expression.[36] Together these results support an important and unique role for the $\alpha 2$-cytoplasmic domain in mediating activation of p38 MAP kinase activation.

Two separate signaling pathways downstream of the $\alpha 2\beta 1$-integrin cytoplasmic domain mediate chemotactic migration and entry into S-phase.[62] Distinct residues within the cytoplasmic domain are responsible for initiating the different signals that lead to either cell migration or cell proliferation. Mutagenesis of the $\alpha 2$ integrin cytoplasmic domain identified the precise residues that are required to mediate chemotactic migration in response to epidermal growth factor (EGF) or entry into the cell cycle on type I collagen matrices in response to insulin. In fact, the seven amino acids following the conserved GFFKR sequence are sufficient to recapitulate the phenotype of the full-length $\alpha 2$ integrin cytoplasmic domain in mammary epithelial cells (Fig. 1). Mutation of the methionine at position 1137 to alanine inhibited both chemotactic migration and entry into S-phase. Mutation of the tyrosine at position 1134 to alanine inhibited migration while it had only a modest effect on the ability of the cells to enter S-phase in the presence of insulin when adherent to type I collagen. Mutation of the lysine at position 1136 to alanine severely inhibited the ability of the mammary epithelial cells to enter S-phase while it had no effect on the ability of the cells to migrate in response to EGF on a type I collagen matrix. These results suggest that two residues separated by only two amino acids control dramatically differing phenotypes via distinct downstream signals.

The two critical residues may interact with different adaptor molecules. Secondary structure prediction indicates that the cytoplasmic domain of the $\alpha 2$ integrin subunit adopts an α-helical conformation across the amino acids FKRKYEKM, which contains all of the residues mutated in the study (Fig. 2).[77,78] Analysis of the mutations revealed that none of them were predicted to disrupt the helical nature of this region of the $\alpha 2$ cytoplasmic domain. Mapping the amino acids FKRKYEKM of the $\alpha 2$ integrin cytoplasmic domain to a helical wheel revealed that the conserved basic residues (K1131, R1132 and K1136) cluster on one face of the helix while the conserved hydrophobic residues (F1130, Y1134 and M1137) are clustered on the other face (Fig. 2). This arrangement suggests a model to explain how mutations at position 1134 can affect migration while mutations at position 1136 can affect entry into S-phase; they are on opposite faces of the helix where they presumably interact with different adapter molecules. The basic amino acids on one face of the helix result in a positive charge on that face that could mediate interactions with other cytoplasmic molecules (Figs. 1 and 2). This detailed structural and mutational analysis defined the multiple signaling pathways downstream of $\alpha 2\beta 1$ integrin ligation.

The $\alpha 2\beta 1$ Integrin and Wound Healing

In vitro studies of human skin and cell lines have suggested that the $\alpha 2\beta 1$ integrin is required for wound healing. In normal skin, the $\alpha 2\beta 1$ integrin is expressed at high levels on the keratinizing stratified squamous epithelium.[32] The high level of expression is most pronounced at the basal cell layer at the site where the keratinocytes interact with the basement membrane. The $\alpha 2\beta 1$ integrin is also expressed in the dermis by fibroblasts and endothelial cells.[32]

The infliction of a wound results in tissue injury that disrupts the epithelium. Wounding of human skin causes a major disruption in the epithelial cell program which stimulates epithelial migration in explant models. Expression of the $\alpha 2\beta 1$ integrin by the keratinocyte is re-oriented from the basal cell area to the forward-basal aspect of migrating keratinocytes where it is concentrated.[79] This redistribution of the keratinocyte $\alpha 2\beta 1$ integrin places the integrin in contact with type I collagen.[15] Keratinocyte migration on complex matrices containing type I collagen requires the catalytic activity of collagenase I (matrix metalloproteinase 1 [MMP-1].[14,80]

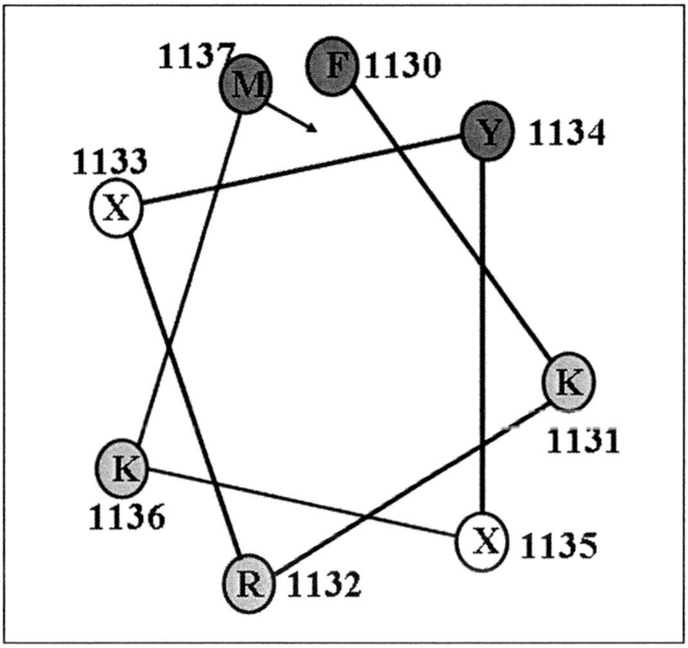

Figure 1. A model: The α2 integrin cytoplasmic domain plays a role in EGF-stimulated chemotactic migration and insulin-dependent G1-progression. Adaptor molecules interact with either a tyrosine or phenylalanine at the 1134 position of the 1139 truncated cytoplasmic domain to initiate a signaling pathway that in conjunction with signals from the EGF receptor results in chemotactic migration on type I collagen. This pathway either activates Rac or acts in a parallel pathway with one containing Rac that culminates in activation of p38 MAPK and migration. On the other side of the α2 cytoplasmic domain a lysine residue at position 1136 interacts with a different adaptor molecule and cooperates with insulin receptor to stimulate sustained phosphorylation of ERK, up-regulation of cyclin E and cdk2 and cell cycle progression.

Expression of MMP-1 is upregulated in basal keratinocytes by the interaction of the α2β1 integrin with type I collagen in the underlying dermis.[15] Keratinocyte migration is inhibited by anti-MMP-1 antibodies and by inhibitory antibodies against the α2β1 integrin.[14] These in vitro studies suggest that α2β1 integrin functions in coordination with MMP-1 to mediate keratinocyte re-epithelialization of cutaneous wounds.

Recent studies have demonstrated that in fact pro-MMP-1 specifically binds to the α2β1 integrin on keratinocytes.[81] Formation of such an α2β1 –pro – MMP-1 complex may provide a mechanism for increased concentration of the enzyme at the correct site to allow epidermal migration. In addition, MMP-1 binds to the α2β1 integrin via the I domain. These observations extend previous studies which have shown localization of matrix metalloproteinases and related enzymes to the cell surface in the appropriate location. Proteinases localized to the cell surface include MMP-2 by TIMP-2,[82,83] MMP-2 by αvβ3 integrin,[84,85] seprase by α3β1 integrin,[86] MMP-9 by CD44,[87,88] and MMP-7 by heparan sulfate proteoglycans.[89] Anchoring proteinases to the cell membrane may afford the cell precise control over both proteinase activity and location. Binding of pro-MMP-1 to a collagen receptor could result in an enzyme that is properly positioned for cleavage of its substrate. In addition, binding of pro-MMP-1 to the α2β1 integrin could result in activation of the enzyme, either by a second proteinase or by a conformational change of MMP-1 itself. In addition to regulating enzyme activity, binding of

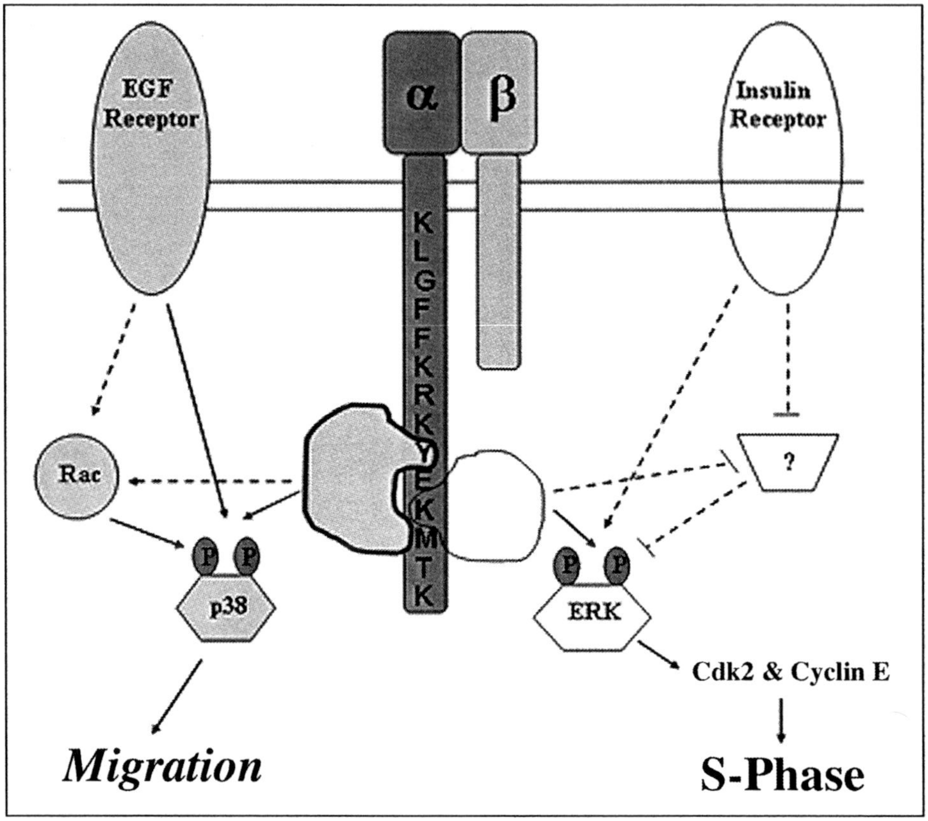

Figure 2. Diagram representing proposed α-helical conformation adopted by the α2 integrin cytoplasmic domain. The conserved residues from the α2 cytoplasmic domain (FKRxYxKM) were mapped on a helical wheel to approximate the position of the amino acids in an α-helix. The dark shaded residues are hydrophobic, the lightly shaded residues are basic, and the unshaded residues marked 'x' are not conserved.

MMP-1 to the α2β1 integrin may allow the cells to carefully control proteinase expression levels in their microenvironment. Ligation of the α2β1 integrin by collagen induces expression of pro-MMP-1 in migrating keratinocytes.[14] When MMP-1 cleaves type I collagen at physiological temperature, the triple helical molecule denatures. Denatured collagen is no longer a high affinity ligand for the α2β1 integrin; consequently, migration and signaling are turned off.[90,91] Thus cells have a mechanism to coordinate and integrate the expression of pro-MMP-1 and the α2β1 integrin, the amount of type I collagen in the matrix, and the activity of MMP-1. Because the enzyme is surface-bound, diffusion of the active enzyme within the extracellular matrix may be low, facilitating focal cleavage of collagen, as well as allowing each cell to sense its microenvironment and adjust its MMP-1 levels independently of its neighbors. Expression of MMP-1 may also be sensitive to α2β1 integrin levels, in terms both of ligated integrin available to signal and of integrin available to orient/activate the enzyme.

In the late phase of wound healing after reepithelialization tissue contraction of collagen fibers results in a strengthened scar. Three-dimensional extracellular matrix culture systems have been used to simulate the in vivo interactions between cells such as fibroblasts and the extracellular matrix environment. Fibroblasts embedded in a lattice consisting mainly of type I collagen can exert forces sufficient to contract the hydrated lattice into a densely organized structure resembling the dermis. The alignment of the collagen fibers that occurs in healing wounds is recapitulated in three-dimensional collagen gels. This process is accompanied by a fundamental reprogramming of fibroblast morphology and metabolism. The interaction of α2β1 integrin on fibroblasts with three-dimensional collagen results in unique signals that lead to altered expression of genes required for tissue remodeling including the attenuation of cellular response to growth factors,[13,92] induction of both MMP-1[93] and the collagen receptor α2 integrin subunit,[94] and modulation of the platelet derived growth factor (PDGF) effect on integrin receptor expression.[13] Therefore the in vitro models suggested critical roles for the α2β1 integrin in wound healing.

The α2β1 Integrin and Regulation of Inflammation

The α2β1 integrin was initially identified as an integrin expressed at very late stages of T cell activation, thus the designation very late activation antigen-2 (VLA-2).[95,96] Recent evidence suggests that the α2β1 integrin is expressed on a variety of cells of the inflammatory and hematopoietic system and may play an important role in leukocyte adhesion and extravasation from the vasculature and into peripheral tissues. Consistent with this initial description, activated T cells which are identified infiltrating in a variety of chronic inflammatory settings express the α2β1 integrin; naïve T cells fail to express the integrin.[97] TCR mediated proliferation and cytokine secretion are enhanced by α2β1-dependent adhesion to collagen.[98] In addition ligand binding of the α2β1 integrin by T cells inhibits fas ligand expression and reduces cell death.[99] In animals, inhibitory monoclonal antibodies directed against the α2β1 integrin significantly inhibit the effector phase of both contact and delayed type hypersensitivity. These results established an important role for the α2β1 integrin in T cell mediated function.[97]

In addition to its expression on activated T cells, the α2β1 integrin but not the α1β1 integrin is expressed at high levels on almost all NK cells, and on subpopulations of monocytes and neutrophils. Arase et al identified the NK cell recognition epitope of the widely used DX5 pan-NK cell monoclonal antibody as CD49b or the α2β1 integrin.[100] These investigators demonstrated that α2β1-expressing and nonexpressing subsets of NK cells are present in the mouse spleen and raised the possibility that α2β1 integrin expression is important in NK cell function. The role of the α2β1 integrin on subsets of neutrophils and monocytes has been evaluated but not firmly established. One study found expression of the α2β1 integrin on extravasated neutrophils in human skin blister chambers and in the rat peritoneal cavity following chemotactic stimulation.[101] These studies, as well as others, suggested that the α2β1 integrin on neutrophils is involved in neutrophil migration from the vasculature into extravascular tissue in response to cytokine induction.

The α2β1 Integrin Deficient Mouse

The previously described in vitro studies as well as the expression of the integrin by diverse cell types implicated the α2β1 integrin in a variety of complex biologic processes. To better define the role of the α2β1 integrin in vivo, a genetically engineered mouse in which expression of the α2β1 integrin was eliminated was developed.[102,103] The α2β1 integrin-null mouse exhibits a multifaceted phenotype, as discussed below.

Mice deficient in the α2β1 integrin are viable and fertile and develop normally. Matings of heterozygous progeny yielded 22% wild type, 58% heterozygous, and 21% homozygous mutant animals, consistent with typical Mendelian inheritance with no excess lethality of homozygotes. α2-null females produced viable litters of the expected sizes and nursed their young.

Although the described in vitro studies suggested that the α2β1 integrin is a critical mediator of platelet adhesion to collagen within the vessel wall following vascular injury, tail vein bleeding times in wild-type, heterozygous, and α2-deficient mice were similar. However, platelets purified from wild-type animals adhered under static conditions to collagen substrates in a

Mg^{2+}-dependent manner, as expected for α2β1 integrin-mediated adhesion (Fig. 3A). In contrast, platelets purified from α2-deficient mice failed to adhere to type I collagen. Platelets from heterozygous littermates exhibited intermediate adhesion. Platelets from all animals adhered to fibronectin in a comparable manner (data not shown), but failed to adhere to BSA.

As shown in Figure 3B, purified platelets from wild-type mice also effectively adhered to type I collagen at shear rates of 150 $seconds^{-1}$, a flow rate where exogenous Von Willebrand factor is not required. In contrast, platelets from α2-deficient mice failed to adhere under shear stress. Platelets from heterozygous mice again adhered to an intermediate extent. As also shown in Figure 3B, platelets from both wild-type and heterozygous mice formed small platelet thrombi on the collagen substrates under conditions of flow.

Platelets from wild-type and heterozygous animals comparably aggregated in response to low concentrations of collagen (2.5 and 5 μg/mL). Although platelets from α2-null animals aggregated in response to type I collagen, they did so with a significantly prolonged lag phase and a decreased rate of aggregation.

The profound deficit in platelet adhesion to collagen observed with platelets from α2 integrin subunit deficient animals is in accord with earlier studies addressing the role of the α2β1 integrin as a collagen receptor on platelets by less direct approaches.[4] The magnitude of the deficit under both static conditions and conditions of flow clearly indicates that the α2β1 integrin is the dominant adhesive receptor for collagen on platelets. Other putative receptors are unable to support significant adhesion in the absence of the α2β1 integrin not only under conditions of flow, but also under the less rigorous static conditions. The dramatic reduction in adhesion of heterozygous platelets under both static and flow conditions is consistent with a gene dosage effect and suggests that the level of α2β1 integrin expression profoundly influences the adhesive functions of platelets on collagenous substrates. The adhesive decrement in heterozygotes suggests that the receptor is not present in significant excess.

The results of the platelet studies are largely consistent with the two-step, two-site model of collagen-induced platelet activation, discussed earlier.[17] Under the extremely low shear conditions of the aggregation assay, as opposed to the more stringent conditions of the adhesion assays, our data suggest that GPVI is sufficient to support activation and aggregation although at a somewhat reduced level. This model is also consistent with the findings of studies with human platelets deficient in GPVI. GPVI deficient platelets exhibit markedly impaired collagen-induced aggregation and activation, little or no defect in adhesion per se, and a modest decrement in total platelet deposition (adhesion plus aggregation) on collagen substrates measured under conditions of flow that was attributable to the loss of platelet-platelet interactions.[55,104] Collagen is only one of several physiologically relevant activators of platelets. The presence of other potent activators, such as thrombin, at the sites of vascular injury in vivo, likely accounts for the relatively normal tail vein bleeding times observed and the absence of an overt bleeding diathesis.

The findings of the α2 integrin subunit null mouse differ from the α2β1 integrin-deficient patient originally described by Nieuwenhuis. In addition to profoundly deficient adhesion to collagen, the patient's platelets[41,42] also exhibited markedly impaired (essentially undetectable) collagen-induced platelet aggregation. It seems likely that the platelets of the patient described by Nieuwenhuis et al[41,42] exhibit an additional defect resulting in the absence of collagen-induced platelet aggregation. Alternatively, the relative contributions of the α2β1 integrin, glycoprotein VI, and perhaps other receptors and signaling pathways differ in human and murine platelets. The mouse model provides an important model to evaluate the molecular mechanisms of platelet adhesion and to evaluate the role in which allelic differences in the α2 integrin gene and different levels of α2β1 integrin expression are associated with increased risk of myocardial infarction, diabetic retinopathy and stroke.[65-67]

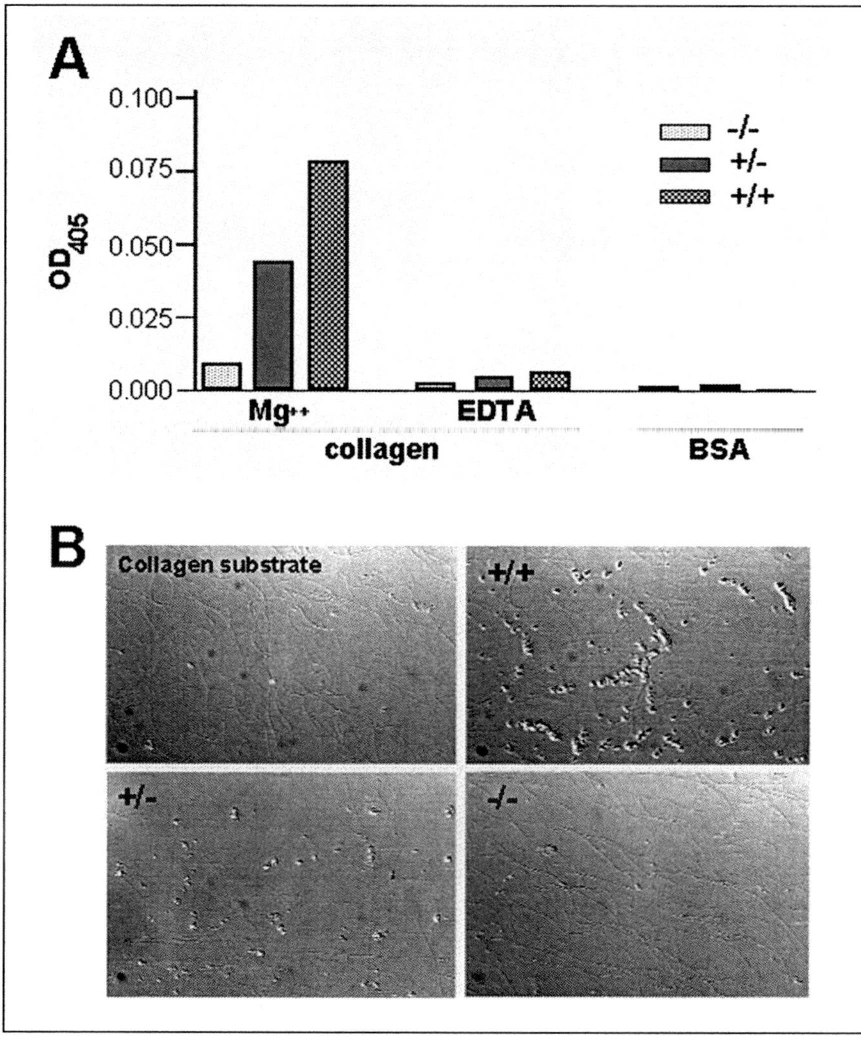

Figure 3. Platelet adhesion and aggregation. (A) Platelets purified from wild-type (+/+), heterozygous (+/-) and α2-deficient (+/-) animals were assayed for adhesion to type I collagen. The assays were carried out for 1 hour in PBS with 2mM Mg^{2+} or EDTA. (B) Platelet adhesion to type I collagen at shear rates of $150s^{-1}$ was assayed in a parallel-plate flow chamber apparatus. Micrographs of the collagen substrate and the deposition of platelets from wild-type (+/+), heterozygous (+/-), and α2-deficient (-/-) mice are depicted.

In contrast to in vitro and explant studies of wound healing, α2-deficient mice demonstrated no defect in wound repair compared to wild-type animals and all animals displayed a similar temporal pattern of wound healing. The morphology of the wounds failed to demonstrate any difference in keratinocyte migration over exposed dermis or angiogenesis into the wound site, suggesting that α2β1 integrin does not play an obligatory role in wound healing. There are at least two possible explanations for the discrepancy. One arises from the use of two different organisms that may not be mechanistically equivalent as model systems. Second,

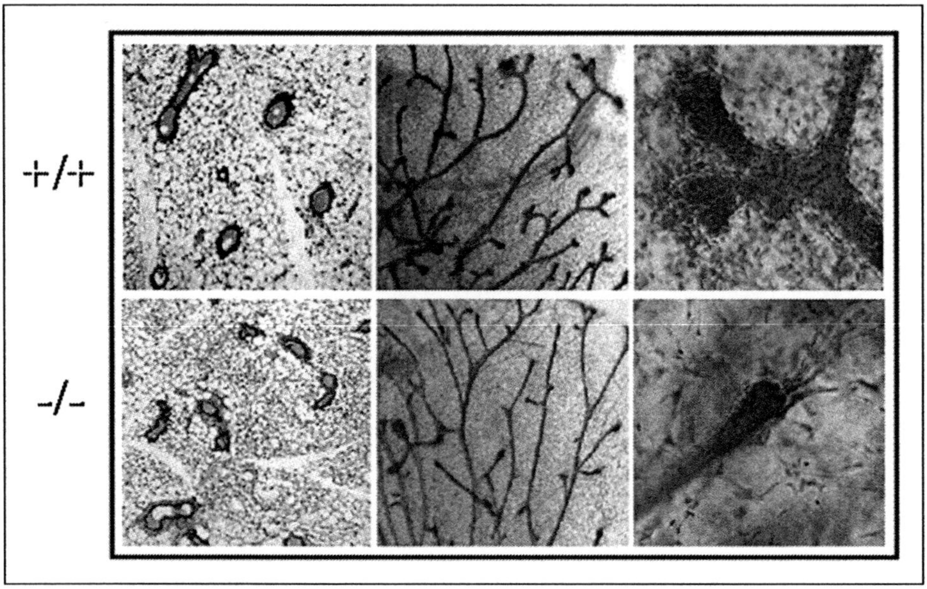

Figure 4. Branching morphogenesis of the mammary gland. Histologic sections of mammary glands of wild-type (+/+) and α2-deficient (-/-) animals 8 to12 weeks of age were evaluated (left panel). Whole mount preparations of 8 to10 old week virgin wild-type (+/+) and α2-deficient (-/-) mammary glands were stained with carmine solution. Terminal ducts were photographed at 100X (middle panel) and 1000X (right panel).

upon binding the integrin, inhibitory antibodies use in the in vitro studies may stimulate signals downstream of the integrin that are responsible for the observed phenotype. Thus, antibodies could produce effects that differ from those due to lack of integrin expression.

The expression of the α2β1 integrin by diverse epithelial cell types and the role of the α2β1 integrin in branching morphogenesis in vitro raised many questions regarding the role of the α2β1 integrin on epithelial cells in vivo. The histology of a number of organs including the heart, lung, gastrointestinal tract, liver, kidney, pancreas, reproductive tracts and mammary gland was entirely normal. However, whole-mount preparations of virgin female mammary glands revealed marked differences in branching complexity. As shown in Figure 4, the branching was more extensive throughout the gland in the wild-type mammary gland than in the α2 -null gland. In fact the number of branch points per terminal duct was significantly greater in the wild-type gland than in the α2 -null gland. These results support the in vitro studies of many investigators that suggested a role for the α2β1 integrin in branching morphogenesis in vivo.

The results of the in vivo studies, although striking, were less dramatic than the complete absence of branching morphogenesis observed in the absence or inhibition of the α2β1 integrin in earlier in vitro studies in three-dimensional collagen gels. The in vitro models were designed to study a single integrin interaction with a single matrix molecule. Mammary epithelial cells express a wide variety of integrin and nonintegrin adhesion receptors that interact with multiple components of the extracellular matrix in vivo and may mitigate the consequences of α2β1 integrin-deficiency.

The mechanism by which branching is impaired in the α2-deficient mouse is not yet understood. One possibility is defective proliferation of α2β1 integrin-deficient mammary epithelial cells. Recent studies have established that the integrin is an important determinant of growth factor-dependent cell cycle progression of mammary epithelial cells.[52,53] Whether the differences seen at weeks 9-11 are due to defective branching per se, or rather to a delay in development, is uncertain. Our earlier studies established a strong correlation between diminished α2β1 integrin expression and a less differentiated phenotype in breast cancer.[68,105] The α2β1 integrin-deficient mouse should afford a system to further assess the relationship between α2β1 integrin expression and the development and progression of breast cancer.

Summary and Conclusions

In summary, in some cell types, such as platelets and mammary epithelial cells, the phenotype of the α2-null mouse was consistent with predictions based upon the expression pattern of the integrin and consistent with the in vitro data. In other instances, the in vivo findings imply that other integrin or nonintegrin receptors can compensate for the loss of the α2β1 integrin. In these cases there are many possible alternative mechanisms for biological processes such as wound healing. The importance of the α2β1 integrin to the pathogenesis of disease processes such as tumorigenesis and metastasis, cardiovascular disease, and the immune surveillance represent exciting new avenues of investigation.

References

1. Staatz WD, Rajpara SM, Wayner EA et al. The membrane glycoprotein Ia-IIa (VLA-2) complex mediates the Mg++-dependent adhesion of platelets to collagen. J Cell Biol 1989; 108(5):1917-1924.
2. Elices MJ, Hemler ME. The human integrin VLA-2 is a collagen receptor on some cells and a collagen/laminin receptor on others. Proc Natl Acad Sci USA 1989; 86(24):9906-9910.
3. Kirchhofer D, Languino LR, Ruoslahti E et al. Alpha 2 beta 1 integrins from different cell types show different binding specificities. J Biol Chem 1990; 265(2):615-618.
4. Santoro SA, Zutter MM. The alpha 2 beta 1 integrin: a collagen receptor on platelets and other cells. Thromb Haemost 1995; 74(3):813-821.
5. Kehrel B. Platelet receptors for collagens. Platelets 1995; 6:11.
6. Moroi M, Jung, SM. Platelet receptors for collagen. Thromb Haemost 1997; 78:439.
7. Sixma J, van Zanten GH, Huizinga EG et al. Platelet adhesion to collagen: an update. Thromb Haemost 1997; 78(434).
8. Chan B, Hemler, ME. Multiple forms of the integrin VLA-2 can be derived from a single alpha 2 cDNA clone: interconversion of forms induced by anti-beta 1 antibody. J Cell Biol 1993; 120(537-543).
9. Berdichevsky F, Alford D, D'Souza B et al. Branching morphogenesis of human mammary epithelial cells in collagen gels. J Cell Sci 1994; 107(Pt 12):3557-3568.
10. Saelman EU, Keely PJ, Santoro SA. Loss of MDCK cell alpha 2 beta 1 integrin expression results in reduced cyst formation, failure of hepatocyte growth factor/scatter factor-induced branching morphogenesis, and increased apoptosis. J Cell Sci 1995; 108 (Pt 11):3531-3540.
11. Zutter MM, Santoro SA, Staatz WD et al. Re-expression of the alpha 2 beta 1 integrin abrogates the malignant phenotype of breast carcinoma cells. Proc Natl Acad Sci USA 1995; 92(16):7411-7415.
12. Zutter MM, Sun H, Santoro SA. Altered integrin expression and the malignant phenotype: the contribution of multiple integrated integrin receptors. J Mammary Gland Biol Neoplasia 1998; 3(2):191-200.
13. Xu J, Clark RA. A three-dimensional collagen lattice induces protein kinase C-zeta activity: role in alpha2 integrin and collagenase mRNA expression. J Cell Biol 1997; 136(2):473-483.
14. Pilcher BK, Dumin JA, Sudbeck BD et al. The activity of collagenase-1 is required for keratinocyte migration on a type I collagen matrix. J Cell Biol 1997; 137(6):1445-1457.
15. Dumin JA, Dickeson SK, Stricker TP et al. Pro-collagenase-1 (matrix metalloproteinase-1) binds the alpha(2)beta(1) integrin upon release from keratinocytes migrating on type I collagen. J Biol Chem 2001; 276(31):29368-29374.
16. Santoro SA. Identification of a 160,000 dalton platelet membrane protein that mediates the initial divalent cation-dependent adhesion of platelets to collagen. Cell 1986; 46(6):913-920.

17. Santoro SA, Walsh JJ, Staatz WD et al. Distinct determinants on collagen support alpha 2 beta 1 integrin-mediated platelet adhesion and platelet activation. Cell Regul 1991; 2(11):905-913.
18. Hall PA, Coates P, Lemoine NR et al. Characterization of integrin chains in normal and neoplastic human pancreas. J Pathol 1991; 165(1):33-41.
19. Ignatius MJ, Large TH, Houde M et al. Molecular cloning of the rat integrin alpha 1-subunit: a receptor for laminin and collagen. J Cell Biol 1990; 111(2):709-720.
20. Ignatius MJ, Reichardt LF. Identification of a neuronal laminin receptor: an Mr 200K/120K integrin heterodimer that binds laminin in a divalent cation-dependent manner. Neuron 1988; 1(8):713-725.
21. Elices MJ, Urry LA, Hemler ME. Receptor functions for the integrin VLA-3: fibronectin, collagen, and laminin binding are differentially influenced by Arg-Gly-Asp peptide and by divalent cations. J Cell Biol 1991; 112(1):169-181.
22. Takada Y, Murphy E, Pil P et al. Molecular cloning and expression of the cDNA for alpha 3 subunit of human alpha 3 beta 1 (VLA-3), an integrin receptor for fibronectin, laminin, and collagen. J Cell Biol 1991; 115(1):257-266.
23. Camper L, Holmvall K, Wangnerud C et al. Distribution of the collagen-binding integrin alpha10beta1 during mouse development. Cell Tissue Res 2001; 306(1):107-116.
24. Camper L, Hellman U, Lundgren-Akerlund E. Isolation, cloning, and sequence analysis of the integrin subunit alpha10, a beta1-associated collagen binding integrin expressed on chondrocytes. J Biol Chem 1998; 273(32):20383-20389.
25. Velling T, Risteli J, Wennerberg K et al. Polymerization of Type I and III Collagens Is Dependent On Fibronectin and Enhanced By Integrins alpha 11beta 1 and alpha 2beta 1. J Biol Chem 2002; 277(40):37377-37381.
26. Velling T, Kusche-Gullberg M, Sejersen T et al. cDNA cloning and chromosomal localization of human alpha(11) integrin. A collagen-binding, I domain-containing, beta(1)-associated integrin alpha-chain present in muscle tissues. J Biol Chem 1999; 274(36):25735-25742.
27. Dickeson SK, Santoro SA. Ligand recognition by the I domain-containing integrins. Cell Mol Life Sci 1998; 54(6):556-566.
28. Kern A, Eble J, Golbik R et al. Interaction of type IV collagen with the isolated integrins alpha 1 beta 1 and alpha 2 beta 1. Eur J Biochem 1993; 215(1):151-159.
29. Calderwood DA, Tuckwell DS, Eble J et al. The integrin alpha1 A-domain is a ligand binding site for collagens and laminin. J Biol Chem 1997; 272(19):12311-12317.
30. Tawil NJ, Houde M, Blacher R et al. Alpha 1 beta 1 integrin heterodimer functions as a dual laminin/collagen receptor in neural cells. Biochemistry 1990; 29(27):6540-6544.
31. Reichardt LF, Tomaselli KJ. Extracellular matrix molecules and their receptors: functions in neural development. Annu Rev Neurosci 1991; 14:531-570.
32. Zutter MM, Santoro SA. Widespread histologic distribution of the alpha 2 beta 1 integrin cell-surface collagen receptor. Am J Pathol 1990; 137(1):113-120.
33. Langholz O, Rockel D, Mauch C et al. Collagen and collagenase gene expression in three-dimensional collagen lattices are differentially regulated by alpha 1 beta 1 and alpha 2 beta 1 integrins. J Cell Biol 1995; 131(6 Pt2):1903-1915.
34. Zutter MM, Santoro SA, Wu JE et al. Collagen receptor control of epithelial morphogenesis and cell cycle progression. Am J Pathol 1999; 155(3):927-940.
35. Riikonen T, Westermarck J, Koivisto L et al. Integrin alpha 2 beta 1 is a positive regulator of collagenase (MMP-1) and collagen alpha 1(I) gene expression. J Biol Chem 1995; 270(22):13548-13552.
36. Ivaska J, Reunanen H, Westermarck J et al. Integrin alpha2beta1 mediates isoform-specific activation of p38 and upregulation of collagen gene transcription by a mechanism involving the alpha2 cytoplasmic tail. J Cell Biol 1999; 147(2):401-416.
37. Gotwals PJ, Chi-Rosso G, Lindner V et al. The alpha1beta1 integrin is expressed during neointima formation in rat arteries and mediates collagen matrix reorganization. J Clin Invest 1996; 97(11):2469-2477.
38. Racine-Samson L, Rockey DC, Bissell DM. The role of alpha1beta1 integrin in wound contraction. A quantitative analysis of liver myofibroblasts in vivo and in primary culture. J Biol Chem 1997; 272(49):30911-30917.
39. Wary KK, Mainiero F, Isakoff SJ et al. The adaptor protein Shc couples a class of integrins to the control of cell cycle progression. Cell 1996; 87(4):733-743.

40. Pozzi A, Wary KK, Giancotti FG et al. Integrin alpha1beta1 mediates a unique collagen-dependent proliferation pathway in vivo. J Cell Biol 1998; 142(2):587-594.
41. Nieuwenhuis HK, Akkerman JW, Houdijk WP et al. Human blood platelets showing no response to collagen fail to express surface glycoprotein Ia. Nature 1985; 318(6045):470-472.
42. Nieuwenhuis HK, Sakariassen KS, Houdijk WP et al. Deficiency of platelet membrane glycoprotein Ia associated with a decreased platelet adhesion to subendothelium: a defect in platelet spreading. Blood 1986; 68(3):692-695.
43. Watson SP, Asazuma N, Atkinson B et al. The role of ITAM- and ITIM-coupled receptors in platelet activation by collagen. Thromb Haemost 2001; 86(1):276-288.
44. Clemetson KJ, Clemetson JM. Platelet collagen receptors. Thromb Haemost 2001; 86(1):189-197.
45. Asselin J, Gibbins JM, Achison M et al. A collagen-like peptide stimulates tyrosine phosphorylation of syk and phospholipase C gamma2 in platelets independent of the integrin alpha2beta1. Blood 1997; 89(4):1235-1242.
46. Polgar J, Clemetson JM, Kehrel BE et al. Platelet activation and signal transduction by convulxin, a C-type lectin from Crotalus durissus terrificus (tropical rattlesnake) venom via the p62/GPVI collagen receptor. J Biol Chem 1997; 272(21):13576-13583.
47. Jandrot-Perrus M, Lagrue AH, Okuma M et al. Adhesion and activation of human platelets induced by convulxin involve glycoprotein VI and integrin alpha2beta1. J Biol Chem 1997; 272(43):27035-27041.
48. Schulte V, Snell D, Bergmeier W et al. Evidence for two distinct epitopes within collagen for activation of murine platelets. J Biol Chem 2001; 276(1):364-368.
49. Clemetson JM, Polgar J, Magnenat E et al. The platelet collagen receptor glycoprotein VI is a member of the immunoglobulin superfamily closely related to FcalphaR and the natural killer receptors. J Biol Chem 1999; 274(41):29019-29024.
50. Jandrot-Perrus M, Busfield S, Lagrue AH et al. Cloning, characterization, and functional studies of human and mouse glycoprotein VI: a platelet-specific collagen receptor from the immunoglobulin superfamily. Blood 2000; 96(5):1798-1807.
51. Keely PJ, Parise LV. The alpha2beta1 integrin is a necessary co-receptor for collagen-induced activation of Syk and the subsequent phosphorylation of phospholipase Cgamma2 in platelets. J Biol Chem 1996; 271(43):26668-26676.
52. Kamiguti AS, Markland FS, Zhou Q et al. Proteolytic cleavage of the beta1 subunit of platelet alpha2beta1 integrin by the metalloproteinase jararhagin compromises collagen-stimulated phosphorylation of pp72. J Biol Chem 1997; 272(51):32599-32605.
53. Poole A, Gibbins JM, Turner M et al. The Fc receptor gamma-chain and the tyrosine kinase Syk are essential for activation of mouse platelets by collagen. Embo J 1997; 16(9):2333-2341.
54. Inoue K, Ozaki Y, Satoh K et al. Signal transduction pathways mediated by glycoprotein Ia/IIa in human platelets: comparison with those of glycoprotein VI. Biochem Biophys Res Commun 1999; 256(1):114-120.
55. Ichinohe T, Takayama H, Ezumi Y et al. Collagen-stimulated activation of Syk but not c-Src is severely compromised in human platelets lacking membrane glycoprotein VI. J Biol Chem 1997; 272(1):63-68.
56. Suzuki-Inoue K, Ozaki Y, Kainoh M et al. Rhodocytin induces platelet aggregation by interacting with glycoprotein Ia/IIa (GPIa/IIa, Integrin alpha 2beta 1). Involvement of GPIa/IIa-associated src and protein tyrosine phosphorylation. J Biol Chem 2001; 276(2):1643-1652.
57. Polanowska-Grabowska R, Geanacopoulos M et al. Platelet adhesion to collagen via the alpha 2 beta 1 integrin under arterial flow conditions causes rapid tyrosine phosphorylation of pp125FAK. Biochem J 1993; 296(Pt3):543-547.
58. Hers I, Berlanga O, Tiekstra MJ et al. Evidence against a direct role of the integrin alpha2beta1 in collagen-induced tyrosine phosphorylation in human platelets. Eur J Biochem 2000; 267(7):2088-2097.
59. Kamiguti AS, Theakston RD, Watson SP et al. Distinct contributions of glycoprotein VI and alpha(2)beta(1) integrin to the induction of platelet protein tyrosine phosphorylation and aggregation. Arch Biochem Biophys 2000; 374(2):356-362.
60. Klekotka PA, Santoro SA, Ho A et al. Mammary epithelial cell-cycle progression via the alpha(2)beta(1) integrin: unique and synergistic roles of the alpha(2) cytoplasmic domain. Am J Pathol 2001; 159(3):983-992.

61. Klekotka PA, Santoro SA, Zutter MM. alpha 2 integrin subunit cytoplasmic domain-dependent cellular migration requires p38 MAPK. J Biol Chem 2001; 276(12):9503-9511.
62. Klekotka PA, Santoro SA, Wang H et al. Specific residues within the alpha 2 integrin subunit cytoplasmic domain regulate migration and cell cycle progression via distinct MAPK pathways. J Biol Chem 2001; 276(34):32353-32361.
63. Kunicki TJ, Orchekowski R, Annis D et al. Variability of integrin alpha 2 beta 1 activity on human platelets. Blood 1993; 82(9):2693-2703.
64. Kunicki TJ, Kritzik M, Annis DS et al. Hereditary variation in platelet integrin alpha 2 beta 1 density is associated with two silent polymorphisms in the alpha 2 gene coding sequence. Blood 1997; 89(6):1939-1943.
65. Kritzik M, Savage B, Nugent DJ et al. Nucleotide polymorphisms in the alpha2 gene define multiple alleles that are associated with differences in platelet alpha2 beta1 density. Blood 1998; 92(7):2382-2388.
66. Santoso S, Kunicki TJ, Kroll H et al. Association of the platelet glycoprotein Ia C807T gene polymorphism with nonfatal myocardial infarction in younger patients. Blood 1999; 93(8):2449-2453.
67. Matsubara Y, Murata M, Maruyama T et al. Association between diabetic retinopathy and genetic variations in alpha2beta1 integrin, a platelet receptor for collagen. Blood 2000; 95(5):1560-1564.
68. Zutter MM, Mazoujian G, Santoro SA. Decreased expression of integrin adhesive protein receptors in adenocarcinoma of the breast. Am J Pathol 1990; 137(4):863-870.
69. Stallmach A, von Lampe B, Matthes H et al. Diminished expression of integrin adhesion molecules on human colonic epithelial cells during the benign to malign tumour transformation. Gut 1992; 33(3):342-346.
70. Koukoulis GK, Virtanen I, Korhonen M et al. Immunohistochemical localization of integrins in the normal, hyperplastic, and neoplastic breast. Correlations with their functions as receptors and cell adhesion molecules. Am J Pathol 1991; 139(4):787-799.
71. Pignatelli M, Cardillo MR, Hanby A et al. Integrins and their accessory adhesion molecules in mammary carcinomas: loss of polarization in poorly differentiated tumors. Hum Pathol 1992; 23(10):1159-1166.
72. Pignatelli M, Hanby AM, Stamp GW. Low expression of beta 1, alpha 2 and alpha 3 subunits of VLA integrins in malignant mammary tumours. J Pathol 1991; 165(1):25-32.
73. Pignatelli M, Smith ME, Bodmer WF. Low expression of collagen receptors in moderate and poorly differentiated colorectal adenocarcinomas. Br J Cancer 1990; 61(4):636-638.
74. Albelda SM. Role of integrins and other cell adhesion molecules in tumor progression and metastasis. Lab Invest 1993; 68(1):4-17.
75. Keely PJ, Fong AM, Zutter MM et al. Alteration of collagen-dependent adhesion, motility, and morphogenesis by the expression of antisense alpha 2 integrin mRNA in mammary cells. J Cell Sci 1995; 108(Pt2):595-607.
76. D'Souza B, Berdichevsky F, Kyprianou N et al. Collagen-induced morphogenesis and expression of the alpha 2-integrin subunit is inhibited in c-erbB2-transfected human mammary epithelial cells. Oncogene 1993; 8(7):1797-1806.
77. O'Toole TE, Katagiri Y, Faull RJ et al. Integrin cytoplasmic domains mediate inside-out signal transduction. J Cell Biol 1994; 124(6):1047-1059.
78. Deleage G, Blanchet C, Geourjon C. Protein structure prediction. Implications for the biologist. Biochimie 1997; 79(11):681-686.
79. Saarialho-Kere UK, Kovacs SO, Pentland AP et al. Cell-matrix interactions modulate interstitial collagenase expression by human keratinocytes actively involved in wound healing. J Clin Invest 1993; 92(6):2858-2866.
80. Pilcher BK, Sudbeck BD, Dumin JA et al. Collagenase-1 and collagen in epidermal repair. Arch Dermatol Res 1998; 290 Suppl:S37-46.
81. Stricker TP, Dumin JA, Dickeson SK et al. Structural analysis of the alpha(2) integrin I domain/procollagenase-1 (matrix metalloproteinase-1) interaction. J Biol Chem 2001; 276(31):29375-29381.
82. Wang Z, Juttermann R, Soloway PD. TIMP-2 is required for efficient activation of proMMP-2 in vivo. J Biol Chem 2000; 275(34):26411-26415.

83. Caterina JJ, Yamada S, Caterina NC et al. Inactivating mutation of the mouse tissue inhibitor of metalloproteinases-2(Timp-2) gene alters proMMP-2 activation. J Biol Chem 2000; 275(34):26416-26422.
84. Brooks PC, Stromblad S, Sanders LC et al. Localization of matrix metalloproteinase MMP-2 to the surface of invasive cells by interaction with integrin alpha v beta 3. Cell 1996; 85(5):683-693.
85. Brooks PC, Silletti S, von Schalscha TL et al. Disruption of angiogenesis by PEX, a noncatalytic metalloproteinase fragment with integrin binding activity. Cell 1998; 92(3):391-400.
86. Mueller SC, Ghersi G, Akiyama SK et al. A novel protease-docking function of integrin at invadopodia. J Biol Chem 1999; 274(35):24947-24952.
87. Yu Q, Stamenkovic I. Cell surface-localized matrix metalloproteinase-9 proteolytically activates TGF-beta and promotes tumor invasion and angiogenesis. Genes Dev 2000; 14(2):163-176.
88. Yu Q, Stamenkovic I. Localization of matrix metalloproteinase 9 to the cell surface provides a mechanism for CD44-mediated tumor invasion. Genes Dev 1999; 13(1):35-48.
89. Yu WH, Woessner JF Jr. Heparan sulfate proteoglycans as extracellular docking molecules for matrilysin (matrix metalloproteinase 7). J Biol Chem 2000; 275(6):4183-4191.
90. Messent AJ, Tuckwell DS, Knauper V et al. Effects of collagenase-cleavage of type I collagen on alpha2beta1 integrin-mediated cell adhesion. J Cell Sci 1998; 111(Pt8):1127-1135.
91. Parks WC, Meecham RP. Matrix Metalloproteinases. San Diego: Academic Press, Inc., 1998.
92. Lin YC, Grinnell F. Decreased level of PDGF-stimulated receptor autophosphorylation by fibroblasts in mechanically relaxed collagen matrices. J Cell Biol 1993; 122(3):663-672.
93. Unemori EN, Werb Z. Reorganization of polymerized actin: a possible trigger for induction of procollagenase in fibroblasts cultured in and on collagen gels. J Cell Biol 1986; 103(3):1021-1031.
94. Klein CE, Dressel D, Steinmayer T et al. Integrin alpha 2 beta 1 is upregulated in fibroblasts and highly aggressive melanoma cells in three-dimensional collagen lattices and mediates the reorganization of collagen I fibrils. J Cell Biol 1991; 115(5):1427-1436.
95. Hemler ME, Jacobson JG, Brenner MB et al. VLA-1: a T cell surface antigen which defines a novel late stage of human T cell activation. Eur J Immunol 1985; 15(5):502-508.
96. Hemler ME. VLA proteins in the integrin family: structures, functions, and their role on leukocytes. Annu Rev Immunol 1990; 8:365-400.
97. de Fougerolles AR, Sprague AG, Nickerson-Nutter CL et al. Regulation of inflammation by collagen-binding integrins alpha1beta1 and alpha2beta1 in models of hypersensitivity and arthritis. J Clin Invest 2000; 105(6):721-729.
98. Rao WH, Hales JM, Camp RD. Potent costimulation of effector T lymphocytes by human collagen type I. J Immunol 2000; 165(9):4935-4940.
99. Aoudjit F, Vuori K. Engagement of the alpha2beta1 integrin inhibits Fas ligand expression and activation-induced cell death in T cells in a focal adhesion kinase-dependent manner. Blood 2000; 95(6):2044-2051.
100. Arase H, Saito T, Phillips JH et al. Cutting edge: the mouse NK cell-associated antigen recognized by DX5 monoclonal antibody is CD49b (alpha 2 integrin, very late antigen-2). J Immunol 2001; 167(3):1141-1144.
101. Werr J, Johansson J, Eriksson EE et al. Integrin alpha(2)beta(1) (VLA-2) is a principal receptor used by neutrophils for locomotion in extravascular tissue. Blood 2000; 95(5):1804-1809.
102. Chen J, Diacovo TG, Grenache DG et al. The alpha(2) integrin subunit-deficient mouse: a multifaceted phenotype including defects of branching morphogenesis and hemostasis. Am J Pathol 2002; 161(1):337-344.
103. Holtkotter O, Nieswandt B, Smyth N et al. Integrin alpha 2-deficient mice develop normally, are fertile, but display partially defective platelet interaction with collagen. J Biol Chem 2002; 277(13):10789-10794.
104. Kehrel B, Wierwille S, Clemetson KJ et al. Glycoprotein VI is a major collagen receptor for platelet activation: it recognizes the platelet-activating quaternary structure of collagen, whereas CD36, glycoprotein IIb/IIIa, and von Willebrand factor do not. Blood 1998; 91(2):491-499.
105. Zutter MM, Santoro SA. The ups and downs of alpha 2 beta 1-integrin expression: contributions to epithelial cell differentiation and the malignant phenotype. Curr Top Microbiol Immunol 1998; 231:167-185.

Chapter 4

Structure Distribution and Function of α10β1 Integrin

Evy Lundgren-Åkerlund and Max Schneller

Introduction

The integrin α10β1 was originally identified as a collagen type II binding integrin on chondrocytes.[1] The bovine integrin subunit α10 was isolated from chondrocyte lysates by affinity chromatography on collagen II and the human homologue was characterized using a human chondrocyte library. Distribution analysis has demonstrated that integrin α10β1 is mainly expressed in cartilage and that it is a major collagen-binding integrin both on fetal and adult chondrocytes.[2] It is also present on specialized fibroblasts in heart valves, in the fascia of skeletal muscle and in tendons/ligaments.[2]

The following chapter details the information currently known concerning the structure, distribution and function of the integrin α10β1.

Structural Features

The α10 subunit has a M_r of around 160 kDa under reducing conditions as determined by SDS-PAGE. When compared to the other collagen-binding integrins on cultured chondrocytes, the integrin subunit α10 is clearly smaller than α1, slightly smaller than α11 and similar in size to the α2 subunit under reducing conditions (Fig. 1).

Sequence analysis of the four collagen-binding integrins shows that α10 has the highest identity with α11, with a 43% identity to α11 and a 33% and 31% identity to α1 and α2 respectively.

I-Domain

α10 integrin contains a von Willebrand factor A-domain, also called I-domain (for inserted domain). The α10 I-domain is inserted in the N-terminal region of the extracellular domain between cation binding sites two and three. As for α1, α2 and α11 integrin, the α10 I-domain is encoded by exons 6-9. Thus, the α10 I-domain extends from C^{140}-G^{337} and consists of 198 amino acids (amino acid numbers correspond to the mature form of α10 integrin after signal peptide cleavage).[1] Three cysteine residues are present within the α10 I-domain, compared to one cysteine in α11, two in α1 and three in α2. It is not known whether these cysteines are involved in disulfide bridge formation, even though this appears likely given the oxidating conditions in the extracellular environment. Indeed, reduction of recombinant α10 I-domain decreases the mobility of the protein in SDS-PAGE compared to the nonreduced form, indicating that the cysteines can form intramolecular disulfide bridges (unpublished

I Domains in Integrins, edited by Donald Gullberg. ©2003 Eurekah.com and Kluwer Academic / Plenum Publishers.

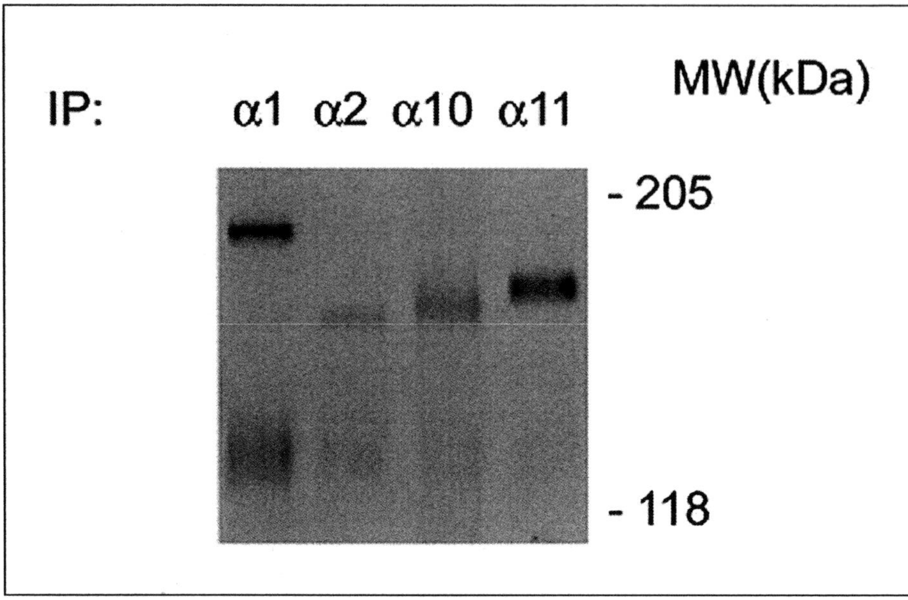

Figure 1. Expression of collagen-binding integrins on human chondrocytes.
Lysates of surface-biotinylated cultured human chondrocytes were immunoprecipitated with antibodies recognizing the integrin subunit α1, α2, α10 and α11 respectively. The precipitated proteins were separated by SDS-PAGE under reducing conditions. The results show that the four collagen-binding integrins are expressed by cultured chondrocytes. The band visible at around 125 kDa corresponds to the integrin β1 subunit that is coprecipitated with the α-subunits

results). It is also possible that the cysteines in the α10 I-domain form disulfide-bridges with other domains of the α10 integrin.

Like the other I-domain-containing collagen-binding integrins, the α10 I-domain contains a so-called MIDAS motif (for metal ion-dependent adhesion site). This motif is believed to have an important role in ligand binding to the I-domain. The overall homology between α10 I-domain and the I-domains of the other collagen-binding integrins is high (see Tuckwell and Johnson, this volume), however, molecular modeling of the α10 I-domain, based on the α2 I-domain crystal structure, has revealed a higher degree of positively charged amino acids in the vicinity of the MIDAS motif[3] compared to the other I-domains. This cluster, which appears not to be present in the other collagen-binding integrin I-domains, may provide α10 with specific functional characteristics.

Transmembrane and Cytoplasmic Domain

The conserved sequence in the transmembrane/cytoplasmic region of α10 is GFFAH (aa 1126-1130) and not GFFKR as in most other integrins α-subunits[1] (see also Gullberg et al this volume). It remains to be elucidated if the amino acids AH provide a function that is unique to α10. It is possible that these amino acids affect activation of the integrin since deletion of the GFFKR motif is known to result in a constitutively active integrin.[4] In a recent study by Armulik et al[5] the border between the transmembrane and cytoplasmic domains was determined using the glycosylation mapping technique.[6] Their results demonstrated that in the absence of interacting ligands the conserved KLGFF sequence in α2 and α5 is buried in the

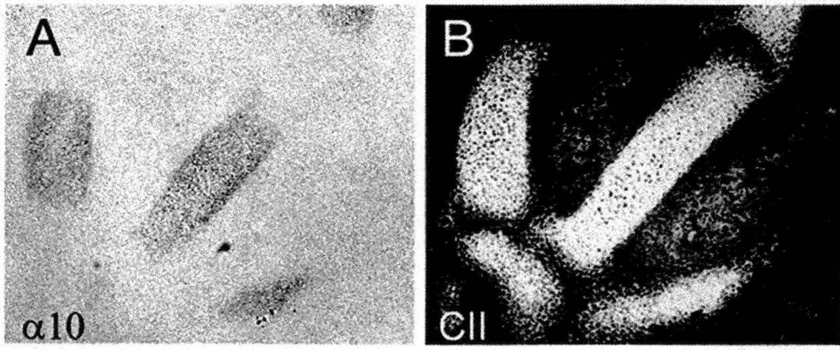

Figure 2. Colocalization of α10 and collagen II (CII) during cartilage development. Cryosections of metacarpal anlage in mouse forelimbs at E13.5 were double stained with antibodies recognizing α10 or CII followed by peroxidase (A)- or Texas red (B)-conjugated secondary antibodies, respectively. α10 was present in the central parts of the developing anlage, while expression of CII was broader.
From Camper et al.[2]

membrane. With this technique the transmembrane region of human α10 was found to extend to the Ala 1129.[7] The cytoplasmic domain of α10 would then consist of 16 amino acids.

Splice Domain

We have recently demonstrated that the human α10 subunit exists as two splice variants due to alternative splicing of exon 25.[8] The spliced domain (between nt 2941 and nt 3055) is located extracellularly, close to the transmembrane region (see also Gullberg et al this volume). In chondrocytes, both forms are expressed as proteins but the smaller form, lacking exon 25, appears not to be transported to the cell surface.[8] However, when we transfected C2C12-cells with the α10 integrin subunit lacking exon 25 we were able to detect the splice variant on the cell surface.[8] In contrast to human α10, mouse exon 25 appears not to be spliced. Instead, mouse exon 26 is alternatively spliced. In the spliced variant, exon 26 containing 144 nt was extended into the intron by 62 nt and resulted in a shift in the reading frame and a premature stop codon.[8] This variant could represent a soluble form of α10. However, we currently do not know if the truncated mRNA is translated into a protein.

Extracellular splice variants have been reported for other α-subunits, although the function of these is not understood.[9] The α-subunits α7 and αIIb, like α10, have alternatively spliced extracellular regions close to the transmembrane spanning domain. In human α7, a 225 nt in-frame-deletion corresponding to 75 amino acids has been found.[10] In the human α-subunit αIIb, a 102 nt sequence encoding a separate exon is spliced in or out.[11]

The α10 Gene

The human α10 gene, which consists of 30 translated exons distributed over a region of 19kb, has been located to chromosome 1, locus q21.[8,12] α10 is the only integrin located on this chromosome.[8] The other human collagen-binding integrin subunits α1 and α2 are located on chromosome 5 while the locus of α11 is on chromosome 15. Several human disease-related genes are located in the α10 region of chromosome 1. Two examples are Gaucher's disease and Pycnodysostosis, in which bone defects are involved.[13,14] In Gaucher's disease, calcification of the aortic and mitral valves has been reported.[15,16] However, it is not clear if Pycnodysostosis and Gaucher's disease are related to abnormalities in matrix or cell-matrix interactions.

The mouse $\alpha 10$ gene has a homology of 90% compared to the human $\alpha 10$, with the highest homology being in the I-domain (97%).[8] Located on chromosome 3, F2.2, the mouse $\alpha 10$ gene consists of 30 translated exons spanning a region of about 18 kb genomic DNA.[8] Primer extension analysis determined that a major transcription start site is located 38 nucleotides upstream from the translation initiation site ATG. The 5'-flanking region of the transcription site at –38 nt lacked a TATA box, as is the case for most other integrin subunits.[8]

When we searched for potential transcription factor binding sites in the TRANSFAC and Matinspector databases, a number of interesting sites for nuclear proteins were found in the 5' sequence of the $\alpha 10$ gene. In agreement with other α-subunits, the $\alpha 10$ gene contained a number of binding sites for Sp1, AP-1 and GATA-boxes, which are expressed ubiquitously. The search also revealed several binding sites for the transcription factors SRY and Sox5, which is also a member of the SRY-family and is the mouse testis determining gene.[17,18] Interestingly, members of the Sox family, including Sox-5, have been shown to be crucial for normal cartilage development.[19]

Distribution

Analysis of mRNA from 50 different human tissues has indicated that expression of $\alpha 10$ mRNA is restricted to cartilage, trachea, lung, heart, aorta and spinal cord.[2,3] It should be noted that expression of $\alpha 10$ in aorta and spinal cord has not been confirmed on the protein level. A BLAST-search of human EST's revealed $\alpha 10$ expression in a number of tumor-derived cell lines, such as melanoma, enchondroma and colon carcinoma.

The restricted distribution of $\alpha 10$ has been confirmed by immunohistochemical analysis of mouse tissues, using an antibody recognizing the cytoplasmic domain of $\alpha 10$.[2] Thus, $\alpha 10$ is mainly expressed in articular cartilage, in the vertebral column, the trachea and in the bronchial cartilage. This expression correlates well with the expression of collagen type II. In the heart, $\alpha 10$ was found in the aortic and atrioventricular valves but not in the muscle. Interestingly, collagen II is expressed in the heart valves during early development but the role for this collagen during heart development is not known.[20] In addition, $\alpha 10\beta 1$ has also been found in the fibrous tissue lining skeletal muscle fibers and tendons/ligaments and in the musculotendon junction.[2] Since we have not been able to detect $\alpha 10\beta 1$ on fibroblasts in other tissues, $\alpha 10\beta 1$ is likely to be expressed only by a specialized subset of fibroblasts. mRNA hybridization of human tissues suggested expression of $\alpha 10$ also in aorta and spinal cord,[3] however, we have not been able to detect the $\alpha 10$ protein in the mouse aorta or spinal cord. This indicates either a species difference between human and mouse or that the $\alpha 10$ mRNA is not translated in these tissues.

Expression of $\alpha 10$ in Cartilage During Development

We have recently shown that $\alpha 10$ is the main collagen-binding integrin during early development of the skeleton.[2] Expression of $\alpha 10$ appears at embryo day 11.5 (E11.5) which coincides with the onset of chondrogenesis as determined by collagen type II expression (Fig. 2). At E13.5, $\alpha 10$ is present throughout the anlage as well as in the perichondrium and in mesenchyme just outside of the perichondrium. $\alpha 1$ was weakly detected in the anlage during development while $\alpha 2$ was undetectable. $\alpha 11$ is found around the growing anlage in the human embryo[21] (see also Gullberg et al in this volume) indicating that $\alpha 10$ and $\alpha 11$ may have overlapping roles in the development of cartilage.

In the articular cartilage of newborn mouse limbs, $\alpha 10\beta 1$ is expressed by all chondrocytes in the growing epiphyseal cartilage and in the hypertrophic region while $\alpha 1\beta 1$ is found on the surface of the epiphyseal cartilage (Fig. 3). Four weeks after birth $\alpha 10$ is prominent both at the articular surface and in the growth plate.[2] At this stage of development $\alpha 1$ is expressed in the upper part of the articular cartilage however its expression in the growth plate is very weak. $\alpha 2$

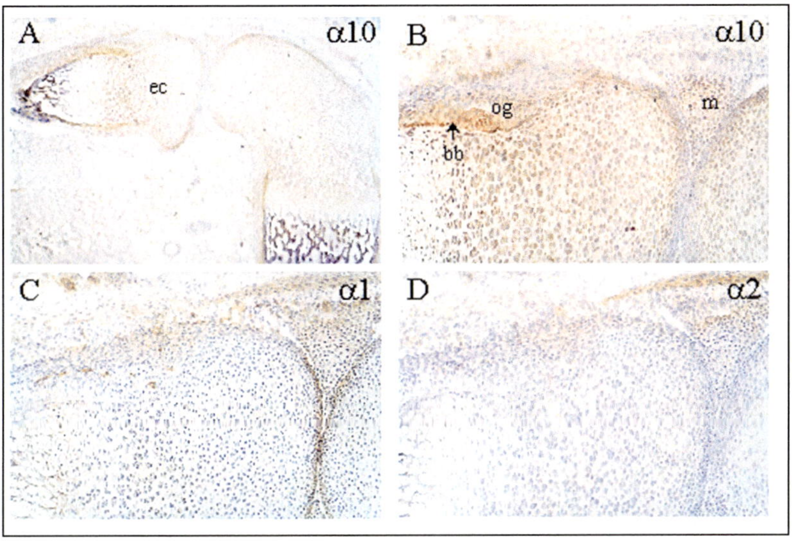

Figure 3, above. Immunohistochemical localization of the integrin subunits α10, α1 and α2 in knee joint. Hind limbs from 3-day-old mice were cryosectioned and stained as indicated with antibodies specific for the integrin subunits α10, α1 and α2. α10 was expressed by all chondrocytes in the growing epiphyseal cartilage (ec), in the ossification groove of Ranvier (*og*), in the bone bark (*bb*) and in the meniscus (*m*). α1 was expressed on the superficial chondrocytes, in the fibrous tissue surrounding the joint and in the meniscus. A few cells in the ossification groove, most likely representing capillaries, also expressed α1. α2 was detected in the fibrous tissue surrounding the joint and in the meniscus but not in the epiphyseal cartilage. From Camper et al.[2]

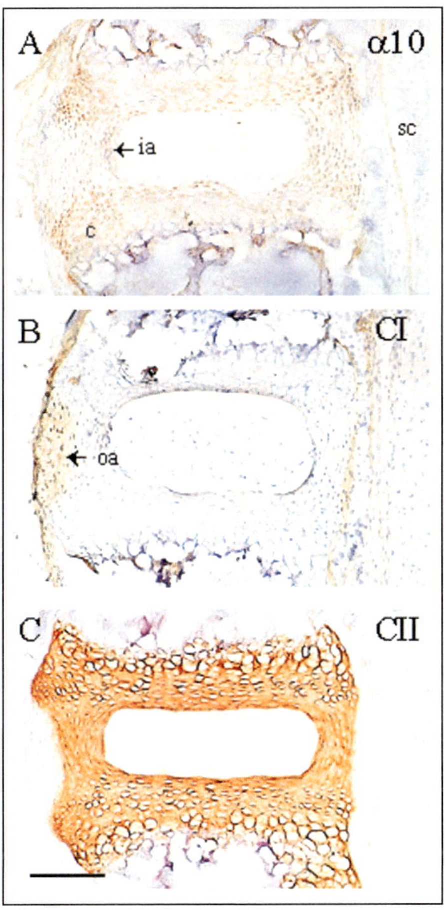

Figure 4, left. Immunohistochemical localization of integrin α10 and the collagen types I and II in the vertebral column. Vertebral columns from 3-day-old mice were cryosectioned and stained with antibodies specific for the integrin subunit α10 and collagen types I and II. (A) α10 expression was found in the inner annulus fibrosus (*ia*) and in the cartilage of the vertebral body (*c*). α10 staining was also detected in the thin lining of the spinal cord (*sc*). (B) CI expression was found in the outermost fibrosus part of the annulus fibrosus (*oa*) and in the lining of the vertebrae. (C) CII was present in the cartilage of the vertebral body and in the inner annulus fibrosus. From Camper et al.[2]

on the on the hand is below detection level. α10 is also detected in the inner cartilaginous region of the meniscus where collagen type II is expressed. Interestingly, α10 is expressed by cells in the ossification groove of Ranvier. The cells in the ossification groove contain precursors for both chondrocytes and osteoblasts and have been suggested to be involved in growth of the bone.[22,23] α10 is also expressed on cells in the perichondrium and periosteum, which points to a possible role for α10 on cells that are involved in the repair of cartilage and bone. Preliminary data have shown that α10 is present during callus formation in fracture healing experiments (unpublished results).

Thus, α10 together with α11 appear to have a central role in cartilage development. During maturation of the cartilage, tissue expression of α1β1 increases and this integrin is highly expressed together with α10β1 in adult cartilage. The involvement of α2, however, is less clear. Our findings that α2 is not present in cartilage or alternatively that expression is below detection level are in agreement with previous studies on tissue sections from fetal and adult cartilage reporting that α2 was not detected in these tissues by immunohistochemistry.[24,25,26] Integrin α2β1 has however been detected in osteoarthritic cartilage[26] and on isolated chondrocytes[25,27,28] (see also Fig. 1). In addition we have found that α2 is upregulated during mechanical stimulation of chondrosarcoma cells[28] indicating that this integrin may be upregulated in cartilage during mechanical stress.

In the vertebral column, α10 was detected in the cartilage of the vertebral body and in the inner annulus fibrosus of the intervertebral discs.[2] In both tissues, α10 expression was colocalized with collagen II (Fig. 4). The outer annulus fibrosus, which contains collagen I as opposed to collagen II, was negative for α10. Interestingly, in contrast to α10, α11 has been detected in the outer annulus fibrosus[21] (see also Gullberg et al this volume). This provides an example of the integrins α10β1 and α11β1 having complementary expression during cartilage development. The integrins α1β1 and α2β1, on the other hand, were not expressed to any detectable degree in the vertebral column or in the intervertebral discs at the 3-day-old stage.[2]

Function

The integrin subunit α10 was originally isolated by affinity purification on collagen type II.[1] Experiments using α10β1-expressing cells have confirmed that α10β1 is indeed a receptor for collagen type II (unpublished results) but that it also interacts with other collagen types. The affinity of α10β1 for different collagens is currently under investigation. A recent study by Tulla et al, using recombinant I-domains, suggested that the α10 I-domain is similar to the α1 I-domain in that it has a preferred affinity for collagen types IV and VI over the fibrillar collagen types I-III.[29] Our findings that α10β1 is also present in tissues that lack collagen type II suggests that α10β1 also in vivo binds to other ligands in addition to collagen type II.

The question as to why mammals express four different types of collagen binding integrins remains to be elucidated. Ligand specificity for the collagen sub-types, as well as other aspects such as intracellular signaling, have to be taken into account. For α10β1 integrin a specific role is suggested by its relatively restricted tissue distribution. It is expressed in tissues that are exposed to high mechanical load such as articular cartilage, vertebral column, heart valves and musculotendon junctions. This points towards a role for α10 integrin in mechanical integrity and/or in mechanical signaling. Our findings that α10 is the major collagen binding integrin on chondrocytes during cartilage development and in the adult stage suggests that it has a specific role both in cartilage formation and cartilage integrity. This, together with the finding that it has a restricted distribution makes α10β1 an interesting target for medical therapy of cartilage diseases. The presence of α10 on cells in perichondrium and periosteum, as well as in the ossification grove of Ranvier, points to a possible role of α10 on cartilage and bone precursor cells that are involved in cartilage growth and repair.

Acknowledgements

We would like to acknowledge the contributions of Lisbet Camper, Karin Holmvall and Therese Bengtsson to the results discussed in this article.

References

1. Camper L, Hellman U, Lundgren-Åkerlund E. Isolation, cloning and sequence analysis of the integrin subunit α10, a β1-associated collagen binding integrin expressed on chondrocytes. J Biol Chem 1998; 273(32):20383-20389.
2. Camper L, Holmvall K, Wängnerud C et al. Distribution of the collagen-binding integrin α10β1 during mouse development. Cell Tissue Res 2001; 306(1):107-116.
3. Gullberg DE, Lundgren-Åkerlund E. Collagen-binding I domain integrins – what do they do? Progr Histochem Cytochem 2002; 37(1):3-54.
4. O'Toole TE, Katagiri Y, Faull RJ et al. Integrin cytoplasmic domains mediate inside-out signal transduction. J Cell Biol 1994; 124(6):1047-1059.
5. Armulik A, Nilsson I, Von Heijne G et al. Determination of the border between the transmembrane and the cytoplasmic domains of human integrin subunits. J Biol Chem 1999; 274(52):37030-37034.
6. Nilsson I, Saaf A, Whitley P et al. Proline-induced disruption of a transmembrane alpha-helix in its natural environment. J Mol Biol 1998; 284(4):1165-1175.
7. Armulik A. Studies on the transmembrane signaling of β1 integrins. Doctoral Thesis (ISBN 91-544-4832-1). Sweden: Uppsala University, 2000.
8. Bengtsson T, Camper L, Schneller M et al. Characterization of the mouse integrin subunit α10 gene and comparison with its human homologue. Genomic structure, chromosomal localization and identification of splice variants. Matrix Biol 2001; 20(8):565-576.
9. De Melker AA, Sonnenberg A. Integrins: alternative splicing as a mechanism to regulate ligand binding and integrin signaling events. Bioassays 1999; 21(6):499-509.
10. Leung E, Lim SP, Berg R et al. A novel extracellular domain variant of the human integrin alpha 7 subunit generated by alternative intron splicing. Biochem Biophys Res Commun 1998; 243(1):317-325.
11. Bray PF, Leung CS, Shuman MA. Human platelets and megakaryocytes contain alternately spliced glycoprotein IIb mRNAs. J Biol Chem 1990; 265(17):9587-9590.
12. Lehnert K, Ni J, Leung E et al. The integrin alpha10 subunit: expression pattern, partial gene structure, and chromosomal localization. Cytogenet. Cell Genet 1999; 87(3-4):238-244.
13. Gelb BD, Shi GP, Chapman HA et al. Pycnodysostosis, a lysosomal disease caused by cathepsin K deficiency. Science 1996; 273(5279):1236-1238.
14. Yamashita DS, Dodds RA. Cathepsin K and the Design of Inhibitors of Cathepsin K. Curr Pharm Des 2000; 6(1):1-24.
15. Chabas A, Cormand B, Grinberg D et al. Unusual expression of Gaucher's disease: cardiovascular calcifications in three sibs homozygous for the D409H mutation. J Med Genet 1995; 32(9):740-742.
16. Veinot JP, Elstein D, Hanania D et al. Gaucher's disease with valve calcification: possible role of Gaucher cells, bone matrix proteins and integrins. Can J Cardiol 1999; 15(2):211-216.
17. Wunderle VM, Critcher R, Ashworth A et al. Cloning and characterization of SOX5, a new member of the human SOX gene family. Genomics 1996; 36(2):354-358.
18. Connor F, Cary PD, Read CM et al. DNA binding and bending properties of the post-meiotically expressed Sry-related protein Sox-5. Nucleic Acids Res 1994; 22(16):3339-3346.
19. de Crombrugghe B, Lefebvre V, Behringer RR et al. Transcriptional mechanisms of chondrocyte differentiation. Matrix Biol 2000; 19(5):389-394.
20. Swiderski RE, Daniels KJ, Jensen KL et al. Type II collagen is transiently expressed during avian cardiac valve morphogenesis. Dev Dyn 1994; 200(4):294-304.
21. Tiger CF, Fougerousse F, Grundström G et al. α11β1 integrin is a receptor for interstitial collagens involved in cell migration and collagen reorganization on mesenchymal nonmuscle cells. Dev Biol 2001; 237(1):116-129.

22. Shapiro F, Holtrop ME, Glimcher MJ. Organization and cellular biology of the perichondrial ossification groove of Ranvier: a morphological study in rabbits. J Bone Joint Surg Am 1977; 59(6):703-723.
23. Schollmeier G, Uhthoff HK, Lewandrowski KU et al. Role of bone bark during growth in width of tubular bones. A study in human fetuses. Clin Orthop 1999; 367:291-299.
24. Salter DM, Hughes DE, Simpson R et al. Integrin expression by human articular chondrocytes. Br J Rheumatol 1992; 31(4):231-234.
25. Woods Jr VL, Schreck PJ, Gesink DS et al. Integrin expression by human articular chondrocytes. Arthritis Rheum 1994; 37(4):537-544.
26. Ostergaard K, Salter DM. Immunohistochemistry in the study of normal and osteoarthritic articular cartilage. Prog Histochem Cytochem 1998; 33(2):93-165.
27. Dürr J, Goodman S, Potocnik A et al. Localization of β1-integrins in human cartilage and their role in chondrocyte adhesion to collagen and fibronectin. Exp Cell Res 1993; 207(2):235-244.
28. Holmvall K, Camper L, Johansson S et al. Chondrocyte and chondrosarcoma cell integrins with affinity for collagen type II and their response to mechanical stress. Exp Cell Res 1995; 221(2):496-503.
29. Tulla M, Pentikäinen OT, Viitasalo T et al. Selective binding of collagen subtypes by integrin α1I, α2I, and α10I domains. J Biol Chem 2001; 276(51):48206-48212.

Chapter 5

Structure and Function of $\alpha 11\beta 1$ Integrin

Donald Gullberg, Svetlana N. Popova and Carl-Fredrik Tiger

Abstract

The $\alpha 11$ integrin chain constitutes the latest addition to the integrin family. $\alpha 11\beta 1$ was originally identified on cultured human muscle cells, but recent studies have shown that it is not expressed on muscle cells in vivo. It remains to be determined if satellite cells in regenerating muscle express $\alpha 11$.

Distribution data indicate that the $\alpha 11$ chain in vivo is expressed on a subset of mesenchymal cells in the developing human embryo. Expression is high in the developing musculoskeletal system in areas of cartilage, bone and in tendon formation. High $\alpha 11$ expression is also seen in mesenchymal tissues characterized by elaborately organized collagen matrices such as the intervertebral disc and the cornea. In agreement with the distribution data, ligand binding studies suggest that $\alpha 11$ prefers collagen I over collagen IV.

We will review the current knowledge about $\alpha 11\beta 1$ and discuss the possible in vivo functions of $\alpha 11\beta 1$ and also address the issue of functional redundancy among collagen-binding integrins.

Introduction

Only some years ago the two major collagen receptors were believed to be $\alpha 1\beta 1$ and $\alpha 2\beta 1$. With the discovery of the nonintegrin tyrosine kinase discoidin receptors DDR1 and DDR2[1] as well as the integrins $\alpha 10\beta 1$ and $\alpha 11\beta 1$,[2] this picture has changed. Mice lacking $\alpha 1\beta 1$,[3] $\alpha 2\beta 1$,[4,5] DDR1[6] or DDR2[7] have been generated, but at this moment in time the role of $\alpha 10\beta 1$ and $\alpha 11\beta 1$ for mouse development and adult tissue integrity has not yet been evaluated. Already now it can be predicted that in order to gain an overall picture of the importance of cell-collagen interactions much work remains and these efforts will involve a careful analysis of redundant/overlapping receptor functions.

Why did the identification of $\alpha 10\beta 1$ and $\alpha 11\beta 1$ remain so elusive?

In the mid 90's many regarded the vertebrate integrin family as being saturated. Degenerate PCR using primers to conserved regions had resulted in the identification of novel integrin chains.[8-10] Yet data existed that indicated the presence of additional integrin chains on chondrocytes and fibroblasts.

Chondrocytes, which only express low levels of the known collagen receptors $\alpha 1\beta 1$ and $\alpha 2\beta 1$, still adhere well to collagen in vitro. Affinity purification of collagen-binding integrins from chondrocytes in the late 90's lead to the isolation of the $\alpha 10$ integrin chain.[11] $\alpha 10$ on SDS-PAGE has a Mw similar to $\alpha 2$ both under reducing and nonreducing conditions,[11] and therefore was easily missed. The restricted expression of $\alpha 10$ to chondrocytes is the most likely reason $\alpha 10$ was not picked up in the degenerate PCR experiments.

I Domains in Integrins, edited by Donald Gullberg. ©2003 Eurekah.com and Kluwer Academic / Plenum Publishers.

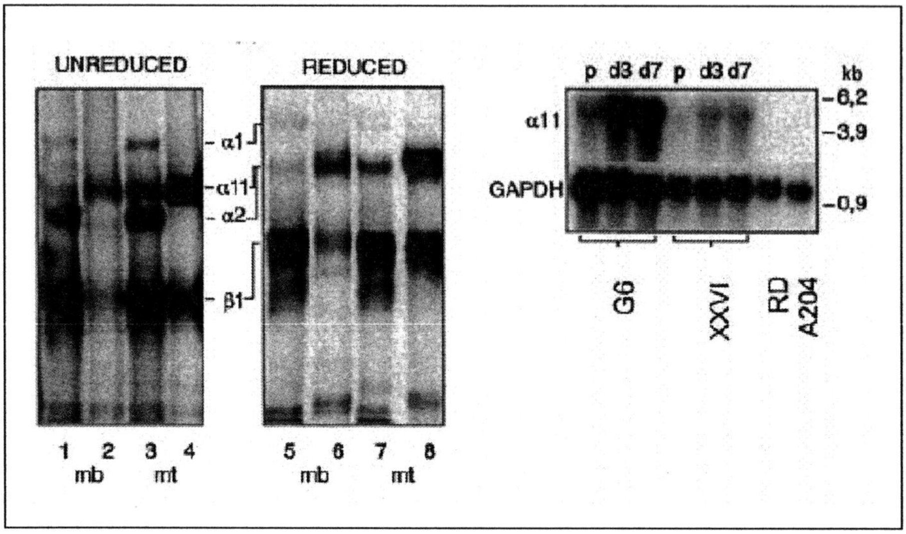

Figure 1. Induction of α11 protein and mRNA during in vitro myogenic differentiation. Left panel: Induction of integrin α11 upon myogenic differentiation in vitro. G6 muscle cells were metabolically labeled with[35] Scysteine/methionine during growth in proliferation medium (mb, proliferating myoblasts) and after 7 days in differentiation medium (mt, myotubes). Integrins were precipitated with antibodies to β1 and α11, and the precipitates were resolved on 7.5% SDS-PAGE gels both under nonreducing (UNREDUCED) and reducing (REDUCED) conditions. Lanes 1, 3, 5, and 7 are immunoprecipitations with the antibody to integrin β1, and lanes 2, 4, 6, and 8 are immunoprecipitations with the antibody to integrin α11. Right panel: Up-regulation of integrin α11 mRNA in differentiated myogenic cells. mRNA was extracted from G6 and XXVI cells grown under proliferating (p) or differentiating (d) conditions for 3 days (d3) or 7 days (d7). Total RNA was isolated from RD and A204 cells. Following separation of RNA on agarose gel and transfer to the membrane, the filter was hybridized with probes to α11 integrin (α11) and glyceraldehyde-3-phosphate dehydrogenase (GAPDH). The size of the bands in the RNA standard (in kb) is marked to the right. From Velling et al.[15]

More recently, α11 has also been found to be present on cultured chondrocytes.[2] With a full panel of functional antibodies to collagen-binding integrins it will be possible to evaluate the relative contribution of α10β1 and α11β1 to chondrocyte adhesion to collagens.

Work with fibroblasts in the early 90's had shown that integrin antibodies to the known collagen receptors could not inhibit collagen gel contraction,[12] and fibroblast spreading on collagen I could not be ascribed to α1β1 and α2β1 alone.[3] It is now known that fibroblasts express α11β1 which can mediate cell adhesion and collagen gel contraction.[13] The α11 chain has a distinct molecular weight on SDS-PAGE[14,15] and under nonreducing conditions α11 migrates as a band in a position in between that of α1 and α2, and is easily detected. The distinct Mw of α11 is most likely due to the relatively high glycosylation of α 11 (see below). High resolution SDS-PAGE and analysis of integrin subunits under both reducing and nonreducing conditions was essential in identifying α11. The reason α11 was not amplified in degenerate PCR experiments was most likely due to sequence variations in the regions used for designing the primers. As in the case for chondrocytes, it will be important to sort out the in vivo and in vitro role of the different collagen-binding integrins on fibroblasts.

In the next chapter we will focus on α11, review the history of α11 and highlight some current topics related to this collagen receptor.

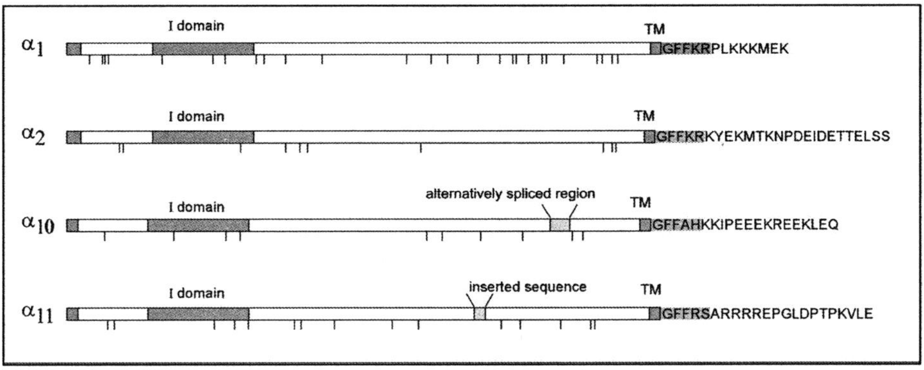

Figure 2. Comparison of the collagen-binding integrin achains. Schematic representation of α1, α2, α10 and α11 integrin chains. Potential N-glycosylation sites are marked with vertical bars. Inserted regions are marked in α1 and α11, as well as the region subjected to alternative splicing in α10. The conserved GFFXX sequence at the transmembrane(TM)/- cytoplasmic interface is shaded. From Gullberg and Lundgren-Åkerlund.[2]

α11 Protein

When studying integrin subunits on human fetal myoblasts and human satellite cells in vitro we noted the induction of a novel integrin chain with a distinct Mw[16] (Fig. 1). Years of attempts to purify enough protein were fruitless, and only after combined PCR and screening of cDNA libraries was the cDNA sequence of the α11 integrin subunit finally obtainable.[15]

α11 Protein Characteristics

Extracellular Domain

As already mentioned the reason that the α11 chain migrates in a position intermediate to that of α1 and α2 chain is most likely due to the relatively high glycoslation level of the α11 chain (Fig. 2). Another distinguishing feature of the α11 chain is the insertion of 22 extra amino acids in the stalk region. With the recent crystallization of the αvβ3 this region can now be mapped to the calf-1 domain (see Fig. 3).

I Domain

When comparing the sequence of α11 with that of other integrin chains it became clear that α11 contained an I domain and was most closely related to the α10 subunit. (Fig. 4). At the I domain level this identity is as high as 60% (see also Tuckwell and Johnson, this volume). Whereas α1 and α2 I domains have been crystallized, α10 and α11 structures have not yet been determined. Modeling of the α11 I domain suggests that the cluster of charged amino acids surrounding the MIDAS is different in α11 compared with the surrounding amino acids in other collagen-binding I domains (Fig. 5). Future studies will determine how these differences influence ligand specificity. Recent modeling of I domains-ligand complexes have suggested that the loop structures surrounding the MIDAS determine ligand specificity.[17]

Cytoplasmic Tail

Similar to α10, the α11 subunit lacks the GFFKR sequence in the plasma membrane / cytoplasmic interface and instead contains the sequence GFFRS. In human α10, the corresponding sequence is GFFAH. Whereas the GFFKR sequence is conserved across different

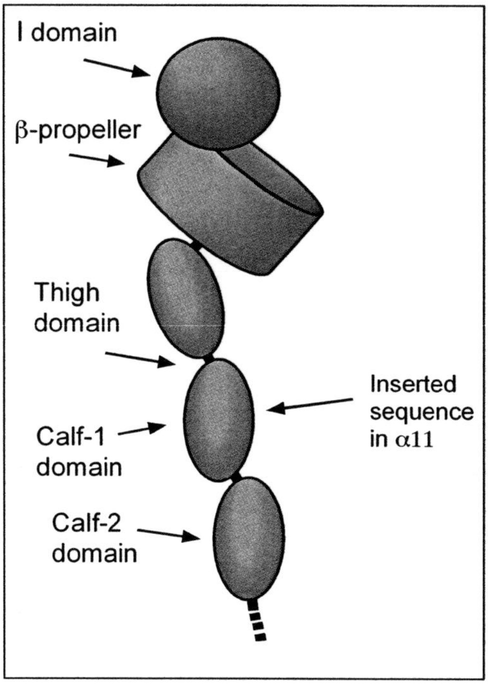

Figure 3. Schematic representation of α11 integrin chain. A schematic representation of α11 chain. β-propeller and calf-1 domain in α11 are marked.

species for most α-chains, this appears not to be the case for the corresponding α10 GFFAH sequence when comparing human and mouse proteins.[18] The sequence variation of α10 and α11 in this region in various species will be informative when identifying the residues that are functionally important.

α11 Expression and Distribution

Probing multiple tissue northern blots with α11 cDNA showed a broad tissue distribution of α11 mRNA with a prominent expression in adult uterus, heart and skeletal muscle.[15] This suggested to us that the mRNA detected in these tissues was derived from the muscle cells. When performing immunohistochemistry and in situ hybridization on human embryonic tissues we, however, failed to detect a signal in muscle cells. Instead α11 protein and mRNA are expressed in non-muscle mesenchymal cells.[13] Maybe the most distinguishing feature is expression throughout the musculoskeletal system adjacent to the perichondrium in areas of cartilage formation (Fig. 6).

Furthermore at this developmental stage the expression overlapped with the expression seen in mouse embryos for the E-box binding transcription factor scleraxis.[19] At later developmental stages scleraxis becomes restricted to tendon cells.[20] To further investigate this issue it will be important to determine the α11 expression and distribution in adult tissues and to establish if α11 and scleraxis are coexpressed at the cellular level.

Based on the embryonic expression pattern it appears that α11 demarcates a subset of mesenchymal cells in the forming musculoskeletal system. At the moment little is known about the sites of the earliest α11 expression in the embryo. To address this issue we have raised antibodies reactive with mouse α11.

The wide expression pattern detected by Northern blotting in adult human tissues might, at the cellular level, be derived from mesenchymal cells in various tissues. Studies of α11 distribution in adult tissues are ongoing.

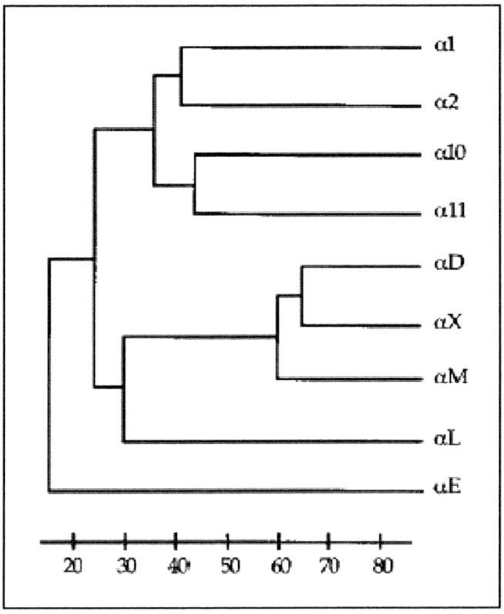

Figure 4. The family of I domain-containing integrins. A distance tree of the I domain-containing α-integrin subfamily members was assembled using ClustalW multiple alignment, based on the SEAVIEW and PHYLOWIN software. A scale at the bottom shows the percentage identity. From Velling et al.[15]

A number of nonmesenchymal cultured cell lines do not express α11, compatible with the idea that mesenchyme-specific transcription factors direct the expression of α11. The default state of expression in vivo thus seems to be in a subset of mesenchymal cells. Why then do muscle cells when cultured in vitro express high levels of α11? It is possible that a transcriptional repression signal active in vivo is released under in vitro conditions. The release of this repression appears to increase α11 expression to a wider set of mesenchymal cells (i.e., chondrocytes and myocytes), but so far α11 has not been detected in cultured nonmesenchymal cells. It remains to be determined if induction of α11 expression can also occur in vivo on mesenchymal cells normally not expressing α11. Since α11 expression on cultured muscle cells is strong in vitro, it will be of particular interest to analyze α11 levels in regenerating muscle where satellite cells are activated and proliferate.[21]

The α11 Integrin Gene

Analysis of the gene structure of integrin α-chains has revealed common themes in the region upstream of the translation start. Integrin promoters generally lack a TATA box and CCAAT box but do contain a conserved consensus initiator promoter element.[22,23] Binding sites contributing to cell- or tissue-specific expression in integrin promoters include a CArG box in the chick IGTA1 promoter,[24] tandem Sp1 elements in the ITGA2 promoter,[25] and E-boxes in the itga7 promoter.[23]

For a number of integrin chains, analysis of the corresponding genomic structure has clarified the mechanism for generating splice variants in the extracellular and intracellular domains.[26,27] This has identified the cleavage site for generating heavy and light chains at a site corresponding to the exon borders.[22] Genomic analysis has furthermore explained disease-causing mutations in the genes encoding α2,[28] α6 and β4,[29] α7[30] and β3[31] integrin chains. The I domain is encoded for by four exons flanked by phase 1 introns, indicative of intergenic acquisition of this domain during evolution[22] and strongly suggests that these four exons define the functional domain.

With the rapid cloning of mouse and human genome, genomic sequences have become available. Assembly of human chromosome 15 genomic sequences has been performed and the human α11 integrin promoter region has been analyzed in silico, and functionally using reporter constructs.[32] Assembly of genomic sequences results in a gene composed of 30 exons[32] (Fig. 7).

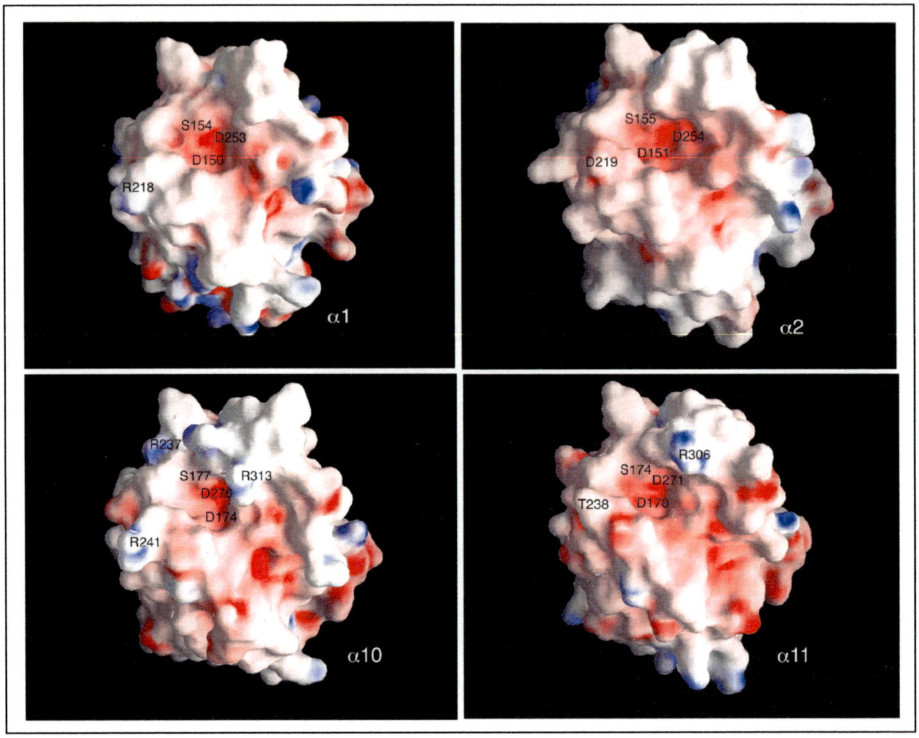

Figure 5. Modeling of integrin I domains. The surface view of the human integrin α1 and α2 I domains crystallized in the absence of ligand are shown. The hypothetical structures of α10 and α11 have been modeled based on the α1 and α2 I domain structures using ProMod and Swiss-Model.[57] Surface charge distributions were calculated using GRASP. Blue, positive charge; red, negative charge; white, neutral. For orientation, residues around MIDAS have been labeled. D219 in α2, R^{218} in α1 suggested to determine collagen specificity have been labeled as well as the corresponding residues in α10 (R^{241}) and α11(T^{238}). From Gullberg and Lundgren-Åkerlund.[2]

In our studies of the ITGA11 gene we have mapped it to a region of chromosome 15 harboring the locus for BBS4.[15] However, performing radiation hybrid mapping we could exclude a role of ITGA11 in BBS4 (data not shown). This agrees with the recent identification of the BBS4 gene as a glucosamine transferase.[33] However, it is not possible to exclude as yet unmapped diseases caused by mutations in ITGA11. Observations from mice lacking functional alleles for integrin α11 gene will be informative when identifying such diseases.

Promoter and Transcription Start

The ITGA11 promoter lacks both TATA and CCAAG boxes. The promoters of other integrin α-chains contain features of a conserved initiator element often associated with an upstream Sp1 site found close to the transcription start.[22,23] Using oligocapping we have identified a transcription start site 30 nucleotides upstream of ATG.[32] The consensus sequence for initiator sequences is pypyANT/Apypy. The experimentally found sequence ACACC in α11 qualifies as an abbreviated initiator sequence. Furthermore an Sp1 binding site is located upstream of the putative initiator sequence, supporting the view that it is functional. Recently the transcription start of the α10 gene has been mapped 38 nucleotides upstream of the translation start.[18]

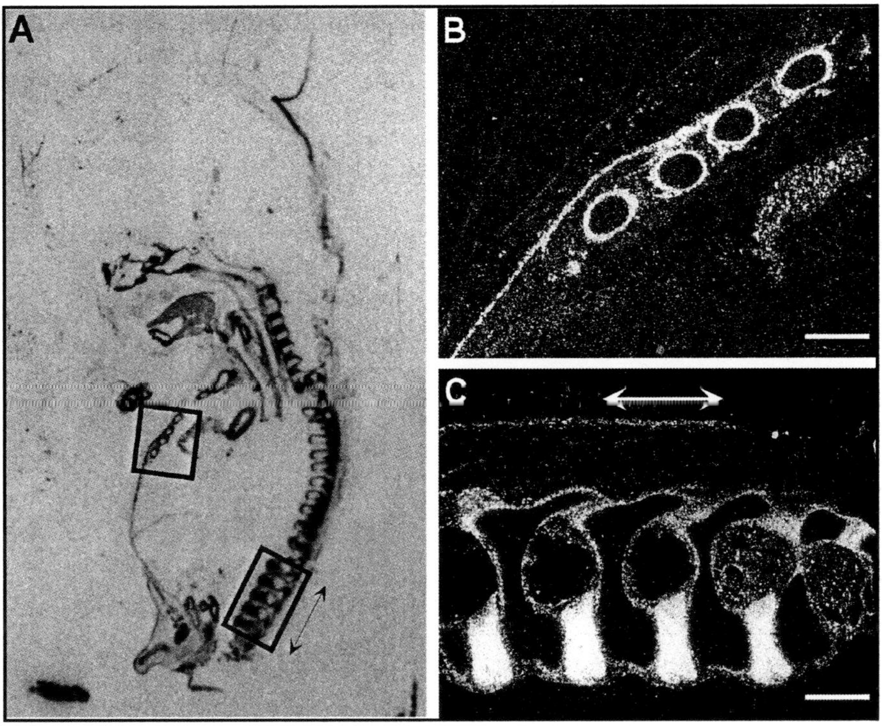

Figure 6. Distribution of α11 mRNA transcripts at 8 weeks of gestation. In situ hybridization of α11 mRNA on sagital sections from a human embryo at 8 weeks of gestation. Low-resolution autoradiogram highlights areas around forming cartilage (A). Boxed areas in (A) are shown in (B) and (C). α11 mRNA is transcribed at high levels around ribs (B), and in intervertebral discs (C). Bars in (B) and (C) represent 800 μm. The arrow depicts the rostro-caudal orientation of the embryo. From Tiger et al.[13]

The result of a transcription start just 30 nucleotides upstream of the translation start was surprising in the light of our previous findings of a 90 nucleotide 5' UTR sequence found in the screening of a uterus cDNA library. Closer examination of this previously obtained cDNA sequence and comparison with genomic sequences revealed that the first 8 nucleotides of the 90 UTR sequence represents vector sequence. The presence of multiple transcription start sites is a common finding in TATA-less promoters and is illustrated by the α2 integrin gene[34] which shows two major transcription start sites at 112 and 129 nt upstream of the ATG.

Analysis of the sequence upstream of the transcription start site predicts the presence of a number of putative transcription factor binding sites. That this region can direct high-level transcription in a cell type-specific manner was determined using luciferase constructs.[32] We could identify a core promoter, a silencer region and a region suggested to contain enhancer activity. This ordered organization of enhancer, silencer and core promoter of the ITGA11 is similar to other integrin promoters including ITGA2[34] and itga7.[23] Studies are currently ongoing to identify the transcription factor binding sites conferring these activities to the different regions in the ITGA11 promoter.

Several E-boxes are present in the proximal promoter region. E-boxes are elements with the consensus sequence CANNTG that bind heterodimers of basic helix-loop- helix transcription

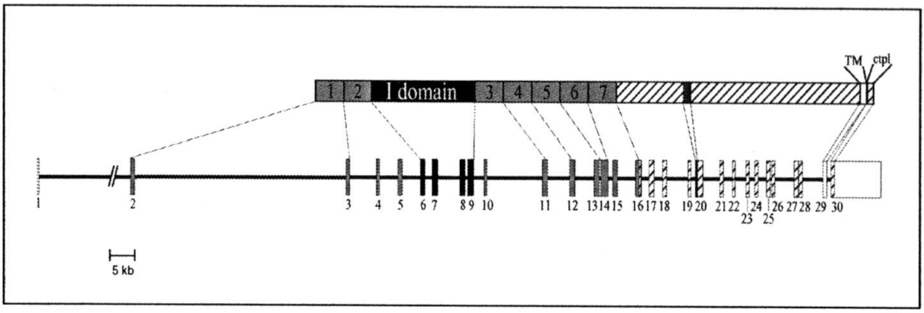

Figure 7. Schematic representation of ITGA11 gene. The top panel shows a schematic representation of human α11 protein, and the lower panel shows a schematic overview of the organization of the ITGA11 gene. For the protein, the I domain (I domain), the 7 FG-GAP repeats (1–7), the transmembrane part (TM) and the cytoplasmic tail (ctpl) are marked. In the gene, exonic sequences representing untranslated regions are marked with open boxes and the interruption of unknown size in intron 1 is marked with -//-. From Gullberg and Lundgren-Åkerlund.[2]

factors, including members of the skeletal muscle-specific MyoD family.[35,36] Closer examination of the E-boxes in the α11 promoter shows that one (CAGCTG) confirms to the MyoD symmetrical core consensus-binding site. The promoter of the muscle α7 integrin chain contains 8 E-boxes, out of which only one contains the MyoD-preferred core sequence.[23] In addition to binding sites for basic helix-loop-helix factors, the ITGA11 promoter also contains one consensus binding site for the myogenic transcription factor MEF-2. We hypothesize that the presence of E-boxes in the α11 promoter explains the up-regulation of the α11 chain during myogenic differentiation in vitro. It is also possible that nonmyogenic basic helix- loop -helix factors such as scleraxis and capsulin bind to the E-boxes in the ITGA11 promoter in vivo. Scleraxis is expressed during mesenchyme/cartilage formation[19,37] and capsulin is expressed in mesothelial cells covering epithelial organs.[38,39] The embryonic expression pattern of scleraxis resembles that of α11.[13] Two putative Sox-5 binding sites were also identified. A Sox-5 variant has recently been found to direct expression of genes in mesenchymal cells in cartilage-forming regions in the embryo.[40] Recent analysis of cells derived from the musculoskeletal system has confirmed the expression of α11 in cultured chondrocytes.[2]

Exon Structure

Signal Peptide

As in other integrin genes, the signal peptide is split by two exons, exon 1 and 2. Unlike some other integrin genes, exon 1 does not solely contain a UTR sequence,[22] but also encodes the major part of the signal peptide. It is interesting to note that in the public domain, preliminary exon assignments can be found for ITGA11 with the locus ID: 2208. However, in this assignment the gene has 29 exons. The size of the final exon is too short to account for the observed 3′UTR, and exon 1 and intron 1 are lacking. In the locus link the signal sequence identified by us is omitted and instead of using M^1 as the translation start, the M^{25} is suggested to be the first amino acid. This would omit one coding exon from the gene structure. Comparison with the closely related α10 integrin gene,[18] shows the presence of a similar signal peptide to α11. If a truncated α11 (starting at M^{25}) is compared with full-length α10, the sequence identity is lower, suggesting that the assignment originally made by us is correct. In further support of our assignment, the experimentally determined transcription start as well as the active promoter region is located just up-stream of the ATG identified by us as the start codon. Amino acid sequencing of the purified integrin will be needed to conclusively resolve this matter.

I Domain

Previous analysis of the exon structure of leukocyte integrins has been very informative in identifying the I domain borders. In the case of α11 the I domain is encoded by 4 exons (exons 6-9) and flanking introns are in phase 1. Based on exon structure the new borders for the α11 I domain are thus Cys^{159} - Gly^{354}. Comparing the exon borders with the borders that we assigned to α11 based on comparison with other I domains, the α11 I -domain is shifted one amino acid C-terminally and includes an unpaired Cys. It is also one amino acid shorter at the N-terminus, retaining the originally suggested size of 195 amino acids.

Inserted Sequence and Alternative Splicing

We have previously identified an inserted region of 22 amino acids that distinguishes α11 from other integrin α chains.[15] More recently we have determined that this sequence is not subject to alternative splicing[32](Fig. 8). It is likely that the inserted region coding for 22 amino acids has arisen by reshuffling of the existing sequence, which has then been inserted at the beginning of exon 20 during evolution. Comparison with ITGA2 exon 20 (AC016619) shows that the homologous sequence starts in a position immediately after the insert observed in exon 20 of ITGA11. Comparison of the α11 sequence with the recently crystallized αv integrin chain[41] places the insertion in a region of the stalk region called the calf-1 domain (Fig. 3). Future studies will reveal the possible importance of this region in cis interactions of the α11 chain.

Careful examination of intron sequences using the program Genscan resulted in the identification of three potential in-frame exon sequences (with the tentative names 10B, 21L and 22B). We, however, failed to detect RNA messages for these variants, and currently no experimental data is thus available to support their existence. In ITGA10 alternative splicing in a region corresponding to exon 25 has been described.[18] PCR amplifications covering this region in α11 suggest that corresponding splicing does not occur in α11.

The cytoplasmic tail in integrin α chains was previously thought to include the conserved sequence GFFKR, which in α11 corresponds to the homologous sequence GFFRS. This sequence is located in a region that can affect receptor latency.[42] It is interesting to note that for those integrin α chains that undergo alternative splicing in the cytoplasmic tail, the alternative exons also encode GFFKR, supporting the view that GFFKR together with the cytoplasmic tail denotes a functional unit. Biochemical analysis of the exact border of the transmembrane domain has recently suggested that GFFK residues are part of the transmembrane domain.[43] Comparison with the gene structure for α11 shows that if the homologous sequence GFFR is considered to be part of the transmembrane domain, the majority of this domain is encoded by exon 29, with the final four residues being encoded by the terminating exon 30, which also encodes the cytoplasmic tail.

Comparison with Other I Domain Integrins

In the locus ID link 2208 the last exon is too short to accommodate the observed 3′UTR sequence. Comparison with other integrin genes, suggests the presence of a last exon larger than 1 kb in size, and it is likely that α11 shows a similar structure.

Lehnert et al has published the partial genomic organization of the human ITGA10.[44] 14 exons were identified and comparisons of these exons reveal conserved exon borders with α11 (not shown). More recently the mouse itga10 gene has been characterized.[18] Comparisons

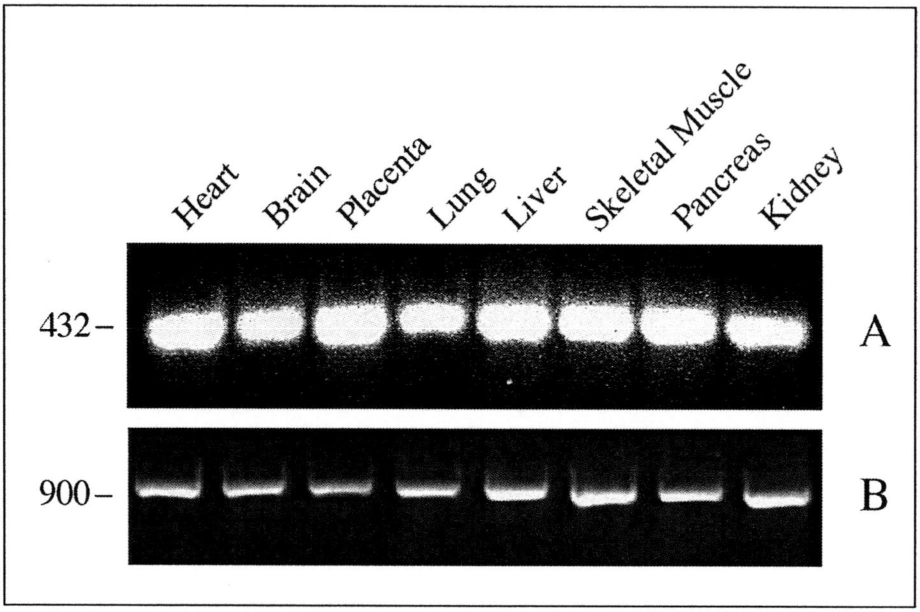

Figure 8. Lack of splicing of exon ITGA11 exon 20. PCR was performed on a panel of cDNAs from adult human tissues. A) The result from amplifications with primers flanking exon 20. B) PCR amplification using G3PDH primers. In A only existing variant of exon 20 is detected.

with these genes lends further support to the view that α10 and α11 have arisen by a recent gene duplication event. Unlike ITGA1 and ITGA2, which are both located on chromosome 5,[45,46] ITGA10 and ITGA11 have been mapped to chromosome 1[18] and 15,[15] respectively. Unlike the ITGA11 gene, which spans at least 100 kb, ITGA10 is much more compact and in the mouse itga10 only spans 18 kb. Comparison of the exon borders among different members of the leukocyte integrin family indicates that exon borders are conserved.[22] Comparison of ITGA11 with the αX exon structure reveals a striking conservation of exon borders, underlining the close evolutionary relationship of integrin I domain α-chain genes.

The complete gene structure including the promoter should be informative when analyzing mutations in integrin α11. The closely related α2 integrin has been shown to display polymorphisms in the promoter region, also identified as risk factors for thrombotic disease.[47,48] The availability of a transgenic mouse model to study α11 integrin function will identify processes likely to be affected by α11 mutations and polymorphisms. Further studies will be instructive in determining the regions in the upstream region that determine the highly localized expression of α11 observed in vivo.

Future Studies of Transcriptional Regulation in Vivo

We have identified different parts of the α11 promoter that are active in cells cultured in vitro. It will be interesting to determine the regions of the promoter that direct the tissue-specificity of the expression. Is it, for example, possible to separate the regions in the promoter that directs expression around forming cartilage, from those elements that direct expression in the fibroblasts in intervertebral disc and cornea? In a recent elegant study lacZ-linked promoter constructs were used to separate the elements in the collagen I promoter that direct expression in tendon fibroblasts from those active in osteoblasts.[49]

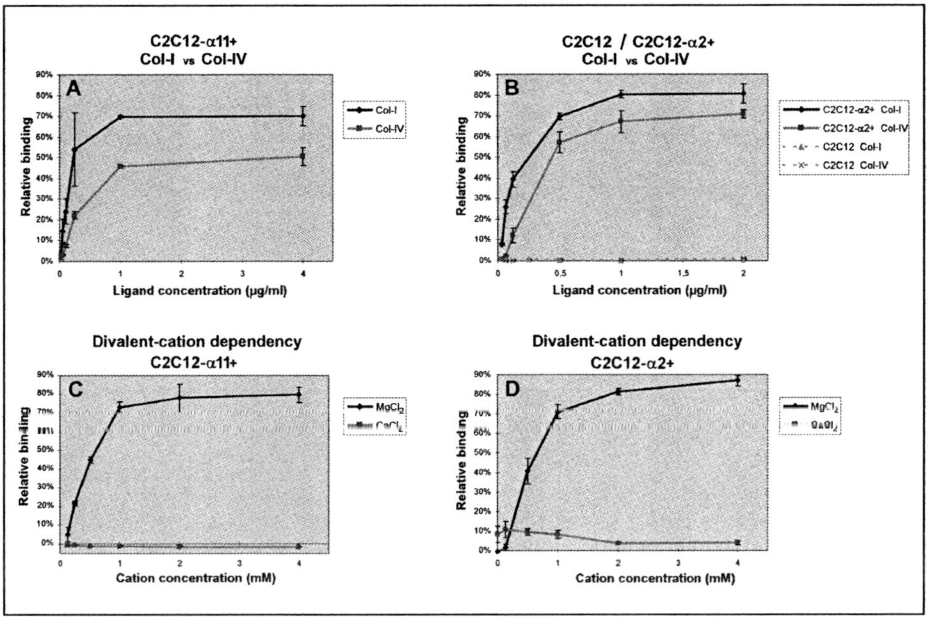

Figure 9. Collagen-specificity and cation-dependence of α11β1-mediated cell adhesion. C2C12 cells transfected with α11 integrin cDNA (C2C12-α11) or α2 integrin cDNA (C2C12-α2) were assayed for attachment to collagen type I and IV (A, B). B) The untransfected C2C12
cells are included as a control. Divalent cation dependence of adhesion was assayed on collagen type I coated at 4 mg/ml using C2C12-α11. (C) and C2C12-α2 (D) cells. Data are presented as relative binding to fibronectin and BSA, where saturated binding to fibronectin represents 100% and background binding to the BSA-blocking represents 0%. Values are given as mean values of triplicate data with error bars denoting standard deviation. From Tiger et al[13]

α11 Function: In Vitro Studies

To be able to study the ligand-binding properties of α11 under controlled conditions, the mouse satellite cell line C2C12 lacking collagen receptors was transfected with α11 cDNA. Upon transfection with α11 the cells acquired the ability to adhere to collagen (Fig. 9), to migrate on collagen and to contract collagen matrices.[13]

Comparison of different ligands suggested that fibrillar collagen I served as a better ligand than the basement membrane collagen IV. The embryonic expression pattern suggests that the in vivo ligands are present in the perichondrial region, which is rich in collagen I and matrilins.[50] In cell migration assays fetal bovine serum (FBS) and PDGF stimulated migration. Interestingly PDGF was more effective in stimulating migration on α11-expressing cells, whereas FBS was more effective on α2-expressing cells, indicating that separate signaling mechanisms may be used by these integrins.

Collagen gel contraction reflects the ability of cells to interact with the collagen matrix. Studies by Harris et al have suggested that traction forces at the surface of fibroblasts are the driving forces for collagen remodeling and tissue morphogenesis.[51] As an extension of these studies a model has recently been put forward where the organization of the collagen meshwork explains how traction forces might act in vivo to organize collagen fibers.[52] The fact that α11β1 is expressed at sites where highly organized collagen matrices are present, indicates that α11β1 might play a role in the organization of the matrix in a subset of mesenchymal tissues.

α11 Function: In Vivo Studies

In Drosophila, integrin α-chains with I domains[53] and interstitial collagens are lacking. Instead, the Drosophila ECM is mainly composed of basement membranes that surrounds the organs and cells.[54] This raises the possibility that some I domain-containing integrin α-chains have, in vertebrates, evolved to fulfil important functions in interstitial collagen matrices. Vertebrate-specific structures including tendon, cartilage and bone all are major building blocks of the musculoskeletal system.

Possible Role in the Musculoskeletal System

The expression of α11 in the perichondrial region is interesting since this region contains mesenchymal cells that will contribute both to forming cartilage and bone.

During endochondral ossification, a number of events take place.[55,56] Mesenchymal precursor cells mature into chondrocytes which subsequently pass through a number of stages including proliferation, hypertrophy and apoptosis. Signals from mesenchymal cells in the perichondrium affect these processes. With vascularization of the hypertrophic cartilage, mesenchymal cells invade the hypertrophic cartilage to form osteogenic cells responsible for ossification. During membranous bone formation, mesenchymal condensates differentiate directly into osteoblasts, and no cartilage forms. Defects in any of these steps may result in musculoskeletal defects.

Functional Redundancy of Collagen-Binding Integrins

To address the issue of functional redundancy among collagen-binding integrins the genes for multiple collagen-binding integrins have to be inactivated. Whereas the genes for α1 and α2 have been ablated , characterization of mice lacking α10 (Bengtsson Lundgen-Åkerlund, unpublished) and α11(Tiger and Gullberg, unpublished) is ongoing. Based solely on distribution overlaps it is hard to predict the phenotypes of the double knockout mice. α1 and α2 integrins are widely expressed on many cell-types including fibroblasts, yet the lack of either of these integrins results in only a mild phenotype.[2] α11 in embryos appears expressed on a subset of fibroblasts. The distribution of α11 in adult tissues is under investigation. α10 during embryogenesis shows a complementary, but not overlapping, expression pattern with α11. We suggest that the simultaneous inactivation of multiple collagen-binding integrins will reveal redundant functions of these integrins for the structural integrity of the musculoskeletal system.

In summary, the elucidation of the in vivo function of $\alpha 11\beta 1$, a major collagen receptor expressed in mesenchymally derived tissues, is likely to significantly contribute to our understanding of the importance of cell-collagen interactions during development and disease. In a longer perspective, careful analysis of mice lacking different set-ups of collagen receptors is likely to be a fruitful approach to further increase our knowledge about cell-collagen interactions.

Acknowledgements

This research was supported by grants from Wenner-Gren foundation, Vetenskapsrådet (DG), Cartela AB (SP) and Konung Gustaf V:s minnesfond (DG).

References

1. Vogel WF. Collagen-receptor signaling in health and disease. Eur J Dermatol 2001; 11(6):506-14.
2. Gullberg D, Lundgren-Åkerlund E. Collagen-binding I domain integrins- what do they do? Prog. Histochem Cytochem 2002; 37:3-54.

3. Gardner H, Kreidberg J, Koteliansky V et al. Deletion of integrin α1 by homologous recombination permits normal murine development but gives rise to a specific deficit in cell adhesion. Dev Biol 1996; 175:301-13.
4. Holtkotter O, Nieswandt B, Smyth N et al. Integrin α2-deficient mice develop normally, are fertile, but display partially defective platelet interaction with collagen. J Biol Chem 2002; 277(13):10789-94.
5. Chen J, Diacovo TG, Grenache DG et al. The α2 integrin subunit-deficient mouse: a multifaceted phenotype including defects of branching morphogenesis and hemostasis. Am J Pathol 2002; 161(1):337-44.
6. Vogel WF, Aszodi A, Alves F et al. Discoidin Domain Receptor 1 Tyrosine Kinase Has an Essential Role in Mammary Gland Development. Mol Cell Biol 2001; 21(8):2906-17.
7. Labrador JP, Azcoitia V, Tuckermann J et al. The collagen receptor DDR2 regulates proliferation and its elimination leads to dwarfism. EMBO Rep 2001; 2(5):446-52.
8. Palmer EL, Ruegg C, Ferrando R et al. Sequence and tissue distribution of the integrin alpha 9 subunit, a novel partner of beta 1 that is widely distributed in epithelia and muscle. J Cell Biol 1993; 123(5):1289-97.
9. Sheppard D, Erle D, Busk M et al. Identification of novel airway epithelial integrins using the homology-based polymerase chain reaction. Chest 1992; 101(3 Suppl):49S.
10. Sheppard D, Rozzo C, Starr L et al. Complete amino acid sequence of a novel integrin beta subunit (beta 6) identified in epithelial cells using the polymerase chain reaction. J Biol Chem 1990; 265(20):11502-7.
11. Camper L, Hellman U, Lundgren-Åkerlund E. Isolation, cloning, and sequence analysis of the integrin subunit α10, α β1-associated collagen binding integrin expressed on chondrocytes. J Biol Chem 1998; 273(32):20383-89.
12. Klein CE, Dressel D, Steinmayer T et al. Integrin α2β1 is upregulated in fibroblasts and highly aggressive melanoma cells in three-dimensional collagen lattices and mediates the reorganization of collagen I fibrils. J Cell Biol 1991; 115(5):1427-36.
13. Tiger C-F, Fougerousse F, Grundström G et al. α11β1 integrin is a receptor for interstitial collagens involved in cell migration and collagen reorganization on mesenchymal nonmuscle cells. Dev Biol 2001; 237:116-29.
14. Gullberg D, Velling T, Sjöberg G et al. Up-regulation of a novel integrin α-chain (αmt) on human fetal myotubes. Dev Dyn 1995b; 204:57-65.
15. Velling T, Kusche-Gullberg M, Sejersen T et al. cloning and chromosomal localization of human α11 integrin. A collagen-binding, I domain-containing, β1-associated integrin α-chain present in muscle tissues. J Biol Chem 1999; 274(36):25735-42.
16. Gullberg D, Sjöberg G, Velling T et al. Analysis of fibronectin and vitronectin receptors during differentiation of human fetal muscle. Exp Cell Res 1995; 220:112-23.
17. Legge GB, Morris GM, Sanner MF et al. Model of the alphaLbeta2 integrin I-domain/ICAM-1 DI interface suggests that subtle changes in loop orientation determine ligand specificity. Proteins 2002; 48(2):151-60.
18. Bengtsson T, Camper L, Schneller M et al. Characterization of the mouse integrin subunit alpha10 gene and comparison with its human homologue. Genomic structure, chromosomal localization and identification of splice variants. Matrix Biol 2001; 20(8):565-76.
19. Cserjesi P, Brown D, Ligon KL et al. Scleraxis: a basic helix-loop-helix protein that prefigures skeletal formation during mouse embryogenesis. Development 1995; 121(4):1099-110.
20. Schweitzer R, Chyung JH, Murtaugh LC et al. Analysis of the tendon cell fate using scleraxis, a specific marker for tendons and ligaments. Development 2001; 128(19):3855-66.
21. Gullberg D, Tiger C-F, Lohikangas L et al. Integrins during muscle development and in muscular dystrophies. Front Biosci 1998; 3:1028-39.
22. Corbi AL, Garcia-Aguilar J, Springer TA. Genomic structure of an integrin α subunit, the leukocyte p150,95 molecule. J Biol Chem 1990; 265(5):2782-88.
23. Ziober BL, Kramer RH. Identification and characterization of the cell type-specific and developmentally regulated α7 integrin gene promoter. J Biol Chem 1996; 271(37):22915-22.
24. Obata H, Hayashi K, Nishida W et al. Smooth muscle cell phenotype-dependent transcriptional regulation of the α1 integrin gene. J Biol Chem 1997; 272(42):26643-51.

25. Zutter MM, Ryan EE, Painter AD. Binding of phosphorylated Sp1 protein to tandem Sp1 binding sites regulates α2 integrin gene core promoter activity. Blood 1997; 90(2):678-89.
26. Graner MW, Bunch TA, Baumgartner S et al. Splice variants of the Drosophila PS2 integrins differentially interact with RGD-containing fragments of the extracellular proteins tiggrin, ten-m, and D-laminin 2. J Biol Chem 1998; 273(29):18235-41.
27. de Melker AA, Sonnenberg A. Integrins: alternative splicing as a mechanism to regulate ligand binding and integrin signaling events. Bioessays 1999; 21(6):499-509.
28. Kritzik M, Savage B, Nugent DJ et al. Nucleotide polymorphisms in the α2 gene define multiple alleles that are associated with differences in platelet α2β1 density. Blood 1998; 92(7):2382-88.
29. Pulkkinen L, Uitto J. Mutation analysis and molecular genetics of epidermolysis bullosa. Matrix Biol 1999; 18(1):29-42.
30. Hayashi YK, Chou F-L, Engvall E et al. Mutations in the integrin α7 gene cause congenital myopathy. Nat Genet 1998; 19:94-97.
31. Bray PF, Shuman MA. Identification of an abnormal gene for the GPIIIa subunit of the platelet fibrinogen receptor resulting in Glanzmann's thrombasthenia. Blood 1990; 75(4):881-88.
32. Zhang W-M, Bergman C, Velling T et al. Analysis of the human integrin α11 gene (ITGA1) and its promoter. Matrix Biol 2002; 21:513-523.
33. Mykytyn K, Braun T, Carmi R et al. Identification of the gene that, when mutated, causes the human obesity syndrome BBS4. Nat Genet 2001; 28(2):188-91.
34. Zutter MM, Santoro SA, Painter AS et al. The human α2 integrin gene promoter. Identification of positive and negative regulatory elements important for cell-type and developmentally restricted gene expression. J Biol Chem 1994; 269(1):463-69.
35. Blackwell TK, Weintraub H. Differences and similarities in DNA-binding preferences of MyoD and E2A protein complexes revealed by binding site selection. Science 1990; 250(4984):1104-10.
36. Kophengnavong T, Michnowicz JE, Blackwell TK. Establishment of distinct MyoD, E2A, and twist DNA binding specificities by different basic region-DNA conformations. Mol Cell Biol 2000; 20(1):261-72.
37. Brown D, Wagner D, Li X et al. Dual role of the basic helix-loop-helix transcription factor scleraxis in mesoderm formation and chondrogenesis during mouse embryogenesis. Development 1999; 126(19):4317-29.
38. Lu J, Richardson JA, Olson EN. Capsulin: a novel bHLH transcription factor expressed in epicardial progenitors and mesenchyme of visceral organs. Mech Dev 1998; 73(1):23-32.
39. Lu J, Chang P, Richardson JA et al. Inaugural article: the basic helix-loop-helix transcription factor capsulin controls spleen organogenesis. Proc Natl Acad Sci USA 2000; 97(17):9525-30.
40. Lefebvre V, Li P, de Crombrugghe B. A new long form of Sox5 (L-Sox5), Sox6 and Sox9 are coexpressed in chondrogenesis and cooperatively activate the type II collagen gene. Embo J 1998; 17(19):5718-33.
41. Xiong JP, Stehle T, Diefenbach B et al. Crystal structure of the extracellular segment of integrin αVβ3. Science 2001; 294:339-45.
42. Briesewitz R, Kern A, Smilenov LB et al. The membrane-cytoplasm interface of integrin alpha subunits is critical for receptor latency. Mol Biol Cell 1996; 7(10):1499-509.
43. Armulik A, Nilsson I, von Heijne G et al. Determination of the border between the transmembrane and cytoplasmic domains of human integrin subunits. J Biol Chem 1999; 274(52):37030-34.
44. Lehnert K, Ni J, Leung E et al. Cloning, sequence analysis, and chromosomal localization of the novel human integrin α11 subunit (ITGA11). Genomics 1999; 60(2):179-87.
45. Jaspers M, Marynen P, Aly MS et al. Localization of the gene encoding the α2 subunit of the human VLA-2 receptor to chromosome 5q23-31. Somat Cell Mol Genet 1991; 17(5):505-11.
46. Seymour AB, Yanak BL, O'Brien EP et al. An integrated genetic map of the pearl locus of mouse chromosome 13. Genome Res 1996; 6(6):538-44.
47. Bray PF. Platelet glycoprotein polymorphisms as risk factors for thrombosis. Curr Opin Hematol 2000; 7(5):284-89.
48. Bray PF. Integrin polymorphisms as risk factors for thrombosis. Thromb Haemost 1999; 82(2):337-44.

49. Terraz C, Brideau G, Ronco P et al. A combination of Cis-acting elements is required to activate the pro- alpha1(I) collagen promoter in tendon fibroblasts of transgenic mice. J Biol Chem 2002; 277(21):19019-26.
50. Klatt AR, Paulsson M, Wagener R. Expression of matrilins during maturation of mouse skeletal tissues. Matrix Biol 2002; 21(3):289-96.
51. Harris AK, Stopak D, P W. Fibroblast traction as a mechanism for collagen morphogenesis. Nature 1981; 290:249-51.
52. Sawhney RK, Howard J. Slow local movements of collagen fibers by fibroblasts drive the rapid global self-organization of collagen gels. J Cell Biol 2002; 157(6):1083-91.
53. Brown NH. Cell-cell adhesion via the ECM: integrin genetics in fly and worm. Matrix Biol 2000; 19(3):191-201.
54. Fessler LI, Nelson RE, Fessler JH. Drosophila extracellular matrix. Methods Enzymol. 1994; 245:271-94.
55. de Crombrugghe B, Lefebvre V, Nakashima K. Regulatory mechanisms in the pathways of cartilage and bone formation. Curr Opin Cell Biol 2001; 13(6):721-27.
56. Wagner EF, Karsenty G. Genetic control of skeletal development. Curr Opin Genet Dev 2001; 11(5):527-32.
57. Peitsch MC. ProMod and Swiss-Model: Internet-based tools for automated comparative protein modelling. Biochem Soc Trans 1996; 24:274-79.

Chapter 6

Structure and Activation of β_2-Integrins

Carl G. Gahmberg and Susanna Fagerholm

Abstract

Cellular adhesion is essential for several leukocyte functions. A number of adhesion molecules are involved in various adhesive events, but perhaps the most important are the leukocyte specific β_2-integrins (CD11/CD18). Recent work has explored integrin structure and the basis for ligand binding in considerable detail. The integrins are normally in an inactive state and need activation for ligand binding. The molecular mechanisms for activation are still incompletely understood, but intracellular signaling by protein kinases and phosphatases are certainly involved in regulation of integrin activity. Therefore we have extensively studied the phosphorylation of the β_2-integrins, mapped the phosphorylation sites and studied the functional significance of phosphorylation. Activated integrins bind to intercellular adhesion molecules (ICAM) and certain soluble proteins. The molecular basis for this is discussed.

Introduction

Most leukocytes circulate in blood and accumulate in lymphoid organs. Upon activation members of different leukocyte subclasses migrate into tissues where they exert a number of functions. The attachment to other cells such as endothelial cells, infected or malignant cells or to other leukocytic cells, needs the cooperation of several members of different families of adhesion molecules.

Perhaps the most important class of leukocyte adhesion molecules is the leukocyte specific β_2-integrin or the CD11/CD18 family.[1-4] Characteristic of leukocyte integrins is that they normally exist in an inactive state, and this is obviously important for circulating cells. It is also noteworthy that mature leukocytes can exist as single cells, which is exceptional because most cells need attachment to other cells or connective tissue components. Otherwise cells die from apoptosis, also called anoikis. The importance of β_2-integrins is further underscored from the fact that individuals who lack or have reduced amounts of functional integrins, due to mutations in the common CD18-polypeptide, develop the leukocyte adhesion deficiency syndrome type I (LAD). The affected children usually die at young age due to repeated infections.[5]

Recent work from a number of laboratories has greatly increased our understanding of integrin structure and function in general, and the current knowledge can perhaps often be applied to the various integrin subclasses. But no integrins are alike and they exert highly specific cellular functions. This is most obvious for the leukocyte β_2-integrins, which exist in highly specialized cells equipped with a number of essential and specific functions.

In the following we discuss recent knowledge of leukocyte integrin structure and how integrins may function. Although much is still poorly understood we have recently witnessed remarkable progress in the field.

I Domains in Integrins, edited by Donald Gullberg. ©2003 Eurekah.com and Kluwer Academic / Plenum Publishers.

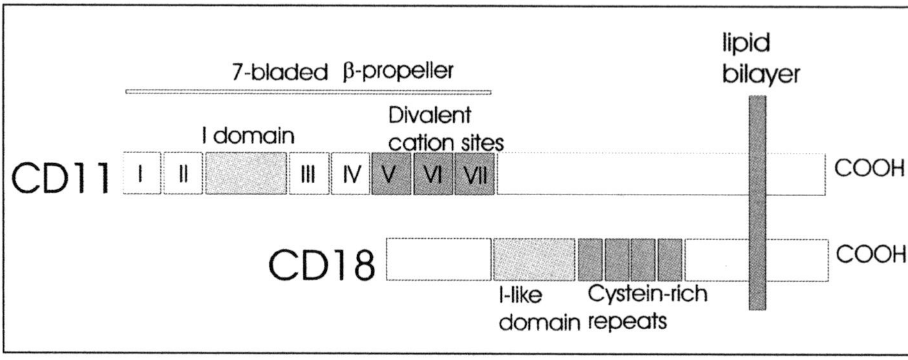

Figure 1. Schematic structure of a β_2-integrin. The CD11 α-chain contains seven divalent cation binding sites which may form a β-propeller structure. The α-I domain is the binding region for ligands, but it may interact with the β-I domain. The β_2-chain is rich in cysteins, which may be involved in integrin regulation. The polypeptides are type I membrane protein, which span the membrane once and parts of the polypeptides are in the cytoplasm interacting with intracellular proteins such as talin, filamin, actin and 14-3-3 proteins.

Structure of Leukocyte β_2-Integrins

Like all integrins the β_2-integrins are heterodimeric type I membrane proteins.[3] (Fig. 1). Most of the integrin structural domains are on the outer surface of the plasma membrane. The polypeptides contain complex N-linked carbohydrate chains, the structures of which have been determined and they may be functionally important.[6,7] Four β_2-integrins have been described, namely CD11a/CD18 (LFA-1, $\alpha_L\beta_2$), CD11b/CD18 (Mac-1, $\alpha_M\beta_2$), CD11c/CD18 (p150/95, $\alpha_X\beta_2$) and CD11d/CD18. The α-chains contain Mg^{++} bound to their I-domains.[8,9] and Ca^{++}-binding repeats.[10-12] The α-chains have been predicted to resemble trimeric G-proteins containing seven-blade propeller structures.[13] The I-domain is located on top of the propeller structure and is directly involved in ligand binding.[14] The I-domains from CD11a and CD11b have been crystallized and their structures determined.[8,9] The CD18 chain contains an I-like domain and several cystein-rich repeats. The cystein-rich region may be important in regulation of integrin activity because several activating monoclonal antibodies bind to this region.[15]

The extracellular part of the β_2-related $\alpha_V\beta_3$ integrin has recently been crystallized and its structure determined at 3.3 Å resolution.[16] The structural determination gave the interesting result that the integrin is flexed back towards the membrane with a "bent knee" structure. The α_V does not have an I-domain but makes contact with the β_3-I -like domain forming the ligand binding site at the external end of the integrin. How similar other integrins, such as β_2-integrins, are to this is not yet known.

The cytoplasmic COOH-terminal portions of the integrins have several interesting structural features. The α-chains contain a membrane-proximal strongly basic sequence, KVGFFKRNLK in CD11a and closely similar sequences in the other CD11 polypeptides,[3] which may be involved in keeping the integrin in an inactive locked structure by interaction with the β-chain cytoplasmic portion. The β_2 cytoplasmic region is essential for the regulation of adhesion and contains several potential phosphorylation sites (Fig. 2), some of which seem functionally important (see below).

The β_2-integrins are leukocyte-specific, but show a characteristic cellular distribution among leukocytes. Thus CD11a/CD18 is enriched on lymphocytes, CD11b/CD18 on granulocytes and CD11c/CD18 and CD11d/CD18 on monocytes and macrophages.[3,4] It should be pointed out that individual leukocytes often contain more than one β_2-integrin.

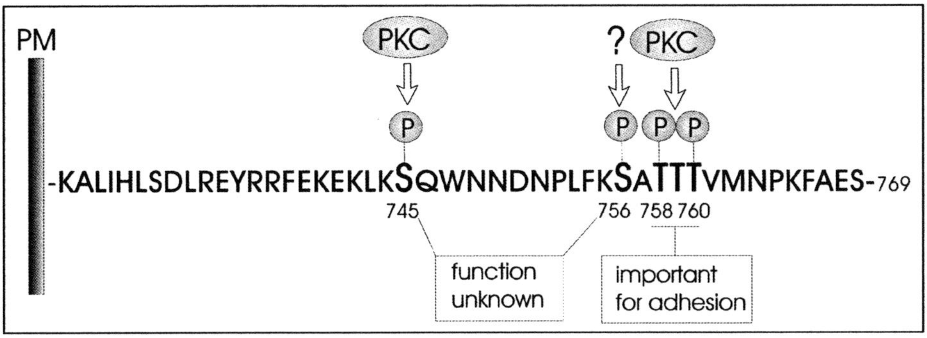

Figure 2. The sequence of the cytoplasmic part of the CD18 (β_2) polypeptide. The phosphorylation sites are marked using bigger letters and PKC isoenzymes phosphorylate Ser-745 and two of the threonines. Ser-756 is phosphorylated by an unknown kinase.

Activation of β_2-Integrins

Although the exact mechanisms of activation are not known, our knowledge has substantially increased during the past few years . It has become evident that integrin activation is unusually complex and may involve several mechanisms and intermediate stages of activation. It should also be emphasized that integrin activation most probably can take place by an increase in *affinity* of individual integrin molecules for their ligands or an increase in *avidity* by clustering of integrins in the plane of the membrane. Both the actin cytoskeleton[17] and lipid rafts[18,19] have been postulated to play a role in the clustering of integrins. Certainly, loss of activity is also of utmost importance and down regulation may occur through mechanisms opposite to activation. How down regulation takes place has not been much studied.

Integrin signaling may occur through outside-in or inside-out activation mechanisms. In outside-in signaling, ligands bind to surface-exposed integrins resulting in intracellular signals, whereas inside-out integrin activation occurs through intracellular signal transduction normally arising through other plasma membrane molecules. Activation of integrins from the outside may occur directly, perhaps by changes in integrin conformation induced by ligands.[3]

Early work showed that leukocyte integrins can be activated by phorbol esters,[20,21] and later it was shown that antibodies to the T cell receptor may activate adhesion.[22,23] How the signaling proceeds is poorly understood and some possibilities are illustrated in (Fig. 3). It is not known if the signaling pathways converge proximal to the integrins. The actin cytoskeleton may be a common downstream target of the signaling pathways. In addition affinity and avidity regulation could well use different molecular mechanisms.

Importantly, many new signaling elements involved in integrin regulation have been revealed using transgene technology, where signaling components have been "knocked out" and the effects of this on integrin-mediated adhesion has been studied. This is certainly a line of investigation, which will become increasingly important.

Initial protein tyrosine kinase activation after T cell receptor engagement leads to the activation of multiple downstream signaling pathways. The protein tyrosine kinase Lck has recently been shown to be important for CD11a/CD18-integrin activation both in response to T cell receptor and phorbol ester triggering of T cells.[24] Downstream of protein tyrosine kinases, adapter proteins are recruited to the plasma membrane to initiate further signaling. SLAP-130/Fyb/ADAP is an adapter protein that is involved in the signaling pathway leading to clustering of CD11a/CD18-integrins on the T cell surface.[25] Also Rap-1, a small GTP-binding protein of the Ras-family, has been implicated in this process,[26-28] but whether there is a link between these proteins is still unknown.

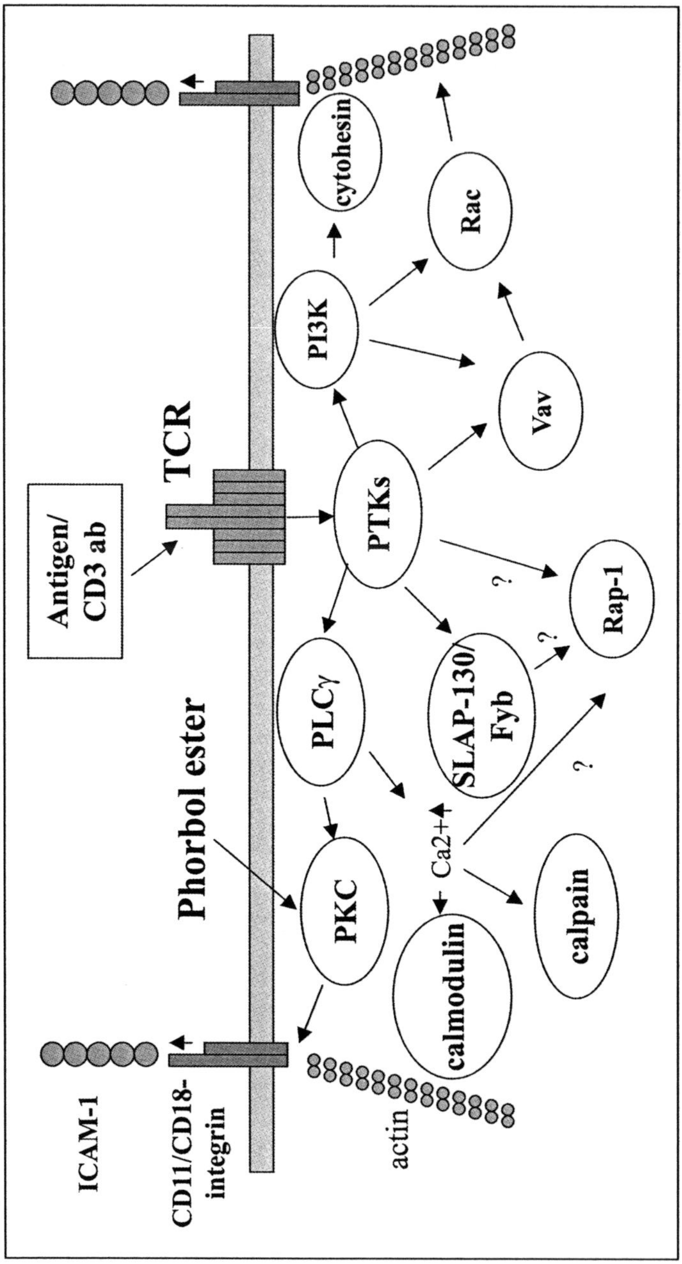

Figure 3. Inside-out signaling resulting in integrin activation. Some of the implicated signaling molecules and pathways are shown.

Another important enzyme in T cell signaling is the lipid kinase, PI3-kinase, the activation of which leads to the recruitment of PH-domain containing proteins to the plasma membrane. PI3-kinase has been implicated in integrin activation,[26,29] and possible downstream effectors of this pathway is Vav[30] and cytohesin,[29,31] which are guanine nucleotide exchange factors for small GTP-binding proteins implicated in actin cytoskeleton rearrangements.

Protein tyrosine kinase activity also leads to the activation of phospholipase Cγ, that generates IP_3 and diacylglycerol. These second messengers go on to activate calcium-signaling and protein kinase C (PKC), respectively. The calcium-activated protease calpain[32] and the calcium-binding protein calmodulin[33] have been implicated in regulating T cell adhesion through CD11a/CD18-integrins. Recently, PKC isoforms have been implicated in CD11a/CD18-integrin regulation using transfection systems.[29] Interestingly, phorbol esters, as well as other stimuli, cause phosphorylation of the integrin CD18-chain, while the CD11-chain is constitutively phosphorylated.[34-37] It was thus postulated that induced integrin phosphorylation may play a role in integrin activation.

Serine-756 was the major observed phorbol ester-induced phosphorylated residue in CD18.[38] Mutation of this serine to alanine did not, however, affect adhesion. This finding may be the reason that interest in studying integrin phosphorylation decreased. Interestingly mutation of threonines 758-760 strongly decreased adhesion, but no or very weak threonine phosphorylation was observed.[38] Valmu and Gahmberg then showed that by using the phosphatase inhibitor, okadaic acid, strong threonine phosphorylation occurred but in a transient manner.[39] Stimulation through the T cell receptor in the presence of okadaic acid gave a similar result.

Fagerholm et al recently carefully mapped the phosphorylation sites in CD18.[40] Using the cytoplasmic peptide of CD18 as substrate, phosphorylating enzymes were purified from leukocytes and identified as PKC isoenzymes.[40] Using in vitro assays it was found that PKCδ and PKCβI/II phosphorylated a serine residue in CD18 but surprisingly this was not the previously identified Ser-756 but Ser-745. In addition, threonine phosphorylation was obtained. Also PKCα and PKC0 phosphorylated these residues in vitro. In contrast, activation of T cells in vivo using phorbol esters resulted in phosphorylation of Ser-756 and Ser-745 as well as threonine phosphorylation. Importantly, activation through the T cell receptor gave no phosphorylation of Ser-756. Threonine-phosphorylation, on the other hand, occurs after stimulation through the T cell receptor. As indicated in (Fig. 2) PKC,s can phosphorylate Ser-745 and probably two threonine residues,[41] whereas the phosphorylation of Ser-756 occurs through a presently unidentified kinase (activated by phorbols).

Further work showed that the phosphorylated β_2-integrins preferentially associated with the cytoskeleton.[42] A continuous association/disassociation release cycle could explain the movement of the integrins along the plane of the membrane resulting in clustering and increased avidity. The β_2-polypeptide binds to the cytoplasmic cytoskeletal proteins talin and filamin but no effect on binding to these proteins by phosphorylation has been seen.[42] Interestingly, the threonine phosphorylated β_2-peptide was found to bind 14-3-3 proteins,[40] which are known to be Ser/Thr-specific cytoplasmic adapter proteins.

It should be pointed out that inside-out activation may occur through numerous other pathways involving other cell surface molecules. Thus antibodies for example to CD43, and CD44 activate integrins.[43] Whether the activation follows similar intracellular pathways, including integrin phosphorylation, as described above is not known. Particularly important may be various chemokines, which are efficient integrin activators.[44] Also here, little is known about the mechanisms involved.

Activation from the outside was first observed using certain anti-β_2-integrin monoclonal antibodies.[45-47] Integrin ligation by antibodies or ligand leads to the activation of numerous signaling pathways in T cells, like the Syk-family tyrosine kinase Zap-70,[48] PI 3-kinase[49] and the MAP-kinase pathways.[49,50] Integrin ligation in T cells also leads to actin remodeling,[51] and therefore, outside-in signaling may be involved in T cell costimulation, integrin crosstalk or induction/strengthening of adhesion through the same integrin. Of course, antibodies are seldom if ever physiological activators of adhesion, but it has been found that ICAM-derived reagents are powerful activators. A peptide from the first immunoglobulin domain of ICAM-2

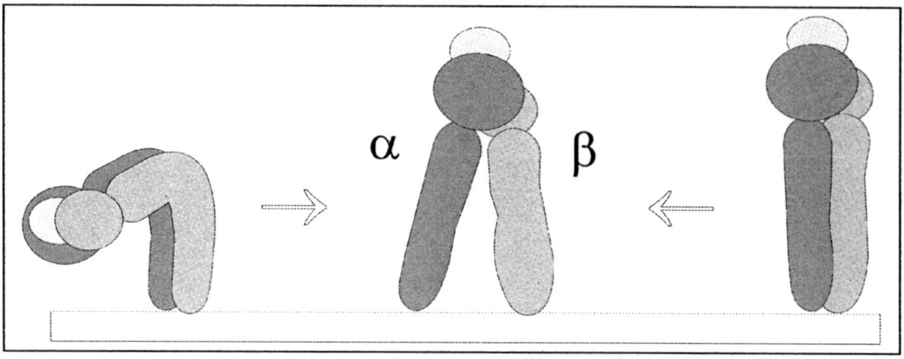

Figure 4. A schematic drawing of an active integrin. To the left a drawing based on the structure of $a_V b_3$. It is possible that its activation does not involve gross conformational changes. Another possibility is that the active conformation is like the integrin figure shown in the middle and the inactive form to the right.

was found to be an effective activator of leukocyte adhesion and mobility.[52,53] Recombinant soluble constructs of ICAMs were subsequently found to activate integrins as well and ICAM-2 seems to be the most powerful one.[54] Importantly, activation by integrin ligands resulted in increased integrin affinity in that sense resembling activation by chemokines.[44] Activation by inside-out out signaling initiated by Ca^{++}-signals, T cell receptor triggering or phorbol esters resulted in increased avidity evidently through integrin clustering in the membrane.[32,55]

Lovastatin and related cholesterol-lowering statins bind to β_2-integrins and the statins seem to bind to integrins at a site separate from the ICAM-binding sites.[56] It is probable that they bind to an activation/inhibition allosteric site and in this way keep the integrins in a closed, inactive state.

There is controversy about how integrin structure changes upon activation. Figure 4 presents this problem. Arnaout and coworkers have determined the structure of the $\alpha_V\beta_3$ soluble recombinant integrin both in free from and bound to an RGD-ligand.[16, 57] The observed changes in structure upon ligand binding were small and primarily involved changes in the ligand binding region. Springer and coworkers, on the other hand, after studying the structure of EGF domains 2 and 3 in β_2 come to the conclusion that integrin activation involves a separation of the "legs" of a relatively straight molecule.[15] The problem has recently been discussed by Liddington.[58]

The Binding Sites in Intercellular Adhesion Molecules

Five ICAM-molecules have been described (Fig. 5), ICAMs 1-5. ICAM-1 and ICAM-3 contain five immunoglobulin domains, ICAM-2 and ICAM-4 two such domains and ICAM-5 is the most complex one with nine domains. ICAM-1, -2 and –3 are expressed in leukocytes, ICAM-1 also on many other cells and ICAM-2 in addition on endothelial cells and platelets.[3] ICAM-4 is red cell specific[59, 60] and ICAM-5 is expressed in brain neurons.[61, 62]

To understand how integrins bind to their ligands, binding sites must be known in detail. Several groups have worked on the identification of the binding regions and CD11a/CD18 has been found to bind to the first immunoglobulin domains of all ICAMs.[63] Mutational analysis combined with structural work has mapped the binding regions in considerable detail.[64] Particular importance has been assigned to an acidic residue, i.e., glu-34 in ICAM-1 and glu-39 in ICAM-2. This claim is mainly based on mutation of the residue to neutral amino acids, which results in loss of adhesion. A problem exists, however, with such an analysis, because the changed

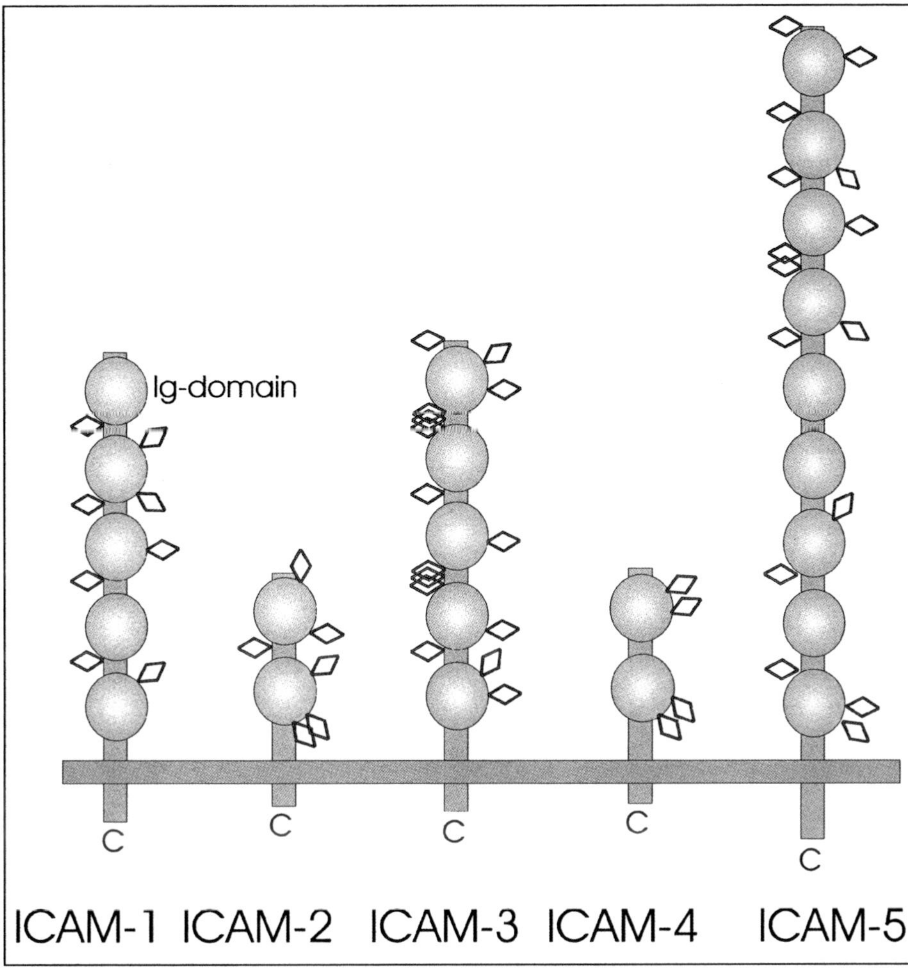

Figure 5. Schematic structures of the known ICAM-molecules. The immunoglobulin domains (O) and the N-glycosylation sites (◊) are shown.

charge may profoundly affect ICAM structure. Structural analysis also indicates that an acidic residue is coordinated to the Mg^{++} -ion in the I-domain binding site.[8,9]

To map the binding sites in more detail we have used phage-peptide libraries to explore the binding sites and we recently described a peptide, which was identified by panning on immobilized β_2-integrins.[65] This peptide called LLG-C4 with the sequence CPCFLLGCC and all cysteines crosslinked efficiently inhibited leukocyte adhesion. The S-S bonds were between residues 1-8 and 3-9 in the active compound. The isomer with the disulfide bonds between cysteines 1-9 and 3-8 was much less active. The active peptide does not contain an acidic amino acid and the finding indicates that the binding region in fact may be rather small and relatively hydrophobic in nature.

It would now be important to obtain the crystal structure of the binding region of β_2-integrins with the bound peptide. Such information would obviously be important for a detailed understanding of integrin-ICAM interactions.

Concluding Remarks

A number of major diseases result from dysfunction of leukocytes. Examples are diseases resulting in loss of such leukocyte functions as immunoglobulin synthesis, cytotoxicity, phagocytosis etc. A dramatic example is the genetic LAD I disease resulting in the malfunctioning of all β_2-integrins. The genetic diseases are rare but much more prevalent are major diseases due to chronic inflammation. Such examples are rheumatoid arthritis, glomerulonephritis, psoriasis and probably diabetes. Here the leukocytes are hyperactive destroying normal tissues. Another example is the reperfusion syndrome which arises after cardiac infarction where leukocytes accumulate in the damaged heart and destroy it further. Further examples are transplantation rejections where the activated T cells destroy transplanted tissues.

On the other hand, increased natural killer and T cell cytotoxicity would be advantageous under some circumstances such as in the combat of malignant diseases and life-threatening infections. Also the development of leukemic cells in the bone marrow may need highly active integrins to keep the cells in place thus avoiding apoptosis. In addition some integrins function as bacterial and viral receptors.[66] Because of this it is understandable that the pharmaceutical industry is heavily involved in developing drugs that affect leukocyte adhesion. Such drugs could in principle interfere with leukocyte-target cell interactions by competing with the binding sites or affect regulatory regions in integrins. Other drugs could be developed to activate integrins to make adhesion more effective. The P1 peptide from ICAM-2 gives hints that this could be possible.

Another major line of development would be to make drugs that interfere with intracellular signaling pathways involving β_2-integrins. Such drugs should be able to penetrate the plasma membrane and be specific for signaling pathways involved in regulation of integrin activity. This may be more difficult to achieve, but not impossible. Specific kinase inhibitors could prove useful as for example the newly discovered inhibitor of the abl and kit tyrosine kinases.[67] The ST1571 tyrosine kinase inhibitor has proved remarkably effective in the treatment of chronic myeloid leukemia and gastrointestinal stromal tumors.[68] The development of corresponding integrin-specific drugs may take time but we are optimistic that eventually useful drugs will be developed.

Acknowledgments

The original work from the authors' laboratories has been funded by the Academy of Finland, the Sigrid Jusélius Foundation, the Finnish Cancer Society, the Magnus Ehrnrooth Foundation and the Oskari Huttunen Foundation. We thank Yvonne Heinilä for secretarial assistance.

References

1. Springer TA. Adhesion receptors of the immune system. Nature 1990; 346:425-34.
2. Arnaout MA. Structure and function of the leukocyte adhesion molecules CD11/CD18. Blood 1990; 75:1037-50.
3. Gahmberg CG, Tolvanen M, Kotovuori P. Leukocyte adhesion. Structure and function of human leukocyte β2-integrins and their cellular ligands. Eur J Biochem 1997; 245:215-32.
4. Gahmberg CG. Leukocyte adhesion. CD11/CD18 integrins and intercellular adhesion molecules. Curr Opin Cell Biol 1997; 9:643-50.
5. Arnaout A, Dana N, Pitt J et al. Deficiency of two human leukocyte surface membrane glycoproteins (Mol and LFA-1). Fed Proc 1985; 44:2664-70.
6. Asada M, Furukawa K, Kantor C et al. Structural study of the sugar chains of human leukocyte cell adhesion molecules CD11/CD18. Biochemistry 1991; 30:1561-71.
7. Kotovuori P, Tontti E, Pigott R et al. The vascular E-selectin binds to the leukocyte integrins CD11/CD18. Glycobiology 1993; 3:131-36.

8. Lee J-O, Rieu P, Arnaout MA et al. Crystal structure of the A domain from the α subunit of integrin CR3 (CD11b/CD18). Cell 1995b; 80:631-38.
9. Qu A, Leahy DJ. Crystal structure of the I-domain from the CD11a/CD18 (LFA-1,αLβ2) integrin. Proc Natl Acad Sci USA 1995; 92:10277-81.
10. Gahmberg CG, Nortamo P, Kotovuori P et al. Activation and binding of leukocyte integrins to their ligands. In: Gahmberg CG, Mandrup-Poulsen T, Wogensen Bach L, Hökfelt B, eds. Leukocyte Adhesion. Basic and Clinical Aspects. Excerpta Medica, International Congress Series 1010. Amsterdam : Elsevier Science Publishers B. V., 1992:205-19.
11. Leitinger B, McDowall A, Stanley P et al. The regulation of integrin function by Ca^{2+}. Biochim Biophys Acta 2000; 1498:91-98.
12. van Kooyk Y, Weder P, Heije K et al. Extracellular Ca2+ modulates leukocyte function-associated antigen-1 cell surface distribution on T lymphocytes and consequently affects cell adhesion. J Cell Biol 1994; 124:1061-70.
13. Springer TA. Folding of the N-terminal, ligand-binding region of integrin α-subunits into a β-propeller domain. Proc Natl Acad Sci USA 1997; 94:65-72.
14. Li R, Rieu P, Griffith DL et al. Two functional states of the CD11b A-domain: Correlations with key features of two Mn^{2+}-complexed crystal structures. J Cell Biol 1998; 143:1523-34.
15. Beglova N, Blacklow SC, Takagi J et al. Cysteine-rich module structure reveals a fulcrum for integrin rearrangement upon activation. Nature Struct Biol 2002; 9:282-87.
16. Xiong J-P, Stehle T, Diefenbach B et al. Crystal structure of the extracellular segment of integrin αvβ3. Science 2001; 294:339-45.
17. van Kooyk Y, Figdor CG. Avidity regulation of integrins: the driving force in leukocyte adhesion. Curr Opin Cell Biol 2000; 12:542-47.
18. Krauss K, Altevogt P. Integrin leukocyte function-associated antigen-1-mediated cell binding can be activated by clustering of membrane rafts. J Biol Chem 1999; 274:36921-27.
19. Leitinger B, Hogg N. The involvement of lipid rafts in the regulation of integrin function. J Cell Sci 2002; 115:963-72.
20. Patarroyo M, Beatty PG, Fabre JW et al. Identification of a cell surface protein complex mediating phorbol ester-induced adhesion (binding) among human mononuclear leukocytes. Scand J Immunol 1985; 22:171-82.
21. Rothlein R, Springer TA. The requirement for lymphocyte function-associated antigen 1 in homotypic leukocyte adhesion stimulated by phorbol ester. J Exp Med 1986; 163:1132-49.
22. Dustin ML, Springer TA. T-cell receptor cross-linking transiently stimulates adhesiveness through LFA-1. Nature 1989; 341:619-24.
23. van Kooyk Y, van de Wiel van de Kemenade P, Weder P et al. Enhancement of LFA-1-mediated cell adhesion by triggering through CD2 or CD3 on T lymphocytes. Nature 1989; 342:811-13.
24. Fagerholm S, Hilden T, Gahmberg CG. Lck tyrosine kinase is important for activation of the CD11a/CD18-integrins in human T lymphocytes. Eur J Immunol 2002; 32:1670-78.
25. Peterson EJ, Woods ML, Dmowski SA et al. Coupling of the TCR to integrin activation by SLAP-130/Fyb. Science 2001; 293:2263-65.
26. Katagiri K, Hattori M, Minato N et al. Rap-1 is a potent activation signal for leukocyte function-associated antigen 1 distinct from protein kinase C and phosphatidylinositol-3-OH kinase. Mol Cell Biol 2000; 20:1956-69.
27. Katagiri K, Hattori M, Minato N et al. Rap1 functions as a key regulator of T-cell and antigen-presenting cell interactions and modulates T-cell responses. Mol Cell Biol 2002; 22:1001-15.
28. Sebzda E, Bracke M, Tugal T et al. Rap1A positively regulates T cells via integrin activation rather than inhibiting lymphocyte signaling. Nature Immunol 2002; 3:251-58.
29. Nagel W, Zeitlmann L, Schlicher P et al. Phosphoinositide 3-OH kinase activates the β2-integrin adhesion pathway and induces membrane recruitment of cytohesin-1. J Biol Chem 1998; 273:14853-61.
30. Krawczyk C, Oliveira-dos-Santos A, Sasaki T et al. Vav1 controls integrin clustering and MHC/peptide-specific cell adhesion to antigen-presenting cells. Immunity 2002; 16:331-43.
31. Kolanus W, Nagel W, Schiller B et al. αLβ2 integrin/LFA-1 binding to ICAM-1 induced by cytohesin-1, a cytoplasmic regulatory molecule. Cell 1996; 86:233-42.

32. Stewart MP, McDowall A, Hogg N. LFA-1-mediated adhesion is regulated by cytoskeletal restraint and by a Ca2+ -dependent protease, calpain. J Cell Biol 1998; 140:699-707.
33. Fagerholm S, Prescott A, Cohen P et al. An essential role for calmodulin in regulating human T cell aggregation. FEBS Lett 2001; 491:131-36.
34. Hara T, Fu SM. Phosphorylation of α, β subunits of 180/100-Kd polypeptides (LFA-1) and related antigens. In: Reinherz EL, Haynes BF, Nadler LM, Bernstein ID, eds. Leukocyte Typing II, Vol. 3. New York: Springer-Verlag, 1986:77-84.
35. Chatila TA, Geha RS, Arnaout MA. Constitutive and stimulus-induced phosphorylation of CD11/CD18 leukocyte adhesion molecules. J Cell Biol 1989; 109:3435-44.
36. Buyon JP, Slade SG, Reibman J et al. Constitutive and induced phosphorylation of the α- and β-chains of the CD11/CD18 leukocyte integrin family. J Immunol 1990; 144:191-97.
37. Valmu L, Autero M, Siljander P et al. Phosphorylation of the β-subunit of CD11/CD18 integrins by protein kinase C correlates with leukocyte adhesion. Eur J Immunol 1991; 21:2857-62.
38. Hibbs ML, Jakes S, Stacker SA et al. The cytoplasmic domain of the integrin lymphocyte function-associated antigen 1 β subunit: Sites required for binding to intercellular adhesion molecule 1 and the phorbol ester-stimulated phosphorylation site. J Exp Med 1991; 174:1227-38.
39. Valmu L, Gahmberg CG. Treatment with okadaic acid reveals strong threonine phosphorylation of CD18 after activation of CD11/CD18 leukocyte integrins with phorbol esters or CD3 antibodies. J Immunol 1995; 155:1175-83.
40. Fagerholm S, Morrice N, Gahmberg CG et al. Phosphorylation of the cytoplasmic domain of the integrin CD18 chain by protein kinase C isoforms in leukocytes. J Biol Chem 2002; 277:1728-38.
41. Valmu L, Hilden TJ, van Willigen G et al. Characterization of β2 (CD18) integrin phosphorylation in phorbol ester-activated T lymphocytes. Biochem J 1999; 339:119-25.
42. Valmu L, Fagerholm S, Suila H et al. The cytoskeletal association of CD11/CD18 leukocyte integrins in phorbol ester-activated cells correlates with CD18 phosphorylation. Eur J Immunol 1999; 29:2107-18.
43. Vermot-Desroches C, Wijdenes J, Valmu L et al. A CD44 monoclonal antibody differentially regulates CD11a/CD18 binding to intercellular adhesion molecules CD54, CD102 and CD50. Eur J Immunol 1995; 25:2460-64.
44. Constantin G, Majeed M, Giagulli C et al. Chemokines trigger immediate β2 integrin affinity and mobility changes: Differential regulation and roles in lymphocyte arrest under flow. Immunity 2000; 13:759-69.
45. Robinson MK, Andrew D, Rosen H et al. Antibody against the Leu-CAM β-chain (CD18) promotes both LFA-1- and CD3-dependent adhesion events. J Immunol 1992; 148:1080-85.
46. Andrew D, Shock A, Ball E et al. KIM185, a monoclonal antibody to CD18 which induces a change in the conformation of CD18 and promotes both LFA-1- and CR3-dependent adhesion. Eur J Immunol 1993; 23:2217-22.
47. Ortlepp S, Stephens PE, Hogg N et al. Antibodies that activate β2 integrins can generate different ligand binding states. Eur J Immunol 1995; 25:637-43.
48. Soede RDM, Driessens MHE, Ruuls-Van Stalle L et al. LFA-1 to LFA-1 signals involve ξ-associated protein-70 (ZAP-70) tyrosine kinase: Relevance for invasion and migration of a T cell hybridoma. J Immunol 1999; 163:4253-61.
49. Ni H-T, Deeths MJ, Li W et al. Signaling pathways activated by leukocyte function-associated Ag-1-dependent costimulation. J Immunol 1999; 162:5183-89.
50. Geginat J, Bossi G, Bender JR et al. Achorage dependence of mitogen-induced G1 to S transition in primary T lymphocytes. J Immunol 1999; 162:5085-93.
51. Porter JC, Bracke M, Smith A et al. Signaling through integrin LFA-1 leads to filamentous actin polymerization and remodeling, resulting in enhanced T cell adhesion. J Immunol 2002; 168:6330-35.
52. Li R, Nortamo P, Kantor C et al. A leukocyte integrin binding peptide from intercellular adhesion molecule-2 stimulates T cell adhesion and natural killer cell activity. J Biol Chem 1993; 268:21474-77.

53. Li R, Xie J, Kantor C et al. A peptide derived from the intercellular adhesion molecule-2 regulates the avidity of the leukocyte integrins CD11b/CD18 and CD11c/CD18. J Cell Biol 1995; 129:1143-53.
54. Kotovuori A, Pessa-Morikawa T, Kotovuori P et al. ICAM-2 and a peptide from its binding domain are efficient activators of leukocyte adhesion and integrin affinity. J Immunol 1999; 162:6613-20.
55. Stewart MP, Cabañas C, Hogg N. T cell adhesion to intercellular adhesion molecule-1 (ICAM-1) is controlled by cell spreading and the activation of integrin LFA-1. J Immunol 1996; 156:1810-17.
56. Weitz-Schmidt G, Welzenbach K, Brinkmann V et al. Statins selectively inhibit leukocyte function antigen-1 by binding to a novel regulatory integrin site. Nature Med 2001; 7(6):687-92.
57. Xiong J-P, Stehle T, Zhang R et al. Crystal structure of the extracellular segment of integrin αVβ3 in complex with an Arg-Gly-Asp ligand. Science 2002; 296:151-55.
58. Liddington RC. Will the real integrin please stand up? Structure 2002; 10:605-07.
59. Bailly P, Tontti E, Hermand P et al. The red cell LW blood group protein is an intercellular adhesion molecule which binds to CD11/CD18 leukocyte integrins. Eur J Immunol 1995; 25:3316-20.
60. Hermand P, Huet M, Callebaut I et al. Binding sites of leukocyte β2 integrins (LFA-1, Mac-1) on the human ICAM-4/LW blood group protein. J Biol Chem 2000; 275:26002-10.
61. Tian L, Yoshihara Y, Mizuno T et al. The neuronal glycoprotein telenchephalin is a cellular ligand for the CD11a/CD18 leukocyte integrin. J Immunol 1997; 158:928-36.
62. Tian L, Nyman H, Kilgannon P et al. Intercellular adhesion molecule-5 induces dendritic outgrowth by homophilic adhesion. J Cell Biol 2000; 150:243-52.
63. Gahmberg CG, Valmu L, Kotovuori A et al. Leukocyte adhesion - an integrated molecular process at the leukocyte plasma membrane. Biosci Rep 1999; 19:273-81.
64. Staunton DE, Dustin ML, Erickson HP et al. The arrangement of the immunoglobulin-like domains of ICAM-1 and the binding sites for LFA-1 and rhinovirus. Cell 1990; 61:243-54.
65. Koivunen E, Ranta T-M, Annila A et al. Inhibition of β_2 integrin-mediated leukocyte cell adhesion by leucine-leucine-glycine motif-containing peptides. J Cell Biol 2001; 153:905-15.
66. Gbarah A, Gahmberg CG, Boner G et al. The leukocyte surface antigens CD11b and CD18 mediate the oxidative burst activation of human peritoneal macrophages induced by type-1 fimbriated Escherichia-coli. J Leuk Biol 1993; 54:111-13.
67. Kantarjian H, Sawyers CL, Hochhaus A et al. Hematologic and cytogenetic responses to imatinib mesylate in chronic myelogenous leukemia. New Engl J Med 2002; 346:645-52.
68. Joensuu H, Roberts P, Sarlomo-Rikala M et al. Effect of the tyrosine kinase inhibirot ST1571 in a patient with a metastatic gastrointestinal stromal tumor. New Engl J Med 2001; 344:1052-56.

CHAPTER 7

Integrin αEβ7: Molecular Features and Functional Significance in the Immune System

Peter J. Kilshaw and Jonathan M.G. Higgins

Abstract

Alpha E beta 7 is an α-I domain-containing integrin that is expressed mainly on T cells, dendritic cells and mast cells in mucosal tissues. Expression depends largely or solely on TGF-β. The best characterized ligand for αEβ7 is E-cadherin on epithelial cells. An exposed acidic residue on the distal aspect of E-cadherin domain 1 interacts with the MIDAS site in the αE α-I domain. By binding to E-cadherin, αEβ7 contributes to mucosal specific retention of T cells, dendritic cells and mast cells within epithelia. Studies on αE knockout mice have identified an additional important function for this integrin in allograft rejection and have also indicated that it may have a role in immunoregulation. A recent finding that αEβ7 is expressed by a major subset of regulatory T cells opens new avenues for functional studies.

Introduction

Alpha E beta 7 is, in many respects, an unusual integrin. The αE subunit (CD103) has unique structural features (Fig. 1) and is the only α-I domain-containing integrin chain that pairs with β7. Beta 7, however, can pair with α4 as well as αE. Both heterodimers are expressed exclusively by leukocytes and have special significance for the mucosal immune system. Alpha 4 beta 7 is the principal mucosal homing receptor for leukocytes[1] whereas αEβ7 appears to play a role in retention of these cells within or near epithelia. In this review we shall discuss the tissue distribution and induction of αE and present a molecular perspective on the interaction between αEβ7 and its principal ligand, E-cadherin. The complex organization of the αE gene locus will be described. Finally, we aim to present a balanced account of functional studies on αEβ7 mentioning, in particular, its significance in inflammatory bowel disease and transplantation.

Tissue and Cellular Distribution

Integrin αEβ7 was discovered in the rat, human and mouse by screening panels of monoclonal antibodies for cell surface features that were distinctive for intestinal intraepithelial lymphocytes (IEL).[2-4] The original mAbs that showed such unique specificity, RGL-1, HML-1

I Domains in Integrins, edited by Donald Gullberg. ©2003 Eurekah.com and Kluwer Academic / Plenum Publishers.

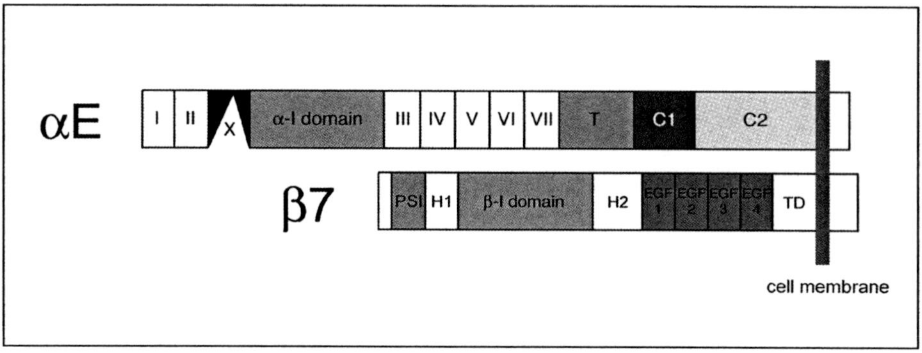

Figure 1. Domain structure of integrin αEβ7. The seven blades of the β-propeller domain of αE are labeled I to VII; the thigh domain, T; the calf-1 and –2 domains, C1 and C2. The extra X-domain, not found in any other integrin chain and containing a post-translational cleavage site, is marked X. In β7, the plexin/semaphorin/integrin domain is labeled PSI; the two components of the hybrid domain, H1 and H2; and the β-tail domain, TD. Note that this figure does not illustrate the relative orientation of the different domains.

and M290 (reactive in rat, man and mouse respectively) were subsequently shown to identify a novel integrin alpha chain now known as αE (CD103).[5-13] A fourth antibody, MRC-OX62, raised against rat lymphatic dendritic cells was later shown also to recognise αE.[14,15] Thus, the main distinguishing feature of αEβ7 is that it is expressed most prominently and abundantly in the gut, particularly on T cells in the epithelium.[2-6,15,16] In other compartments of the immune system and among other lymphoid/myeloid cell lineages expression is confined to sub-populations which may be numerically much smaller but are, nevertheless, functionally important.

Transcription of the αE subunit is stimulated by TGF-β.[5,6,10,12,17] Correspondingly, αEβ7$^+$ T cells are usually found in locations where active TGF-β is abundant. Expression of αEβ7 on T cells is usually skewed towards the CD8 subset,[3,4,16] a phenomenon that is readily seen in mixed T cell cultures stimulated with mitogen in the presence of TGF-β.[18-20] In the gut, almost all IEL and about half the T lymphocyte population in the lamina propria (LPTL) express αEβ7.[3,4,16] Similarly, the integrin is present on T cells in or near other epithelial surfaces, including those of the lung[20] and genital tract.[21,22] In lymphoid tissues, including Peyer's patches and mesenteric lymph nodes and in peripheral blood, the percentage of αEβ7$^+$ T cells is generally small.[3,5,23]

Although αEβ7 is usually considered to be a mucosal T-cell marker, the molecule is also found in other cell lineages. Dendritic antigen-presenting cells (DC) in the gut mucosa, the mesenteric lymph nodes and the epithelium of the airways of rats and mice are αEβ7$^+$ [14,15,24-26] but in lymph nodes which have no mucosal involvement the proportion is considerably smaller[24] and in the spleen αEβ7 expression is confined to the small subset of CD8$^+$ DC.[27] In man expression of αEβ7 by mucosal dendritic cells has not been well documented but one report shows that αEβ7$^+$ DC are present in the dome epithelium of Peyer's patches.[16] In contrast, Langerhans-type DC generated in vitro from hematopoietic stem cells in the presence of TGF-β and other cytokines do not express the integrin.[28] Most studies on the distribution of αEβ7 have failed to detect the molecule on tissue macrophages. There is, however, one exception in which a proportion of macrophages in lung, liver and lymph node sinuses is reported to have stained positively with mAb HML-1[29] Among other cell lineages, the interesting observation was made that mucosal-type mast cells generated in vitro from bone marrow precursors by culturing in the presence of stem cell factor, IL-3, IL-9 and TGF-β expressed αEβ7 strongly.[12,30]

The presence of the integrin on mucosal mast cells in vivo is strongly supported by circumstantial evidence from functional studies described below, but has yet to be demonstrated by direct observation.

Detailed scrutiny of lymphoid subsets for expression of αEβ7 has revealed a complex picture. Early studies showed that the integrin was expressed by few if any B cells in the gut mucosa or elsewhere. However, Csencsits et al[31] recently identified a population of $\alpha E\beta 7^+$ $B220^+$ cells in the intestinal mucosa following intranasal immunisation of mice with cholera toxin. That cells of B lymphocyte lineage can, in certain circumstances, express αEβ7 is supported by the detection of a small population of $CD19^+$ αE^+ B cells in peripheral blood[32] and also by much earlier observations that αEβ7 expression is a diagnostic marker for hairy cell leukemias.[7,33,34]

Studies on αEβ7 expression during thymic ontogeny in the mouse have shown that 3-5% of cells express the integrin and that it is represented in both TCRαβ and γδ lineages, particularly in the late developmental stages.[23,35] The integrin is present on about half the population of thymic precursors of dendritic epidermal T cells and on all mature cells of this subset.[35] In humans, αEβ7 was found to be expressed by a major subpopulation of single positive $CD8^+$ human thymocytes and a smaller proportion of less mature double negative cells.[36,37] It has also been reported that most, if not all, naïve $CD8^+$ that have recently emigrated from the thymus into the circulation express αEβ7.[36] Thus, the frequency of $\alpha E\beta 7^+$ CD8 T cells in the blood with a naïve phenotype appears to be a useful indicator of thymopoiesis. Maintenance of αEβ7 expression by this cell population, and also by splenic and blood $CD8^+$ T cells, has been reported to depend on lymphotoxin alpha (LTα).[38] It is not clear whether LTα regulates synthesis of the αE or the β7 subunit.

Study of T regulatory (Treg) cells (formerly known as suppressor T cells) is currently undergoing a renaissance and their role in immune homeostasis and in the prevention of autoimmune diseases and allograft rejection is being keenly debated. In vivo models of suppression of autoimmunity involving adoptive cell transfer and in vitro studies on suppression of lymphocyte proliferation by spleen or lymph node T cell sub-populations have shown that Treg cells reside within a population that is $CD4^+$ $CD25^+$ $CD45RB^{low}$.[39,40] Four recent studies have shown that αEβ7 is expressed by 20-30% of this T cell subset.[41-44] McHugh et al[43] and Lehmann et al[41] both reported that that the $\alpha E\beta 7^+$ population was more efficient at suppressing anti-CD3 stimulated proliferation of $CD4^+$ $CD25^-$ cells than the $\alpha E\beta 7^-$ subset. Moreover, suppressor function was detected also in $CD4^+$ $CD25^-$ αE^+ cells.[41] In another recent study, regulatory $CD8^+$ T cells generated by co-culture of intestinal epithelial cells and peripheral T cells were shown to express αEβ7.[45] The relative importance of TGF-β and IL-10 in the development and function of Treg cells remains highly controversial but the presence of αEβ7 on the surface of a major subpopulation of Treg cells argues, at least, that these cells have recently been exposed to TGF-β. It is notable that $CD4^+$ $CD25^+$ cells that have been stimulated with anti-CD3 are reported to express TGF-β on the cell surface and to exert TGF-β-dependent suppressive activity in vitro via a cell contact-dependent mechanism[46,47] although this finding has not been reproduced in all experimental systems.[48,49] It is not yet clear whether αEβ7 has any role in the effector function or localization of Treg cells or merely acts as a surface marker reflecting exposure to TGF-β produced by the cell itself or from another source.

Induction of αEβ7

Early studies established that αEβ7 expression is induced by TGF-β[5,6,10,12,17] No other stimulus has consistently been demonstrated to have this effect, although it has been reported that ligation of β1 integrins can act synergistically with TGF-β in αE induction,[20] and that activation of naïve human $CD8^+$ T cells with anti-CD3 in the presence of IL-4 can also increase αEβ7 expression.[50] It believed that αEβ7 expressed by T cells located in the vicinity of epithelia is induced locally by TGF-β produced mainly by epithelial cells. This view is supported by the observation that T cells stimulated in vitro by co-culture with allogeneic kidney epithelial cells, or T cells that migrate into epithelial monolayers, are induced to express αEβ7 and that expression is blocked by anti-TGF-β antibody.[51,52] The results of a study of mucosal T cell memory by Kim et al[53] are also consistent with the idea that αEβ7 is upregulated locally. Ovalbumin-specific transgenic $CD8^+$ T cells were adoptively transferred to recipients that were then infected with recombinant vesicular stomatitis virus expressing ovalbumin (VSV-OVA). Analysis of donor-type memory cells in various lymphoid compartments indicated that αEβ7 was strongly upregulated on IEL over the five week study period. The notion that αEβ7 expression is mainly, if not solely, TGF-β-dependent is supported by a recent study showing that in transgenic mice which express the negative regulator of TGF-β signalling, Smad 7, under an *lck* promoter, 50% of intraepithelial T cells in the gut no longer express αE.[54] Expression of the integrin by the remaining cells probably reflects insufficient expression of the transgene in this population but leaves open the possibility that an alternative signaling pathway could be responsible for αE expression in these circumstances. Although transcription appears to depend on signaling via the Smad pathway, the integrin is expressed normally in *smad3* null/null mice (PJK unpublished). Involvement of other signaling pathways is likely because inhibition of p38 MAP kinase compromises induction of αE by TGF-β, suggesting an essential role for a MAPK pathway (PJK unpublished). Using a T cell line, Robinson et al showed that TGF-β induces αE transcription de novo within 30 min.[17] The speed of induction suggests that synthesis of signaling intermediaries or new transcription factors was probably not required. These authors also looked for transcription control elements in the promoter region of the human αE gene using deletion analysis to examine 4 kb of genomic sequence upstream of the transcription start site. Although the promoter functioned well in reporter assays, it bestowed neither cell lineage specificity nor TGF-β responsiveness. Thus, transcription control mechanisms for αE are likely to be considerably more complex than those of most other integrin α-chain genes, where lineage specificity is generally determined by the proximal promoter.

Gene Structure

Recent studies have revealed considerable complexity at the locus of the integrin αE gene. Schön et al[55] generated a partial map of the murine integrin αE gene, and subsequently the human genome sequencing project has provided more complete information on the human αE gene.[17,56] The human αE gene contains 31 exons spanning approximately 85 kb (Fig 2). Comparison with the closely related αM and αX genes[57-59] reveals a highly conserved gene structure. All the introns are located in similar positions and have the same phase in the three genes, although αE contains an extra exon (exon 6) that encodes the X domain not present in other integrins. The α-I domain is encoded by exons 7-10. Alpha E is found at chromosome 17p13.3 rather than in the αL/αM/αX/αD integrin cluster at chromosome 16p11,[58,60] and is syntenic with the murine αE gene on chromosome 11.[55] Robinson et al[17] analyzed the transcription start site of the human αE gene, and identified two start sites 51 and 44 bp upstream of the start codon, and a third possible initiation site about 40bp further upstream. During these studies, another tightly linked gene was discovered within 6 kb of the 5' end of the αE gene (Fig. 2). This gene, named ELG, contains 9 exons and encodes a predicted protein containing a nuclear localization signal and a highly acidic region. Transcription of ELG is regulated independently of αE (PJK unpublished).

Higgins[56] reported the existence of an alternatively spliced form of the integrin αE gene. The transcript, named human Alpha-E Derived (hAED), originates from a CpG island-associated promoter within intron 26 near the 3' end of the conventional αE gene (Fig. 2). A similar transcript found in mice was named mAED. The hAED transcript does not contain a start codon in the open reading frame of αE, and so probably does not encode a modified form of

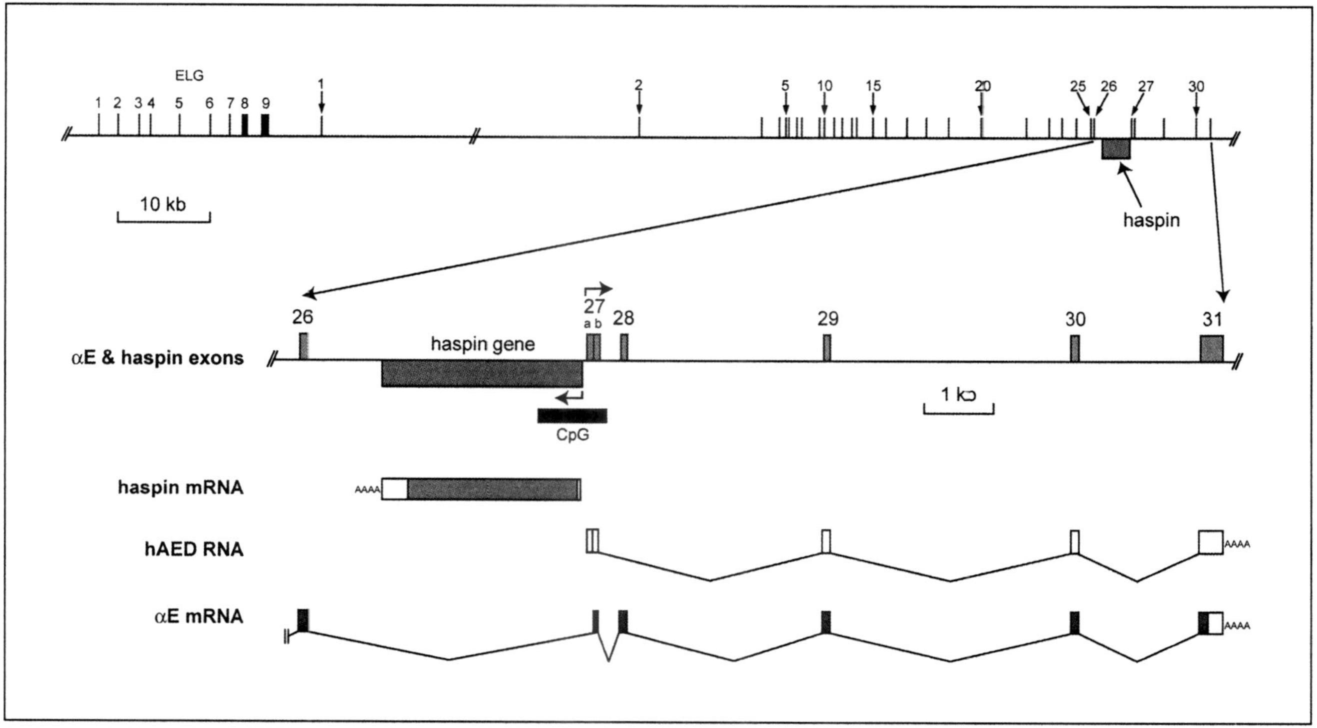

Figure 2. Outline structure of the αE genomic locus. The top line shows the intron-exon structure of the ELG, integrin αE and haspin genes. On the second line, the 3' region of the αE gene containing the haspin gene is shown in more detail. Intronic regions are shown as horizontal lines and exons as boxes. Bent arrows represent transcription start sites, and the thick black line indicates the location of a CpG island. The three lower lines show the three transcribed products of this genomic region. In each case, boxes represent exonic portions of RNA, thin lines indicate introns, dark shading indicates protein coding regions, and AAAA indicates a poly(A) tail.

the αE integrin chain. In hAED, a short reading frame for 30 amino acids is present, but no protein corresponding to this sequence has yet been identified. The corresponding mAED RNA could not encode a similar peptide. The hAED RNA is ubiquitously expressed in human tissues, but is particularly abundant in testis, while mAED can be detected only in the testis. The function, if any, of hAED/mAED is unknown, but it could act as a non-coding RNA. In both humans and mice intron 26 also contains the 2.8 kb intronless gene for the haploid cell specific nuclear protein kinase (haspin/Gsg2).[56,61,62] The hAED and haspin genes are transcribed in opposite directions with less than 70 bp of intervening sequence between them (Fig. 2). It is possible that hAED/mAED is a byproduct of transcription of haspin and could be involved in its regulation. Like hAED, haspin is expressed predominantly in the testis but haspin does not appear always to be co-regulated with hAED or αE.[56] Given the location of the haspin gene and its lack of introns, both Higgins[56] and Yoshimura et al[61] speculated that it might be a retrogene, created by retro-insertion of a spliced mRNA into a pre-existing intron of the αE gene. Haspin is the prototypical member of a distinctive group of serine/threonine kinases that are found throughout the eukaryotes.[63] Murine haspin may play a role in cell cycle control since it is reported to cause cell cycle arrest when overexpressed in HEK-293 cells.[64]

It has been reported that in the rat there are two αE genes, αE1 and αE2.[15] It appears that the two cDNAs were cloned from a rat/mouse T cell heterohybridoma. This difficulty together with the observation that rat αE1 and mouse αE are closely similar casts serious doubt on the provenance of rat αE1.

The human β7 gene comprises 14 exons spanning approximately 10 kb and maps to chromosome 12q13.13,[65,66] syntenic with murine β7 on chromosome 15.[67] The intron-exon structure of the β7 gene is more similar to that of the β1 and β2 genes than the β3, β5 and β6 genes, consistent with a similar sub-grouping derived from analysis of sequence homology.[65]

Ligand Binding

Expression of αEβ7 by T cells closely juxtaposed to epithelial surfaces suggested that this integrin might bind a counter-receptor on the surface of epithelial cells or within the local extracellular matrix (ECM). No ECM ligand has yet been found.[68] It has been reported that αEβ7 can mediate adhesion of cells to synthetic polymeric RGD peptides, but the physiological significance of this is doubtful because adhesion to authentic RGD-containing proteins could not be demonstrated.[69] In 1993 three groups reported that a ligand for αEβ7 was present on epithelial cell lines. Cepek et al,[70] Roberts and Kilshaw[71] and Roberts et al[72] found that αEβ7-expressing T cells adhered to epithelial cells in a manner that was inhibited by antibodies to αEβ7. To identify the epithelial ligand, Cepek et al[73] generated a panel of monoclonal antibodies to epithelial cells and screened them for the ability to block adhesion of IEL to an epithelial monolayer. An antibody with this property was shown to recognize E-cadherin. Transfection of L-cells with a human E-cadherin cDNA construct conferred on them the ability to bind IEL. Conversely, transfection of COS-7 cells with αEβ7 allowed them to bind to E-cadherin transfected, but not mock transfected, L-cells. Karecla et al[74] took an alternative approach, screening monoclonal antibodies to known epithelial cell surface molecules for blocking capacity. Again, E-cadherin was identified as the likely counter-receptor for αEβ7 and transfection of L cells with murine E-cadherin confirmed that it supported increased adhesion of αEβ7-expressing T cells.

The identification of E-cadherin as a likely ligand for αEβ7 was both satisfying and surprising. The expression pattern of E-cadherin makes it an appealing tissue-specific counter-receptor for a leukocyte integrin because E-cadherin has a restricted tissue distribution and is found predominantly on the basolateral aspect of epithelial cells, the site at which IEL are found.[73] On the other hand, the principal function of cadherins is calcium-dependent homophilic cell-to-cell adhesion and E-cadherin has a pivotal role in the formation of adherens junctions between epithelial cells. Cadherins are not therefore obvious candidates for integrin ligands. Given this background, it was important to demonstrate a direct interaction between integrin αEβ7 and E-cadherin at a molecular level. Higgins et al[75] demonstrated that human E-cadherin-Fc fusion proteins could support αEβ7-mediated cell adhesion and, more importantly, that E-cadherin-Fc could bind to αEβ7 solubilised from IEL and to a soluble recombinant form of αEβ7. In common with other integrins, the adhesive capacity of αEβ7 can be regulated. Adhesion of αEβ7-expressing cells to E-cadherin-Fc was found to be dependent on magnesium ions, and could be activated by manganese. In addition, cell stimulation with phorbol ester or by cross-linking the TCR/CD3 complex with antibody increased αEβ7-mediated adhesion to E-cadherin.[75]

The extracellular domains of E-cadherin have an Ig-like fold[76] and several solvent-exposed acidic side chains that, by analogy with other integrin ligands, could be contact residues for integrin binding. Karecla et al[77] were the first to show that, in mouse E-cadherin, the BC-loop of the most distal domain (domain 1) contains a glutamate residue (E31) that is required for adhesion of αEβ7-expressing T cells to E-cadherin transfectants. Mutation of this residue to alanine, but not mutation of other exposed acidic residues, abolished αEβ7-mediated adhesion. Analysis of human E-cadherin confirmed the critical role of E31 and defined a wider αEβ7 binding site formed by the BC and FG loops at the top of domain 1.[78] Mutation of residues N27 and K30 in the BC loop and E89 in the FG loop within the E-cadherin-Fc fusion protein all substantially reduced adhesion of αEβ7-expressing cells. Modeling of human E-cadherin domain 1 on the murine E-cadherin crystal structure suggested that N27, K30 and E31 might directly contact the integrin, while E89 might stabilize the BC-FG loop interface[78] and/or be involved in cadherin cis-dimer formation[79,80] which could be important for integrin binding (see below). It is interesting to compare E-cadherin with three other Ig-like structures in fibronectin, MAdCAM-1 and ICAM-1, that have exposed acidic amino acids and serve as integrin ligands. The loop that displays the critical integrin-binding residue and the contact face of the module involved is different in each case. The integrin binding site on E-cadherin is clearly distinguishable from that involved in homophilic E-cadherin interactions since the mutation of E31 has no influence on the localization of E-cadherin to cell-cell junctions or the adhesion of E-cadherin expressing cells to E-cadherin-Fc.[77,78] This is an important finding given the interest in developing small molecule inhibitors of E-cadherin binding to αEβ7 that do not affect E-cadherin homophilic interactions.

As described elsewhere in this volume, for those integrins that contain an α-I domain, this structure is critically involved in ligand binding. In each case, the critical acidic residue in the ligand is thought to co-ordinate a magnesium ion in the MIDAS site of the α-I domain. Given the distinctive nature of the interaction, it was attractive to test a similar model for αEβ7 binding to E-cadherin. Higgins et al[81] found that the mutation D190A in the DxSxS motif of the MIDAS in the αE α-I domain did indeed abolish E-cadherin binding. A model of the αE α-I domain based on the known structure of the most homologous α-I domain, from αM, allowed identification of other surface residues that might be involved in E-cadherin binding. Mutation of F298, a residue predicted to be prominently exposed on the α-I domain surface near the MIDAS cleft, also completely abolished E-cadherin binding. Furthermore, monoclonal antibodies that block αEβ7 binding to E-cadherin mapped to the surface of the α-I domain surrounding the MIDAS site. A docking model of the α-I domain with E-cadherin domain 1 indicated that co-ordination of the αE MIDAS metal ion by E-cadherin residue E31 and a novel projection of F298 into a hydrophobic pocket between the BC and FG loops on E-cadherin might provide the basis for the interaction (Fig. 3). The N27, K30 and E89 residues of E-cadherin are all found at the predicted binding interface and could form bonds to residues within the α-I domain. The location of the binding site on the αE α-I domain resembles that on other integrins, but its structure appears distinctive and particularly adapted to

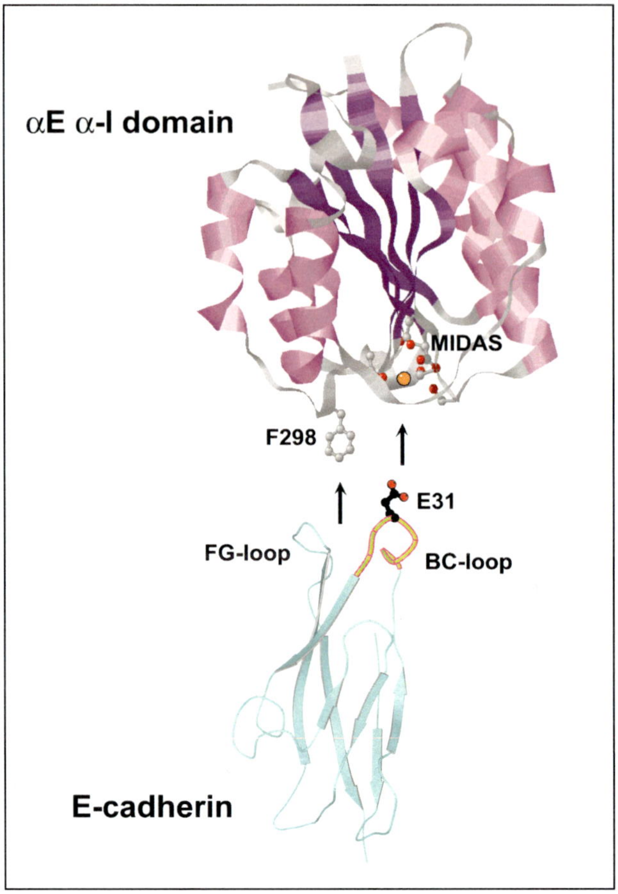

Figure 3. A model of the αE integrin α-I domain docked onto E-cadherin domain 1. Residue E31 in the BC-loop of E-cadherin is predicted to coordinate the MIDAS magnesium ion in the α-I domain and F298 of αE is predicted to project into a hydrophobic pocket between the BC and FG loops of E-cadherin. For details see ref. 81. Reproduced with permission from Agace WW, Higgins JMG, Sadasivan B, Brenner MB, Parker CM. Curr. Opin. Cell Biol. 2000; 12:563-568 (Copyright 2000, Elsevier Science).

recognize the tip of E-cadherin.[81] Crystal structures of the α-I domain of the related integrin αM[82] and of α2 integrin,[83] strongly suggest that co-ordination of an acidic residue from the ligand by the MIDAS cleft requires a major conformational change involving displacement of the C-terminal α7 helix of the α-I domain. On the basis of this structural data Lu et al[84] and Shimaoka et al.[85] demonstrated that the α-I domain of αL could be locked in an 'open' high affinity conformation by adjusting the position of the C-terminal α7 helix by introducing two cysteine substitutions which would form a disulphide bond. In these circumstances αLβ2 showed enhanced affinity for ICAM-1. A similar approach with the αE α-I domain also greatly increased ligand binding by the αEβ7 heterodimer and allowed the αI domain in isolation to bind strongly to E-cadherin, suggesting that affinity is regulated by a similar conformational change (PJK unpublished).

Surprisingly, no direct role for the X-domain of αE in ligand binding has been established. Removal of the cleavage site, deletion of a single residue within its charged region, and inclusion of a polyclonal antiserum to the X-domain all failed to prevent αEβ7 binding to E-cadherin.[81] Cleavage within the X domain could, in principal, optimize the orientation of the α-I domain for ligand binding by loosening constraints on its positioning relative to the β-propeller and to the integrin heterodimer as a whole. X-domains of differing lengths are present in murine, rat and feline αE chains,[12,15,86] perhaps consistent with a role in refining ligand binding.

Although the model described above for binding to E-cadherin via the α-I domain is highly plausible, it remains unclear whether the interaction of the integrin α-I domain with the tip of E-cadherin domain 1 is the only contact between these two molecules. Mutation of the MIDAS motif in the β-I domain of β7 (D140A) prevents the binding of both αEβ7 to E-cadherin and α4β7 to MAdCAM-1.[81] This might suggest that both the alpha and beta chains contact ligand in these two cases. Indeed, it seems clear that in non α-I domain integrins like α4β7, both chains are involved directly in ligand binding. For example, portions of both the β-propeller of αV and the β-I domain of β3 comprise the binding site on αVβ3 for an RGD peptide. The MIDAS of the β-I domain in the β3 subunit directly contacts the critical acidic residue of the RGD sequence.[87] In addition to this primary binding site possessing an exposed acidic residue, "accessory" or "synergy" binding sites have been localized on ligands of non α-I domain integrins.[88-92] It seems likely that integrin residues that interact with these accessory sites are located primarily on the alpha subunit.[93] For α-I domain containing integrins that have MIDAS motifs in both the alpha and beta chains, it might be hypothesized that two acidic residues in the ligand are vital for ligand binding. In general however, only one such residue has been identified. It is conceivable that the necessary pairing of acidic residues might be achieved by dimerization of ligand. Indeed, both ICAM-1 and E-cadherin are thought to form dimers on the cell surface[94-96] However, the two critical E34 residues in an ICAM-1 dimer appear too far apart to bind simultaneously a single αLβ2 molecule,[97] and recent studies have shown that a single ICAM-1 monomer provides a fully competent αLβ2-binding surface. Dimerization of ICAM-1 probably enhances αLβ2 binding simply by providing two αLβ2 binding sites.[98] Moreover, studies of an isolated αL α-I domain locked in the active conformation suggest that this domain alone contains the sole binding site for ICAM-1.[84,85,99] It has been suggested that for α-I domain integrins, the β-I domain serves to regulate the affinity of the α-I domain for ligand,[82-84,99,100] possibly via interaction with an acidic residue in the alpha chain[101] This contrasts with non-α-I-domain integrins where, as discussed above, both alpha and beta chains participate directly in ligand binding.

It remains to be seen how the interaction between αEβ7 and E-cadherin fits these proposed binding mechanisms. Studies by Corps et al[102] on cadherin dimerization showed that when the potential mechanisms for E-cadherin dimer formation were disrupted by point mutations, interaction with αEβ7 was inhibited. The propensity of E-cadherin to dimerize to form a functional unit was demonstrated using antibodies to epitopes that are hidden in the dimer but exposed in the monomer. Cadherin dimerization is thought to depend on Ca^{2+} binding sites at the base of domain 1[76,103,104] and on a tryptophan residue near the N-terminus.[79,80] The NMR and crystal structures of E-cadherin suggest that the point mutations that were made in these positions are unlikely to directly affect the integrin binding site on the distal surface of domain 1.[76,103,104] These structures also suggest that, for steric reasons, a cadherin dimer is unlikely to bind to two integrin holoreceptors. Thus, if dimerization plays a part in facilitating the interaction with αEβ7, it implies that contact sites on each component of the two cadherin molecules may be required for stable adhesion to occur. However, more recent work has established that single E-cadherin molecules can support adhesion of Mn^{2+}-activated $\alpha E\beta 7^{+}$ cells (PJK unpublished). It is conceivable that a secondary ligand contact provided by the cadherin dimer could regulate affinity of the α-I domain. The true significance of cadherin dimerization in this situation is not yet clear.

A number of studies have shown that αEβ7 is specific for E-cadherin. The integrin is unable to recognise MAdCAM-1, VCAM-1 and fibronectin, known ligands for the related integrin α4β7; nor ICAM-1, the ligand for αLβ2.[75,105] A search for αEβ7 ligands in the cadherin family showed that the integrin interacts weakly with human P-cadherin[75] but not with a panel of other cadherins (PJK unpublished). More than one hundred fold more P-cadherin-Fc than E-cadherin-Fc is required to support the same level of adhesion. While human P-cadherin has an acidic residue at position 31 in domain one, and may bind αEβ7 as does E-cadherin, murine P-cadherin lacks such a residue and no adhesion of αEβ7-expressing cells to murine P-cadherin transfectants could be detected.[77]

Notwithstanding this narrow specificity, more recent reports have provided preliminary evidence for the existence of at least one further ligand for αEβ7. Brown et al[106] found that human peripheral blood T cells activated with phytohemagglutinin and TGF-β (to up-regulate αEβ7) adhered to an oral keratinocyte cell line, H376, that lacks E-cadherin expression. Adhesion was reduced in the presence of an antibody to αEβ7. The identity of the putative receptor remains unknown. A different perspective comes from the work of Pauls et al[107] who reported that adhesion of similarly activated T cells to human primary skin keratinocytes and to psoriatic epidermis was dependent on both αEβ7 and E-cadherin. Evidence for a second ligand has also been reported by Strauch et al[108] who found that cultured human IEL could adhere to primary human endothelial cells derived from intestinal lamina propria microvessels (HIMEC), in a manner that could be inhibited by antibodies to αEβ7 but not E-cadherin. Indeed, the HIMEC cells lacked detectable E-cadherin expression. The authors showed that αEβ7 transfectants had increased adherence to HIMEC cells. Further, HIMEC cells adhered to an αEβ7-Fc fusion protein coated on plastic, but not to an αEβ7-Fc protein containing the D190 to alanine MIDAS mutation of the αE α-I domain previously shown to prevent E-cadherin binding. Lastly, αEβ7-dependent adhesion of cells to lamina propria blood vessels in jejunal tissue sections was observed. Again, the identity of this potential counter-receptor remains unknown.

Functions

Early in vitro studies identified a number of potential roles for the αEβ7 integrin. Most obviously, many workers have shown that αEβ7 mediates lymphocyte adhesion to epithelial cells through E-cadherin.[70-74,105,107,109,110] As described above, the avidity of αEβ7 for its counter-receptor increases upon triggering of the TCR/CD3 complex.[75] Thus, as proposed for αLβ2 binding to ICAM-1,[111] αEβ7 could act as a "servomotor" to allow stable interactions between a lymphocyte and a cell presenting an activating antigen. This could be important in arresting lymphocytes within the epithelium when specific antigen is recognized. Since IEL contact more than one cell when resident within the epithelium, localized upregulation of αEβ7 avidity within an IEL could aid in polarizing lymphocyte interactions toward the appropriate cell and contribute to the formation of an immunological synapse.[112] Because some dendritic cells express E-cadherin, including epidermal Langerhans cells,[113] the interaction with αEβ7 could also be important during recognition of antigen on such professional antigen presenting cells.

Once the binding of αEβ7 to its ligand is established, the interaction has the potential to provide co-stimulation to the T cell. In the presence of a sub-optimal level of anti-CD3 antibodies, immobilized anti-αEβ7 antibodies or E-cadherin-Fc increase T cell proliferation[68,69,114] and surface expression of markers including CD5,[115] CD25 and CD71.[114] Further, treatment of IEL with certain αE antibodies can increase the killing of P815 target cells in an anti-CD3 mediated redirected lysis assay[71] and one anti-αEβ7 monoclonal antibody, 2E7, can cause redirected lysis in the absence of anti-CD3 antibodies.[35] Consistent with the idea that αEβ7 can provide co-stimulation for effector T cell function, it has been shown that αEβ7 blocking antibodies inhibit the cytotoxicity of IEL or αEβ7-expressing T cells toward E-cadherin$^+$ target cells.[71,110,116,117] In these circumstances the integrin may provide both adhesive and signaling functions. We envisage that co-stimulatory effects are especially important for the effector function of IEL. Because ICAM-1, -2 and -3 are, in general, not expressed on the intestinal epithelium,[118-121] the interaction of αEβ7 with E-cadherin may provide a tissue-specific alternative to αLβ2 binding to ICAM family members. In fact, preliminary studies suggest that the signaling events triggered by αEβ7 and αLβ2 in cultured IEL are similar (JMGH and MB Brenner, unpublished). An unusual feature of αEβ7 and α4β7 that is not found in β1 and β2 integrins is that the cytoplasmic domains of both α and β subunits contain a membrane-proximal, nine amino acid, tyrosine-containing motif that interacts with a novel WD repeat protein, WAIT-1.[122] This molecule is homologous to mouse EED, a transcription regulator of homeobox genes. The functional significance of this interaction for inside-out or outside-in signaling via αEβ7 has not yet been investigated.

The expression pattern of αEβ7 suggests that it might be particularly important for localization of T cells to the intestinal mucosa, and potentially also to the bronchial and genital mucosae. Most authors have been careful to refer to the role of αEβ7 in "retention" rather than "homing" of leukocytes. This reflects the fact that E-cadherin is generally not expressed on endothelial cells, and thus a classical homing receptor/addressin function for this pair of adhesion molecules seems unlikely. Indeed, the interaction of α4β7 with MAdCAM-1 is considered critical for normal homing of lymphocytes to Peyer's patches and inflamed lamina propria.[1] In contrast, it has been proposed that, in vivo, αEβ7 expression is only induced on T cells after arrival in the TGF-β-rich environment of the mucosa (see above).

Analysis of mice with a "knockout" of the αE gene has provided a direct way to test αEβ7 function in vivo. Disruption of the αE gene results in reduced numbers of intestinal and vaginal IEL consistent with a role of αEβ7 in epithelial retention or expansion of IEL.[55] Mice lacking αE also have reduced numbers of dendritic epidermal T cells in the skin.[123] Interestingly, no difference was seen in the number of peribronchial T cells, suggesting that other interactions may contribute to T cell localization to this site. The extent of the intestinal IEL deficit was influenced by the environment and/or genetic background of the mice. This is perhaps not surprising given the proximity of the IEL compartment to the gut flora. Indeed, the deficiency in αE led to a proportionally greater loss of TCRαβ$^+$ than TCRγδ$^+$ IEL, and the reduction in IEL number relative to wild-type was not apparent until approximately four weeks after birth. This coincides with the time of arrival in the neonatal intestine of TCRαβ$^+$ cells driven by the intestinal flora.[124] Disruption of TGF-β signaling in mature T cells (see above) results in decreased αEβ7 expression on IEL and a reduction in IEL numbers, broadly consistent with these findings.[54] However, it seems that αEβ7 interaction with E-cadherin is not always critical for epithelial localization of T cells. Mice with disrupted E-cadherin function in the epithelium have an increased population of IEL expressing β7-integrin.[125] This can be attributed to the observed disruption of the epithelial barrier and the accompanying infection and inflammatory response that likely results in the upregulation of other homing and adhesive interactions in the intestine. Similarly, analysis of mice infected with VSV-OVA showed that αE deficiency did not prevent ovalbumin-specific TCRαβ$^+$ CD8$^+$ T cell localization to the intestinal epithelium during a primary immune response.[126] Perhaps even more surprisingly, αEβ7 was not required for the retention of the ovalbumin-specific CD8$^+$ IEL in the epithelium up to three weeks after infection. In a similar model, VSV-specific memory cells that express high levels of αEβ7 could be detected among IEL more than 11 weeks post infection,[127] so it would be interesting to see if αE deficiency affected this longer term maintenance of IEL. It would also be valuable to examine the retention of αE-deficient IEL when mice are challenged with a mucosal pathogen rather than systemic VSV infection. Despite the reduction in vaginal T cell numbers observed in αE deficient mice, the majority of T cells isolated from the genital mucosa of *Chlamydia trachomatis* infected mice do not express αEβ7. The expression of other integrins and addressin molecules in the genital, conjunctival and pulmonary mucosae suggests a more important role for α4β1 binding to VCAM-1 and αLβ2 binding to ICAM-1 rather than β7-mediated interactions.[128,129] Somewhat similar conclusions were drawn from a study of *Mycobacterium tuberculosis* in the lung.[130]

Interestingly, the numbers of intestinal lamina propria T lymphocytes (LPTL) in some strains of αE-deficient mice are significantly reduced despite the absence of E-cadherin at this site. The difference is more pronounced in mouse colonies that otherwise have large numbers of LPTL, perhaps due to less stringent housing conditions.[55] Comparing the impact of targeted disruption of integrin αE, β7 and α4 genes in mice, Schön et al concluded that αEβ7 might play a direct role in localization of T cells to the lamina propria as well as the epithelium. In β7-deficient mice, there are very low numbers of lymphocytes in the Peyer's patches, intestinal lamina propria and epithelium.[131] However, in RAG-2-deficient mice reconstituted with bone marrow from α4-deficient fetuses, IEL and LPTL numbers are not reduced, although there is still a deficit of Peyer's patch lymphocytes.[132,133] Together these data indicate that, in the absence of an overt immune response, α4β7 is required for entry into the Peyer's patch but not the lamina propria, while αEβ7 could conceivably participate in the latter process. Lúdvíksson et al[134] also observed a decrease in LPTL numbers upon anti-αEβ7 antibody treatment in a mouse colitis model (see below). Nevertheless, it is possible that the reduction in numbers of LPTL could be secondary to the reduction in IEL. This could occur, for example, if IEL secrete cytokines or chemokines necessary for maintaining LPTL numbers.

Evidence against a role for αEβ7 in homing to the gut has come from several independent studies. Induction of αEβ7 on anti-CD3 activated T cells from lymph nodes by treatment in vitro with TGF-β1 does not cause the cells to home to the intestine when they are injected intravenously into mice.[135] Although this suggests that αEβ7 has no role in homing it is not decisive. Extravasation is a multistep process, requiring a combination of specific chemokine and homing receptors.[1] For example, the majority of human LPL, IEL and αEβ7⁺ CD8⁺ PBL express the chemokine receptor CCR9 [136] that is known to be required for migration of T cells to the intestine in mice.[137] The TGF-β treated splenic T cells used in the Austrup study may have lacked CCR9 or another critical receptor. In another study that should circumvent this criticism, Koseki et al[138] investigated the adhesion of adoptively transferred IEL to the lamina propria microvessels of recipient mice. Antibodies to β7 or MAdCAM-1 partially inhibited the accumulation of IEL in microvessels of intestinal villi but antibodies to neither αEβ7 nor α4 had any effect. In addition, substantial differences were observed when comparing adhesion of $CD4^+$ and $CD4^-$ cells, or $TCR\alpha\beta^+$ and $TCR\gamma\delta^+$ cells, although these populations express similar levels of αEβ7. The authors concluded that α4β7, but not αEβ7, is involved in the initial step of adherence to the endothelial wall. The study emphasises the difficulty of quantifying the contribution of an individual adhesive mechanism when several receptor-ligand pairs are involved. Finally, in the study by Lefrancois et al[126] of antigen-specific $CD8^+$ T cell trafficking during VSV-OVA infection, β7-deficiency caused a marked decrease in T cell migration to the lamina propria. In contrast, αE-deficient T cells were not different from wild type cells in this regard, providing strong evidence that αEβ7 is not required for migration to the lamina propria during this primary immune response.

Taken together, the results discussed above support a role for αEβ7 in the retention and/or expansion of intestinal and vaginal IEL, particularly during the initial population of the epithelium, and in the localization of dendritic epidermal T cells in mice. In other tissues and during some inflammatory responses, however, it seems that localization to the epithelium can be accomplished by other adhesive interactions. The reduction in LPTL seen in αE-deficient mice has lead to a re-evaluation of αEβ7 binding to ligands in the lamina propria, and an endothelial ligand that might fulfill this role has been partially characterized.[108] Currently, we cannot rule out involvement of integrin αEβ7 homing to the gut, but this putative function is clearly dispensable in some circumstances.

A number of studies have examined whether αEβ7 on T cells could play a role in allograft rejection. Hadley et al[139,140] reported that up to 63% of T cells infiltrating renal allografts undergoing a late rejection crisis expressed αEβ7 and that the cells were localized mainly in the tubular epithelium. Similar findings were reported by Robertson et al[141] who observed a correlation between the prevalence of $\alpha E\beta 7^+$ cells in the tubular epithelium, the severity of tubulitis and the levels of TGF-β in the epithelium. Earlier studies established that αEβ7 was induced on $CD8^+$ T cells co-cultured with renal epithelial cells[51] and that αEβ7 provided accessory function for cytotoxic lysis of target epithelial cells.[142] This evidence supports the view that αEβ7 is induced on infiltrating $CD8^+$ cells by TGF-β produced locally in the allograft and causes the cells to accumulate in the graft epithelium by adhesion to E-cadherin expressed by tubular epithelial cells. The integrin/ligand interaction would then provide accessory function for cytotoxic lysis or cytokine production. This interaction may be especially important in rejection when other integrin/ligand interactions, principally αLβ2/ICAM-1, are unavailable. This view is strongly supported by a recent observation that αE null/null mice are unable to reject pancreatic islet allografts.[143,144] Although $CD8^+$ cells accumulate around the graft they do not come into intimate contact with islet cells, which are known to express E-cadherin but not ICAM-1. These observations on the role of αEβ7 in allograft rejection identify a potential opportunity for therapeutic intervention using inhibitors specific for this integrin.

To investigate the function of β7 integrins on mast cells, Artis et al[145] exploited the dramatic rise in intestinal mast cell numbers seen during enteric *Trichinella spiralis* infection. Beta 7-deficient mice showed delayed worm expulsion and reduced mast cell numbers in the duodenum. This did not seem to be a secondary phenomenon caused by reduced T cell localization. The mast cells that did appear in the intestine showed an aberrant distribution and were found in clusters surrounding lamina propria blood vessels suggesting that the lack of αEβ7 might have prevented the mast cells from localizing normally to the epithelium. In a follow-up study, McDermott et al[146] found that treatment with anti-αEβ7, anti-α4 and anti-β7 antibodies all suppressed the increase in mast cell numbers, while T cell numbers remained unaffected. This indicates that both αEβ7 and α4β7 are required for mast cell precursors to enter the gut, or to mature or expand there. Using an assay for mast cell precursors (MCp) in mucosal tissues, Gurish et al[147] clarified the situation by showing unequivocally that β7, but not αE, was required for MCp to home to the small intestine. Taken together, these studies suggest that α4β7 binding to MAdCAM-1 is critical for MCp to extravasate into the lamina propria of the small intestine, and that αEβ7 influences mast cell localization or maturation within the mucosa. Intimate contact between mast cells and epithelial cells via αEβ7 could facilitate the role of the mast cell as a 'gatekeeper' modulating epithelial permeability.[148] Mast cell proteases are known to increase permeability by acting on junctional proteins.[149] Thus, activation of mast cells would allow transient exudation of plasma proteins onto the mucosal surface, which in some circumstances could be beneficial in immunity to parasites.

The role of αEβ7 on dendritic cells (DC) has not been addressed in the published literature, but invites speculation. As described above, DC expressing αEβ7 have been identified in the epithelium and underlying lamina propria of the airways and intestine[14,16,24-26,150] and in thoracic duct lymph.[14,151] Binding of αEβ7 to E-cadherin is likely to be important for DC localization to the epithelium in these tissues. Huang et al[150] reported that in the rat a major subpopulation of $\alpha E\beta 7^+$ DC in the lamina propria, thoracic duct lymph and mesenteric lymph nodes contain fragments of apoptotic epithelial cells, apparently sampled from the villus epithelium. This population, which is $CD4^-$ $CD41^-$, was previously described as being relatively inefficient in antigen presentation compared to $CD4^+$ $CD41^+$ DC from the same tissues.[151] It has been suggested that this mechanism of epithelial sampling could contribute to the induction of peripheral tolerance to self antigens or food antigens.[150] Binding of αEβ7 to E-cadherin on the epithelial cell may assist this process. Furthermore, endocytosis of the epithelial cell membrane by the DC could be a mechanism to facilitate release of DCs from contact with the epithelium so that they can migrate to the afferent lymphatics. An interesting recent report shows that DCs are capable of opening the junctions between intestinal epithelial cells to directly sample bacteria in the lumen.[152] It is possible that αEβ7, as well as the junction proteins E-cadherin, occludin and ZO-1, also found on DC, might contribute to the process of DC penetration through the intestinal or bronchial epithelium.

Schön et al[153] made the surprising observation that αE deficient mice develop inflammatory skin lesions when the αE null/null gene modification is bred onto a 129/Sv x Balb/c genetic background. Moreover, similar lesions resulted when unfractionated spleen cells from αE null/null mice were transferred to MHC-matched *scid* mice. The mechanism is unknown but the observations would be consistent with several possible explanations, including the following: Development of αEβ7$^-$ effector T cells in the gut and failure to retain them in that tissue could allow abnormal dispersal to other sites, including the skin. Inappropriate secretion of cytokines there could trigger the pathology. Alternatively, signaling through αEβ7 on DC or Treg cells, or via E-cadherin on epithelial cells, could, in normal mice, have an immunomodulatory effect which is lost in the αE knockout. The deficiency could tip the balance towards inflammation when genetic and environmental factors are favourable.

Although an E-cadherin antibody that prevents αEβ7 binding had no effect on the acquisition of CD3 by maturing thymocytes in fetal thymic reaggregate cultures,[154] a more recent study has suggested a role for αEβ7 in the human thymus.[37] E-cadherin is present on most thymic epithelial cells[154,155] and was found in the new study to be strongly expressed by medullary epithelium. In vitro, αEβ7$^+$ thymocytes adhered to thymic epithelial cells and proliferated in a αEβ7-dependent manner. This indicates that αEβ7 may well play a significant role in the proliferative phases of human thymocyte development.

Integrin αEβ7 in Disease

Immuno-staining for αEβ7 in inflammatory bowel disease has given an inconsistent picture. In Crohn's disease, expression of the integrin on IEL in the ileum is diminished[156] a finding seemingly consistent with the recent observation that elevated expression of Smad7 in gut T cells in inflammatory bowel disease blocks TGF-β signaling.[157] Conversely, in a mouse model of colitis, αEβ7 has been reported to be expressed strongly by T cells infiltrating the epithelium. In this case, colitis was evoked by injection of CD4$^+$ CD45RB^{high} spleen cells into *scid* mice and 73% of the IEL in the colon expressed αEβ7; a similar proportion was seen in control mice that received CD45RB^{low} cells and did not develop colitis.[158] Similarly, examination of the gut mucosa of C3H/HeJ mice with heritable colitis showed normal numbers of αEβ7$^+$ cells in the large and small intestine.[117] Attempts have been made to influence the severity of colitis by blocking or inhibiting expression of αEβ7. In colitis induced in IL-2 null/null mice by immunisation with TNP-ovalbumin, administration of αE blocking antibody reduced the number of CD4$^+$ T cells in the lamina propria and diminished the severity of disease.[134] This treatment was found also to ameliorate established, ongoing, colitis. The effect was attributed to inhibition of the adhesive mechanism retaining αEβ7$^+$ T cells in the inflamed gut. However, the role of αEβ7 in inflammatory bowel disease is not yet fully resolved as a more recent report shows that β7 integrins are not necessary for the development of colitis in IL-2 null/null mice or after transfer of CD4$^+$ CD45RB^{high} T cells into *scid* mice.[159]

Integrin αEβ7 is also expressed by inflammatory T cells infiltrating other epithelial tissues. In Sjogren's sydrome, a T cell-mediated autoimmune disease of salivary and lachrimal glands, CD8$^+$ effector T cells found in close association with ductal and acinar epithelium express αE.[160] Similarly, as discussed above, a major subset of CD8$^+$ T cells infiltrating renal tubules during acute renal allograft rejection or invading pancreatic islet allografts undergoing rejection express αEβ7. Recent studies of patients with psoriasis have demonstrated that 80% of CD8$^+$ T cells and 30% of CD4$^+$ cells in the epidermis in psoriatic lesions are αEβ7$^+$.[50,107] In all these circumstances the integrin plays a role in retention of T cells close to the epithelium and may, in addition, provide co-stimulatory signals for effector function against the epithelial target cells.

Certain inflammatory diseases of the lung also cause an abnormal pattern of αEβ7 expression. In normal human bronchiolar lavage fluid most CD8$^+$ cells but fewer CD4$^+$ cells express αEβ7.[20,161] This picture is changed in hypersensitivity pneumonitis, sarcoidosis and idiopathic pulmonary fibrosis where the proportion of CD4$^+$ cells expressing αEβ7 is greatly increased.[20] It is not known whether αEβ7 has any role in the pathogenesis of these conditions or merely reflects the presence of high levels of TGF-β associated with the fibrosis.

Particular categories of T and B cell lymphomas have been reported to express αEβ7. Early stage cutaneous T cell lymphomas are αEβ7 positive and expression has been correlated with epidermotropism of the cells.[162,163] More invasive stages of the lymphoma that develop later are associated with loss of the integrin. Similarly, nearly all primary T cell lymphomas of the intestine are αEβ7$^+$ whereas mucosal B cell lymphomas and peripheral T cell lymphomas do not express αE.[164,165] This integrin has long been known to be a diagnostic marker of hairy cell leukemias.[7] Extensive phenotypic analysis of these cells has shown them to express a distinctive and unusual gamut of surface molecules including the αEβ7 integrin, several B cell markers as well as CD11c/αX and CD25.[166]

Conclusions

Integrin αEβ7 has proved to be enigmatic and tantalising. Considerable efforts to define its true significance in vivo have met with mixed fortunes. Whilst this integrin undoubtedly contributes to mucosal specific retention of T cells and probably DCs and mast cells also, there are valid grounds in future to seek deeper significance in its signaling capacity, especially in relation to cross-talk with the epithelium. The possibility that αEβ7 binding to E-cadherin might trigger signaling in epithelial cells remains to be explored. Studies of αE knockout mice have clearly identified an important role for this integrin in allograft rejection[143] and also have provided a glimpse of its possible significance in immunoregulation.[153] Resonance with the recent finding that αEβ7 is expressed by a major subset of Treg cells[41-44] is striking and the functional relationships between these observations provide fertile ground for investigation. In this context, TGF-β signaling to both the Treg cell and its target, and the significance of cell surface-bound TGF-β, merit particular attention. In mice, the αE integrin gene locus is sandwiched between the Th2 cytokine gene cluster (IL-4, IL-5 and IL-13) and a cluster of chemokine genes (eotaxin, TCA 3, MCP 1, 3, 5, MIP 1α and 1β, RANTES). In future studies to address the role of αEβ7 in immunoregulation it will be essential to utilize αE null/null and control mice that are congenic at the Th2 and chemokine loci. Current crystallographic studies on the αE α-I domain and further investigations into αE activation and the requirement for cadherin dimerization in the recognition process will resolve the separate roles of the αE and β7 subunits and clarify the relationship between αEβ7 and other α-I-domain-containing integrins.

References

1. Butcher EC, Williams M, Youngman K et al. Lymphocyte Trafficking and Regional Immunity. Adv Immunol 1999; 72:209-253.
2. Cerf-Bensussan N, Guy-Grand D, Lisowska-Grospierre B et al. A monoclonal antibody specific for rat intestinal lymphocytes. J Immunol 1986; 136(1):76-82.
3. Cerf-Bensussan N, Jarry A, Brousse N et al. A monoclonal antibody (HML-1) defining a novel membrane molecule present on human intestinal lymphocytes. Eur J Immunol 1987; 17(9):1279-1285.

4. Kilshaw PJ, Baker KC. A unique surface antigen on intraepithelial lymphocytes in the mouse. Immunol Lett 1988; 18(2):149-154.
5. Kilshaw PJ, Murant SJ. A new surface antigen on intraepithelial lymphocytes in the intestine. Eur J Immunol 1990; 20(10):2201-2207.
6. Kilshaw PJ, Murant SJ. Expression and regulation of β7 (βp) integrins on mouse lymphocytes: relevance to the mucosal immune system. Eur J Immunol 1991; 21(10):2591-2597.
7. Micklem KJ, Dong Y, Willis A et al. HML-1 antigen on mucosa-associated T cells, activated cells, and hairy leukemic cells is a new integrin containing the β7 subunit. Am J Pathol 1991; 139(6):1297-1301.
8. Yuan Q, Jiang WM, Hollander D et al. Identity between the novel integrin β7 subunit and an antigen found highly expressed on intraepithelial lymphocytes in the small intestine. Biochem Biophys Res Comm 1991; 176(3):1443-1449.
9. Krissansen GW, Print CG, Prestidge RL et al. Immunologic and structural relatedness of the integrin β7 complex and the human intraepithelial lymphocyte antigen HML-1. FEBS Letters 1992; 296(1):25-28.
10. Parker CM, Cepek KL, Russell GJ et al. A family of β7 integrins on human mucosal lymphocytes. Proc Natl Acad Sci USA 1992; 89(5):1924-1928.
11. Cerf-Bensussan N, Begue B, Gagnon J et al. The human intraepithelial lymphocyte marker HML-1 is an integrin consisting of a β7 subunit associated with a distinctive alpha chain [published erratum appears in Eur J Immunol 1992 Mar; 22(3):885]. Eur J Immunol 1992; 22(1):273-277.
12. Smith TJ, Ducharme LA, Shaw SK et al. Murine M290 integrin expression modulated by mast cell activation. Immunity 1994; 1(5):393-403.
13. Shaw SK, Cepek KL, Murphy EA et al. Molecular cloning of the human mucosal lymphocyte integrin α_E subunit. Unusual structure and restricted RNA distribution. J Biol Chem 1994; 269(8):6016-6025.
14. Brenan M, Puklavec M. The MRC OX-62 antigen: a useful marker in the purification of rat veiled cells with the biochemical characteristics of an integrin. J Exp Med 1992; 175(6):1457-1465.
15. Brenan M, Rees DJG. Sequence analysis of rat integrin α_{E1} and α_{E2} subunits: tissue expression reveals phenotypic similarities between intraepithelial lymphocytes and dendritic cells in lymph. Eur J Immunol 1997; 27(11):3070-3079.
16. Farstad IN, Halstensen TS, Lien B et al. Distribution of β_7 integrins in human intestinal mucosa and organized gut-associated lymphoid tissue [published erratum appears in Immunology 1997 Jun; 91(2):322]. Immunology 1996; 89(2):227-237.
17. Robinson PW, Green SJ, Carter C et al. Studies on transcriptional regulation of the mucosal T-cell integrin αEβ7 (CD103). Immunology 2001; 103(2):146-154.
18. Schieferdecker HL, Ullrich R, Weiss-Breckwoldt AN et al. The HML-1 antigen of intestinal lymphocytes is an activation antigen. J Immunol 1990; 144(7):2541-2549.
19. Brew R, West DC, Burthem J et al. Expression of the human mucosal lymphocyte antigen, HML-1, by T cells activated with mitogen or specific antigen in vitro. Scand J Immunol 1995; 41(6):553-562.
20. Rihs S, Walker C, Virchow JC, Jr. et al. Differential expression of αEβ7 integrins on bronchoalveolar lavage T lymphocyte subsets: regulation by α4β1-integrin crosslinking and TGF-β. Am J Resp Cell Mol Biol 1996; 15(5):600-610.
21. Cresswell J, Robertson H, Neal DE et al. Distribution of lymphocytes of the αEβ7 phenotype and E-cadherin in normal human urothelium and bladder carcinomas. Clin Exp Immunol 2001; 126(3):397-402.
22. Quayle AJ, Pudney J, Munoz DE et al. Characterization of T lymphocytes and antigen-presenting cells in the murine male urethra. Biol Reprod 1994; 51(5):809-820.
23. Andrew DP, Rott LS, Kilshaw PJ et al. Distribution of α4β7 and αEβ7 integrins on thymocytes, intestinal epithelial lymphocytes and peripheral lymphocytes. Eur J Immunol 1996; 26(4):897-905.
24. Kilshaw PJ. Expression of the mucosal T cell integrin $\alpha_{M290}\beta_7$ by a major subpopulation of dendritic cells in mice. Eur J Immunol 1993; 23(12):3365-3368.
25. Nelson DJ, McMenamin C, McWilliam AS et al. Development of the airway intraepithelial dendritic cell network in the rat from class II major histocompatibility (Ia)-negative precursors: differential regulation of Ia expression at different levels of the respiratory tract. J Exp Med 1994; 179(1):203-212.

26. Maric I, Holt PG, Perdue MH et al. Class II MHC antigen (Ia)-bearing dendritic cells in the epithelium of the rat intestine. J Immunol 1996; 156(4):1408-1414.
27. McLellan AD, Kapp M, Eggert A et al. Anatomic location and T-cell stimulatory functions of mouse dendritic cell subsets defined by CD4 and CD8 expression. Blood 2002; 99(6):2084-2093.
28. Riedl E, Stockl J, Majdic O et al. Functional involvement of E-cadherin in TGF-β1-induced cell cluster formation of in vitro developing human Langerhans-type dendritic cells. J Immunol 2000; 165(3):1381-1386.
29. Tiisala S, Paavonen T, Renkonen R. $\alpha_E\beta_7$ and $\alpha_4\beta_7$ integrins associated with intraepithelial and mucosal homing, are expressed on macrophages. Eur J Immunol 1995; 25(2):411-417.
30. Wright SH, Brown J, Knight PA et al. Transforming growth factor-β1 mediates coexpression of the integrin subunit αE and the chymase mouse mast cell protease-1 during the early differentiation of bone marrow-derived mucosal mast cell homologues. Clin Exp Allergy 2002; 32(2):315-324.
31. Csencsits KL, Walters N, Pascual DW. Cutting edge: dichotomy of homing receptor dependence by mucosal effector B cells: αE versus L-selectin. J Immunol 2001; 167(5):2441-2445.
32. Hoffkes HG, Schmidtke G, Uppenkamp M et al. Multiparametric immunophenotyping of B cells in peripheral blood of healthy adults by flow cytometry. Clin Diagn Lab Immunol 1996; 3(1):30-36.
33. Moldenhauer G, Mielke B, Dorken B et al. Identity of HML-1 antigen on intestinal intraepithelial T cells and of B-ly7 antigen on hairy cell leukaemia. Scand J Immunol 1990; 32(2):77-82.
34. Möller P, Mielke B, Moldenhauer G. Monoclonal antibody HML-1, a marker for intraepithelial T cells and lymphomas derived thereof, also recognizes hairy cell leukaemia and some B-cell lymphomas. Am J Pathol 1990; 136(3):509-512.
35. Lefrancois L, Barrett TA, Havran WL et al. Developmental expression of the $\alpha_{IEL}\beta_7$ integrin on T cell receptor γδ and T cell receptor αβ cells. Eur J Immunol 1994; 24(3):635-640.
36. McFarland RD, Douek DC, Koup RA et al. Identification of a human recent thymic emigrant phenotype. Proc Natl Acad Sci USA 2000; 97(8):4215-4220.
37. Kutlesa S, Wessels JT, Speiser A et al. E-cadherin-mediated interactions of thymic epithelial cells with $CD103^+$ thymocytes lead to enhanced thymocyte cell proliferation. J Cell Sci 2002; 115(23):4505-4515.
38. Gabor MJ, Sedgwick JD, Lemckert FA et al. Lymphotoxin controls αEβ7-integrin expression by peripheral $CD8^+$ T cells. Immunol Cell Biol 2001; 79(4):323-331.
39. Chatenoud L, Salomon B, Bluestone JA. Suppressor T cells—they're back and critical for regulation of autoimmunity! Immunol Rev 2001; 182:149-163.
40. Maloy KJ, Powrie F. Regulatory T cells in the control of immune pathology. Nat Immunol 2001; 2(9):816-822.
41. Lehmann J, Huhn J, de la Rosa M et al. Expression of the integrin αEβ7 identifies new potent subsets of regulatory $CD4^+$ T cells. Proc Natl Acad Sci USA 2002; 99(20):13031-13036.
42. Zelenika D, Adams E, Humm S et al. Regulatory T cells overexpress a subset of Th2 gene transcripts. J Immunol 2002; 168(3):1069-1079.
43. McHugh RS, Whitters MJ, Piccirillo CA et al. $CD4^+CD25^+$ immunoregulatory T cells: gene expression analysis reveals a functional role for the glucocorticoid-induced TNF receptor. Immunity 2002, 16(2).311-323.
44. Gavin MA, Clarke SR, Negrou E et al. Homeostasis and anergy of $CD4^+CD25^+$ suppressor T cells in vivo. Nat Immunol 2002; 3(1):33-41.
45. Allez M, Brimnes J, Dotan I et al. Expansion of $CD8^+$ T cells with regulatory function after interaction with intestinal epithelial cells. Gastroenterology 2002; 123(5):1516-1526.
46. Nakamura K, Kitani A, Strober W. Cell contact-dependent immunosuppression by $CD4^+CD25^+$ regulatory T cells is mediated by cell surface-bound transforming growth factor β. J Exp Med 2001; 194(5):629-644.
47. Annunziato F, Cosmi L, Liotta F et al. Phenotype, Localization, and Mechanism of Suppression of $CD4^+$ $CD25^+$ Human Thymocytes. J Exp Med 2002; 196(3):379-387.
48. Piccirillo CA, Letterio JJ, Thornton AM et al. $CD4^+$ $CD25^+$ Regulatory T Cells Can Mediate Suppressor Function in the Absence of Transforming Growth Factor β1 Production and Responsiveness. J Exp Med 2002; 196(2):237-245.
49. Jonuleit H, Schmitt E, Kakirman H et al. Infectious Tolerance: Human $CD25^+$ Regulatory T Cells Convey Suppressor Activity to Conventional $CD4^+$ T Helper Cells. J Exp Med 2002; 196(2):255-260.

50. Teraki Y, Shiohara T. Preferential expression of $\alpha E\beta 7$ integrin (CD103) on CD8+ T cells in the psoriatic epidermis: regulation by interleukins 4 and 12 and transforming growth factor-beta. Br J Dermatol 2002; 147(6):1118-1126.
51. Hadley GA, Bartlett ST, Via CS et al. The Epithelial Cell-Specific Integrin, CD103 (αE Integrin), Defines a Novel Subset of Alloreactive CD8+ CTL. J Immunol 1997; 159(8):3748-3756.
52. Shibara T, Si-Tahar M, Shaw SK et al. Adhesion molecules expressed on homing lymphocytes in model intestinal epithelia. Gastroenterology 2000; 118(2):289-298.
53. Kim SK, Schluns KS, Lefrancois L. Induction and visualization of mucosal memory CD8 T cells following systemic virus infection. J Immunol 1999; 163(8):4125-4132.
54. Suzuki R, Nakao A, Kanamaru Y et al. Localization of intestinal intraepithelial T lymphocytes involves regulation of $\alpha E\beta 7$ expression by transforming growth factor-β. Int Immunol 2002; 14(4):339-345.
55. Schön MP, Arya A, Murphy EA et al. Mucosal T lymphocyte numbers are selectively reduced in integrin α_E (CD103)-deficient mice. J Immunol 1999; 162(11):6641-6649.
56. Higgins JMG. The Haspin gene: location in an intron of the Integrin αE gene, associated transcription of an Integrin αE-derived RNA and expression in diploid as well as haploid cells. Gene 2001; 267(1):55-69.
57. Fleming JC, Pahl HK, Gonzalez DA et al. Structural Analysis of the CD11b Gene and Phylogenetic Analysis of the α-Integrin Gene Family Demonstrate Remarkable Conservation of Genomic Organization and Suggest Early Diversification during Evolution. J Immunol 1993; 150(2):480-490.
58. Corbi AL, Garcia-Aguilar J, Springer TA. Genomic structure of an integrin alpha subunit, the leukocyte p150,95 molecule [published erratum appears in J Biol Chem 1990 Jul 25; 265(21):12750-1]. J Biol Chem 1990; 265(5):2782-2788.
59. Noti JD, Gordon M, Hall RE. Human p150,95 α-Subunit: Genomic Organization and Analysis of the 5'-Flanking Region. DNA Cell Biol 1992; 11(2):123-138.
60. Wong DA, Davis EM, LeBeau M et al. Cloning and chromosomal localization of a novel gene-encoding a human β2-integrin alpha subunit. Gene 1996; 171(2):291-294.
61. Yoshimura Y, Tanaka H, Nozaki M et al. Nested structure of haploid germ cell specifc haspin gene. Gene 2001; 267(1):49-54.
62. Tanaka H, Iguchi N, Nakamura Y et al. Cloning and characterization of human *haspin* gene encoding haploid germ cell-specific nuclear protein kinase. Mol Hum Reprod 2001; 7(3):211-218.
63. Higgins JMG. Haspin-like proteins: A new family of evolutionarily conserved putative eukaryotic protein kinases. Prot Sci 2001; 10(8):1677-1684.
64. Tanaka H, Yoshimura Y, Nozaki M et al. Identification and characterization of a Haploid Germ Cell-specific Nuclear Protein Kinase (Haspin) in spermatid nuclei and its effects on somatic cells. J Biol Chem 1999; 274(24):17049-17057.
65. Jiang WM, Jenkins D, Yuan Q et al. The gene organization of the human β_7 subunit, the common β subunit of the leukocyte integrins HML-1 and LPAM-1. Int Immunol 1992; 4(9):1031-1040.
66. Baker E, Sutherland GR, Jiang WM et al. Mapping of the human integrin β_7 gene (ITG β7) to 12q13.13 by non-isotopic in situ hybridization. Mamm Genome 1992; 2(4):272-273.
67. Yuan Q, Kozak CA, Jiang WM et al. Genetic mapping of the gene coding for the integrin β7 subunit to the distal part of mouse chromosome 15. Immunogenetics 1992; 35(6):403-407.
68. Russell GJ, Parker CM, Cepek KL et al. Distinct structural and functional epitopes of the $\alpha_E\beta_7$ integrin. Eur J Immunol 1994; 24(11):2832-2841.
69. Berg RW, Yang Y, Lehnert K et al. Mouse M290 is the functional homologue of the human mucosal lymphocyte integrin HML-1: antagonism between the integrin ligands E-cadherin and RGD tripeptide. Immunol Cell Biol 1999; 77(4):337-344.
70. Cepek KL, Parker CM, Madara JL et al. Integrin $\alpha_E\beta_7$ mediates adhesion of T lymphocytes to epithelial cells. J Immunol 1993; 150(8):3459-3470.
71. Roberts K, Kilshaw PJ. The mucosal T cell integrin α_{M290} β_7 recognizes a ligand on mucosal epithelial cell lines. Eur J Immunol 1993; 23(7):1630-1635.
72. Roberts AI, O'Connell SM, Ebert EC. Intestinal intraepithelial lymphocytes bind to colon cancer cells by HML-1 and CD11a. Cancer Res 1993; 53(7):1608-1611.
73. Cepek KL, Shaw SK, Parker CM et al. Adhesion between epithelial cells and T lymphocytes mediated by E-cadherin and the $\alpha_E\beta_7$ integrin. Nature 1994; 372(6502):190-193.

74. Karecla PI, Bowden SJ, Green SJ et al. Recognition of E-cadherin on epithelial cells by the mucosal T cell integrin $\alpha_{M290}\beta_7$ ($\alpha_E\beta_7$). Eur J Immunol 1995; 25(3):852-856.
75. Higgins JMG, Mandlebrot DA, Shaw SK et al. Direct and regulated interaction of integrin αEβ7 with E-cadherin. J Cell Biol 1998; 140(1):197-210.
76. Nagar B, Overduin M, Ikura M et al. Structural basis of calcium-induced E-cadherin rigidification and dimerization. Nature 1996; 380(6572):360-364.
77. Karecla PI, Green SJ, Bowden SJ et al. Identification of a binding site for integrin $\alpha_E\beta_7$ in the N-terminal domain of E-cadherin. J Biol Chem 1996; 271(48):30909-30915.
78. Taraszka KS, Higgins JMG, Tan K et al. Molecular Basis of Integrin αEβ7 Recognition of E-cadherin. J Exp Med 2000; 191(9):1555-1567.
79. Shapiro L, Fannon AM, Kwong PD et al. Structural basis of cell-cell adhesion by cadherins. Nature 1995; 374(6520):327-337.
80. Boggon TJ, Murray J, Chappuis-Flament S et al. C-Cadherin Ectodomain Structure and Implications for Cell Adhesion Mechanisms. Science 2002; 296(5571):1308-1313.
81. Higgins JMG, Cernadas M, Tan K et al. The Role of α and β Chains in Ligand Recognition by β7 Integrins. J Biol Chem 2000; 275(33):25652-25664.
82. Lee J-O, Bankston LA, Arnaout MA et al. Two conformations of the integrin A-domain (I-domain): a pathway for activation? Structure 1995; 3(12):1333-1340.
83. Emsley J, Knight CG, Farndale RW et al. Structural Basis of Collagen Recognition by Integrin α2β1. Cell 2000; 101(1):47-56.
84. Lu C, Shimaoka M, Ferzly M et al. An isolated, surface-expressed I domain of the integrin αLβ2 is sufficient for strong adhesive function when locked in the open conformation with a disulfide bond. Proc Natl Acad Sci USA 2001; 98(5):2387-2392.
85. Shimaoka M, Lu C, Palframan RT et al. Reversibly locking a protein fold in an active conformation with a disulfide bond: integrin αL I domains with high affinity and antagonist activity in vivo. Proc Natl Acad Sci USA 2001; 98(11):6009-6014.
86. Woo JC, Roccabianca P, van Stijn A et al. Characterization of a feline homologue of the αE integrin subunit (CD103) reveals high specificity for intra-epithelial lymphocytes. Vet Immunol Immunopathol 2002; 85(1-2):9-22.
87. Xiong JP, Stehle T, Zhang R et al. Crystal structure of the extracellular segment of integrin αVβ3 in complex with an Arg-Gly-Asp ligand. Science 2002; 296(5565):151-155.
88. Aota S, Nomizu M, Yamada KM. The short amino acid sequence Pro-His-Ser-Arg-Asn in human fibronectin enhances cell-adhesive function. J Biol Chem 1994; 269(40):24756-24761.
89. Briskin MJ, Rott L, Butcher EC. Structural requirements for mucosal vascular addressin binding to its lymphocyte receptor $\alpha_4\beta_7$. Common themes among integrin-Ig family interactions. J Immunol 1996; 156(2):719-726.
90. Newham P, Craig SE, Seddon GN et al. α4 Integrin Binding Interfaces on VCAM-1 and MAdCAM-1. J Biol Chem 1997; 272(31):19429-19440.
91. Tan K, Casasnovas JM, Liu JH et al. The structure of immunoglobulin superfamily domains 1 and 2 of MAdCAM-1 reveals novel features important for integrin recognition. Structure 1998; 6(6):793-801.
92. Green N, Rosebrook J, Cochran N et al. Mutational analysis of MAdCAM-1/α4β7 interactions reveals significant binding determinants in both the first and second immunoglobulin domains. Cell Adhes Commun 1999; 7(3):167-181.
93. Mould AP, Askari JA, Aota S et al. Defining the topology of integrin α5β1-fibronectin interactions using inhibitory anti-α5 and anti-β1 monoclonal antibodies. Evidence that the synergy sequence of fibronectin is recognized by the amino-terminal repeats of the α5 subunit. J Biol Chem 1997; 272(28):17283-17292.
94. Miller J, Knorr R, Ferrone M et al. Intercellular adhesion molecule-1 dimerization and its consequences for adhesion mediated by lymphocyte function associated-1. J Exp Med 1995; 182(5):1231-1241.
95. Reilly PL, Woska JRJ, Jeanfavre DD et al. The native structure of intercellular adhesion molecule-1 (ICAM-1) is a dimer. Correlation with binding to LFA-1. J Immunol 1995; 155(2):529-532.
96. Takeda H, Shimoyama Y, Nagafuchi A et al. E-cadherin functions as a cis-dimer at the cell-cell adhesive interface in vivo. Nat Struct Biol 1999; 6(4):310-312.

97. Casasnovas JM, Stehle T, Liu JH et al. A dimeric crystal structure for the N-terminal two domains of intercellular adhesion molecule-1. Proc Natl Acad Sci USA 1998; 95(8):4134-4139.
98. Jun CD, Shimaoka M, Carman CV et al. Dimerization and the effectiveness of ICAM-1 in mediating LFA-1- dependent adhesion. Proc Natl Acad Sci USA 2001; 98(12):6830-6835.
99. Lu C, Shimaoka M, Zang Q et al. Locking in alternate conformations of the integrin αLβ2 I domain with disulfide bonds reveals functional relationships among integrin domains. Proc Natl Acad Sci USA 2001; 98(5):2393-2398.
100. Zang Q, Lu C, Huang C et al. The top of the inserted-like domain of the integrin lymphocyte function- associated antigen-1 β subunit contacts the α subunit β-propeller domain near β-sheet 3. J Biol Chem 2000; 275(29):22202-22212.
101. Alonso JL, Essafi M, Xiong JP et al. Does the Integrin αA Domain Act as a Ligand for its βA Domain? Curr Biol 2002; 12(10):R340-R342.
102. Corps E, Carter C, Karecla P et al. Recognition of E-cadherin by integrin αEβ7: requirement for cadherin dimerization and implications for cadherin and integrin function. J Biol Chem 2001; 276(33):30862-30870.
103. Pertz O, Boxic D, Koch AW et al. A new crystal structure, Ca^{2+} dependence and mutational analysis reveal molecular details of E-cadherin homoassociation. EMBO J 1999; 18(7):1738-1747.
104. Haussinger D, Ahrens T, Sass HJ et al. Calcium-dependent homoassociation of E-cadherin by NMR spectroscopy: changes in mobility, conformation and mapping of contact regions. J Mol Biol 2002; 324(4):823-839.
105. Strauch UG, Lifka A, Gosslar U et al. Distinct binding specificities of integrins α4β7 (LPAM-1), α4β1 (VLA-4), and αIELβ7. Int Immunol 1994; 6(2):263-275.
106. Brown DW, Furness J, Speight PM et al. Mechanisms of binding of cutaneous lymphocyte-associated antigen- positive and $\alpha_e\beta_7$-positive lymphocytes to oral and skin keratinocytes. Immunology 1999; 98(1):9-15.
107. Pauls K, Schön M, Kubitza RC et al. Role of integrin αE(CD103)β7 for tissue-specific epidermal localization of $CD8^+$ T lymphocytes. J Invest Dermatol 2001; 117(3):569-575.
108. Strauch UG, Mueller RC, Li XY et al. Integrin αE(CD103)β7 mediates adhesion to intestinal microvascular endothelial cell lines via an E-cadherin-independent interaction. J Immunol 2001; 166(5):3506-3514.
109. Ni J, Hollander D, Sydora B et al. Adhesion molecule expression and adhesion properties of murine intestinal intraepithelial lymphocyte hybridomas. Cell Immunol 1995; 164(1):156-160.
110. French JJ, Cresswell J, Wong WK et al. T cell adhesion and cytolysis of pancreatic cancer cells: a role for E-cadherin in immunotherapy? Br J Cancer 2002; 87(9):1034-1041.
111. Dustin ML, Springer TA. T-cell receptor cross-linking transiently stimulates adhesiveness through LFA-1. Nature 1989; 341(6243):619-624.
112. Krummel MF, Davis MM. Dynamics of the immunological synapse: finding, establishing and solidifying a connection. Curr Opin Immunol 2002; 14(1):66-74.
113. Tang A, Amagai M, Granger LG et al. Adhesion of epidermal Langerhans cells to keratinocytes mediated by E-cadherin. Nature 1993; 361(6407):82-85.
114. Sarnacki S, Begue B, Buc H et al. Enhancement of CD3-induced activation of human intestinal intraepithelial lymphocytes by stimulation of the β7-containing integrin defined by HML-1 monoclonal antibody. Eur J Immunol 1992; 22(11):2887-2892.
115. Müller S, Jungo M, Aichele P et al. $CD5^-$ CD8 alpha beta intestinal intraepithelial lymphocytes (IEL) are induced to express CD5 upon antigen-specific activation: $CD5^-$ and $CD5^+$ CD8 αβ IEL do not represent separate T cell lineages. Eur J Immunol 1997; 27(7):1756-1761.
116. Ebert EC, Roberts AI. Lymphokine-activated killing by human intestinal lymphocytes. Cell Immunol 1993; 146(1):107-116.
117. Ni J, Chen SF, Hollander D. Immunological abnormality in C3H/HeJ mice with heritable inflammatory bowel disease. Cell Immunol 1996; 169(1):7-15.
118. Smart CJ, Calabrese A, Oakes DJ et al. Expression of the LFA-1 β2 integrin (CD11a/CD18) and ICAM-1 (CD54) in normal and coeliac small bowel mucosa. Scand J Immunol 1991; 34(3):299-305.
119. de Fougerolles AR, Stacker SA, Schwarting R et al. Characterization of ICAM-2 and evidence for a third counter-receptor for LFA-1. J Exp Med 1991; 174(1):253-267.

120. Vazeux R, Hoffman PA, Tomita JK et al. Cloning and characterization of a new intercellular adhesion molecule ICAM-R. Nature 1992; 360(6403):485-488.
121. Bloom S, Simmons D, Jewell DP. Adhesion molecules intercellular adhesion molecule-1 (ICAM-1), ICAM-3 and B7 are not expressed by epithelium in normal or inflamed colon. Clin Exp Immunol 1995; 101(1):157-163.
122. Rietzler M, Bittner M, Kolanus W et al. The human WD repeat protein WAIT-1 specifically interacts with the cytoplasmic tails of β7-integrins. J Biol Chem 1998; 273(42):27459-27466.
123. Schön MP, Schön M, Parker CM et al. Dendritic Epidermal T Cells (DETC) are Diminished in Integrin α_E(CD103)-Deficient Mice. J Invest Dermatol 2002; 119(1):190-193.
124. Guy-Grand D, Cerf-Bensussan N, Malissen B et al. Two Gut Intraepithelial $CD8^+$ Lymphocyte Populations with Different T Cell Receptors: A Role for the Gut Epithelium in T Cell Differentiation. J Exp Med 1991; 173(2):471-481.
125. Hermiston ML, Gordon JI. Inflammatory bowel disease and adenomas in mice expressing a dominant negative N-cadherin. Science 1995; 270(5239):1203-1207.
126. Lefrancois L, Parker CM, Olson S et al. The role of β7 integrins in CD8 T cell trafficking during an antiviral immune response. J Exp Med 1999; 189(10):1631-1638.
127. Masopust D, Vezys V, Marzo AL et al. Preferential localization of effector memory cells in nonlymphoid tissue. Science 2001; 291(5512):2413-2417.
128. Perry LL, Feilzer K, Portis JL et al. Distinct homing pathways direct T lymphocytes to the genital and intestinal mucosae in *Chlamydia*-infected mice. J Immunol 1998; 160(6):2905-2914.
129. Igietseme JU, Portis JL, Perry LL. Inflammation and clearance of *Chlamydia trachomatis* in enteric and nonenteric mucosae. Infect Immun 2001; 69(3):1832-1840.
130. Feng CG, Britton WJ, Palendira U et al. Up-regulation of VCAM-1 and differential expansion of beta integrin- expressing T lymphocytes are associated with immunity to pulmonary *Mycobacterium tuberculosis* infection. J Immunol 2000; 164(9):4853-4860.
131. Wagner N, Lohler J, Kunkel EJ et al. Critical role for β7 integrins in formation of the gut-associated lymphoid tissue. Nature 1996; 382(6589):366-370.
132. Arroyo AG, Yang JT, Rayburn H et al. Differential requirements for α4 integrins during fetal and adult hematopoiesis. Cell 1996; 85(7):997-1008.
133. Arroyo AG, Taverna D, Whittaker CA et al. In vivo roles of integrins during leukocyte development and traffic: insights from the analysis of mice chimeric for α5, αv, and α4 integrins. J Immunol 2000; 165(8):4667-4675.
134. Lúdvíksson BR, Strober W, Nishikomori R et al. Administration of mAb against αEβ7 prevents and ameliorates immunization-induced colitis in $IL\text{-}2^{-/-}$ mice. J Immunol 1999; 162(8):4975-4982.
135. Austrup F, Rebstock S, Kilshaw PJ et al. Transforming growth factor-beta 1-induced expression of the mucosa-related integrin αE on lymphocytes is not associated with mucosa-specific homing. Eur J Immunol 1995; 25(6):1487-1491.
136. Zabel BA, Agace WW, Campbell JJ et al. Human G protein-coupled receptor GPR-9-6/CC chemokine receptor 9 is selectively expressed on intestinal homing T lymphocytes, mucosal lymphocytes, and thymocytes and is required for thymus-expressed chemokine-mediated chemotaxis. J Exp Med 1999; 190(9):1241-1256.
137. Uehara S, Grinberg A, Farber JM et al. A role for CCR9 in T lymphocyte development and migration. J Immunol 2002; 168(6):2811-2819.
138. Koseki S, Miura S, Fujimori H et al. In situ demonstration of intraepithelial lymphocyte adhesion to villus microvessels of the small intestine. Int Immunol 2001; 13(9):1165-1174.
139. Hadley GA, Rostapshova EA, Gomolka DM et al. Regulation of the epithelial cell-specific integrin, CD103, by human $CD8^+$ cytolytic T lymphocytes. Transplantation 1999; 67(11):1418-1425.
140. Hadley GA, Charandee C, Weir MR et al. $CD103^+$ CTL accumulate within the graft epithelium during clinical renal allograft rejection. Transplantation 2001; 72(9):1548-1555.
141. Robertson H, Wong WK, Talbot D et al. Tubulitis after renal transplantation: demonstration of an association between $CD103^+$ T cells, transforming growth factor β1 expression and rejection grade. Transplantation 2001; 71(2):306-313.
142. Rostapshova EA, Burns JM, Bartlett ST et al. Integrin-mediated interactions influence the tissue specifity of $CD8^+$ cytolytic T lymphocytes. Eur J Immunol 1998; 28(10):3031-3039.
143. Feng Y, Wang D, Yuan R et al. CD103 expression is required for destruction of pancreatic islet allografts by $CD8^+$ T cells. J Exp Med 2002; 196(7):877-886.

144. Kilshaw PJ, Higgins JMG. Alpha E: no more rejection? J Exp Med 2002; 196(7):873-875.
145. Artis D, Humphreys NE, Potten CS et al. Beta7 integrin-deficient mice: delayed leukocyte recruitment and attenuated protective immunity in the small intestine during enteric helminth infection. Eur J Immunol 2000; 30(6):1656-1664.
146. McDermott JR, Grencis RK, Else KJ. Leucocyte recruitment during enteric nematode infection. Immunology 2001; 103(4):505-510.
147. Gurish MF, Tao H, Abonia JP et al. Intestinal mast cell progenitors require CD49dβ7 (α4β7 integrin) for tissue-specific homing. J Exp Med 2001; 194(9):1243-1252.
148. Miller HR. Mucosal mast cells and the allergic response against nematode parasites. Vet Immunol Immunopathol 1996; 54(1-4):331-336.
149. Scudamore CL, Jepson MA, Hirst BH et al. The rat mucosal mast cell chymase, RMCP-II, alters epithelial cell monolayer permeability in association with altered distribution of the tight junction proteins ZO-1 and occludin. Eur J Cell Biol 1998; 75(4):321-330.
150. Huang FP, Platt N, Wykes M et al. A discrete subpopulation of dendritic cells transports apoptotic intestinal epithelial cells to T cell areas of mesenteric lymph nodes. J Exp Med 2000; 191(3):435-444.
151. Liu L, Zhang M, Jenkins C et al. Dendritic cell heterogeneity in vivo: two functionally different dendritic cell populations in rat intestinal lymph can be distinguished by CD4 expression. J Immunol 1998; 161(3):1146-1155.
152. Rescigno M, Urbano M, Valzasina B et al. Dendritic cells express tight junction proteins and penetrate gut epithelial monolayers to sample bacteria. Nat Immunol 2001; 2(4):361-367.
153. Schön MP, Schön M, Warren HB et al. Cutaneous inflammatory disorder in integrin αE (CD103)-deficient mice. J Immunol 2000; 165(11):6583-6589.
154. Müller KM, Luedecker CJ, Udey MC et al. Involvement of E-cadherin in thymus organogenesis and thymocyte maturation. Immunity 1997; 6(3):257-264.
155. Lee MG, Sharrow SO, Farr AG et al. Expression of the homotypic adhesion molecule E-cadherin by immature murine thymocytes and thymic epithelial cells. J Immunol 1994; 152(12):5653-5659.
156. Elewaut D, Van Damme N, De Keyser F et al. Altered expression of αEβ7 integrin on intra-epithelial and lamina propria lymphocytes in patients with Crohn's disease. Acta Gastroenterologica Belgica 1998; 61(3):288-294.
157. Monteleone G, Kumberova A, Croft NM et al. Blocking Smad7 restores TGF-β1 signaling in chronic inflammatory bowel disease. J Clin Invest 2001; 108(4):601-609.
158. Aranda R, Sydora BC, McAllister PL et al. Analysis of intestinal lymphocytes in mouse colitis mediated by transfer of $CD4^+$, $CD45RB^{high}$ T cells to SCID recipients. J Immunol 1997; 158(7):3464-3473.
159. Sydora BC, Wagner N, Löhler J et al. β7 integrin expression is not required for the localization of T cells to the intestine and colitis pathogenesis. Clin Exp Immunol 2002; 129(1):35-42.
160. Fujihara T, Fujita H, Tsubota K et al. Preferential localization of $CD8^+$ $\alpha E\beta 7^+$ T cells around acinar epithelial cells with apoptosis in patients with Sjogren's syndrome. J Immunol 1999; 163(4):2226-2235.
161. Erle DJ, Brown T, Christian D et al. Lung epithelial lining fluid T cell subsets defined by distinct patterns of β7 and β1 integrin expression. Am J Resp Cell Mol Biol 1994; 10(3):237-244.
162. Simonitsch I, Volc-Platzer B, Mosberger I et al. Expression of monoclonal antibody HML-1-defined αEβ7 integrin in cutaneous T cell lymphoma. Am J Pathol 1994; 145(5):1148-1158.
163. Schechner JS, Edelson RL, McNiff JM et al. Integrins α4β7 and αEβ7 are expressed on epidermotropic T cells in cutaneous T cell lymphoma and spongiotic dermatitis. Lab Invest 1999; 79(5):601-607.
164. Chott A, Dragosics B, Radaszkiewicz T. Peripheral T-cell lymphomas of the intestine. Am J Pathol 1992; 141(6):1361-1371.
165. Spencer J, Cerf-Bensussan N, Jarry A et al. Enteropathy-associated T cell lymphoma (malignant histiocytosis of the intestine) is recognized by a monoclonal antibody (HML-1) that defines a membrane molecule on human mucosal lymphocytes. Am J Pathol 1988; 132(1):1-5.
166. Babugikova, Tomova A, Kusenda J et al. Flow cytometry of peripheral blood and bone marrow cells from patients with hairy cell leukemia: phenotype of hairy cells, lymphocyte subsets and detection of minimal residual disease after treatment. Neoplasma 2001; 48(5):350-357.

Chapter 8

Structural Aspects of Integrins

Robert C. Liddington

Structural studies on integrins have recently made great strides. In the past three years, crystal structures of the complete extracellular fragment of one integrin, αVβ3,[1] two α-I domains in complex with ligands,[2,3] and an intracellular activating protein in complex with a β tail[4] have been obtained; and an αβ tail complex has been determined by NMR.[5] High resolution EM studies complement these atomic resolution techniques by studying the integrin in different activation states.[6,7] Although we have only a few experimental examples among integrin family members, the high level of sequence homology between integrins means that reliable models can be built of the other members of the integrin family. These structures make sense of a lot of preceding biochemical, biophysical and mutagenesis studies, and generate many new testable hypotheses of integrin function. This chapter emphasizes new structural insights applicable to all integrins, with an emphasis on those integrins that contain an α-I domain. The structural data reinforce the notion of the integrin as a molecule in dynamic equilibrium at the cell surface, regulated by binding both to extracellular and intracellular ligands.

Overall Structure

Integrins are αβ heterodimers, consisting of a head domain from which emerge two legs, one from each subunit, ending in a pair of single-pass transmembrane helices and short cytoplasmic tails (Fig. 1). The integrin "head" comprises a seven-bladed propeller from the α-subunit that makes an intimate contact with a GTP-ase like domain of the β subunit (the β-I domain), in a manner that strongly resembles the heterotrimeric G proteins.[1] Nine α-subunits (α1, α2, α10, α11, αD, αE, αL, αM and αX) contain an additional domain, the α-I domain, that is inserted between two loops on the upper surface of the propeller, where it plays a central role in ligand binding.[8-10] Instead of a catalytic center, the I domain contains an invariant ligand binding site called MIDAS, for Metal Ion-Dependent Adhesion Site, in which a metal ion is coordinated by three loops from the I domain, and a glutamic or aspartic acid from the ligand completes an octahedral coordination sphere around the metal. In those integrins that lack an α-I domain, the β-I domain is a major ligand recognition site; in the α-I domain integrins, the β-I domain plays a regulatory role (see below).

In the absence of ligand, bonds between the legs, tails and head are believed to hold the head in an "inactive" or "resting" conformation that has low affinity for ligand.[5,11,12] Recent structural data suggest that this conformation is severely bent,[7] in contrast to the extended conformation assumed from earlier studies. During "outside-in" signaling, the head binds to ECM proteins or counter-receptors on other cells, triggering conformational changes in the integrin that are propagated down the "legs" and through the plasma membrane, leading to a reorganization of the C-terminal tails that allows them to bind intracellular proteins.[7,13] During

I Domains in Integrins, edited by Donald Gullberg. ©2003 Eurekah.com and Kluwer Academic / Plenum Publishers.

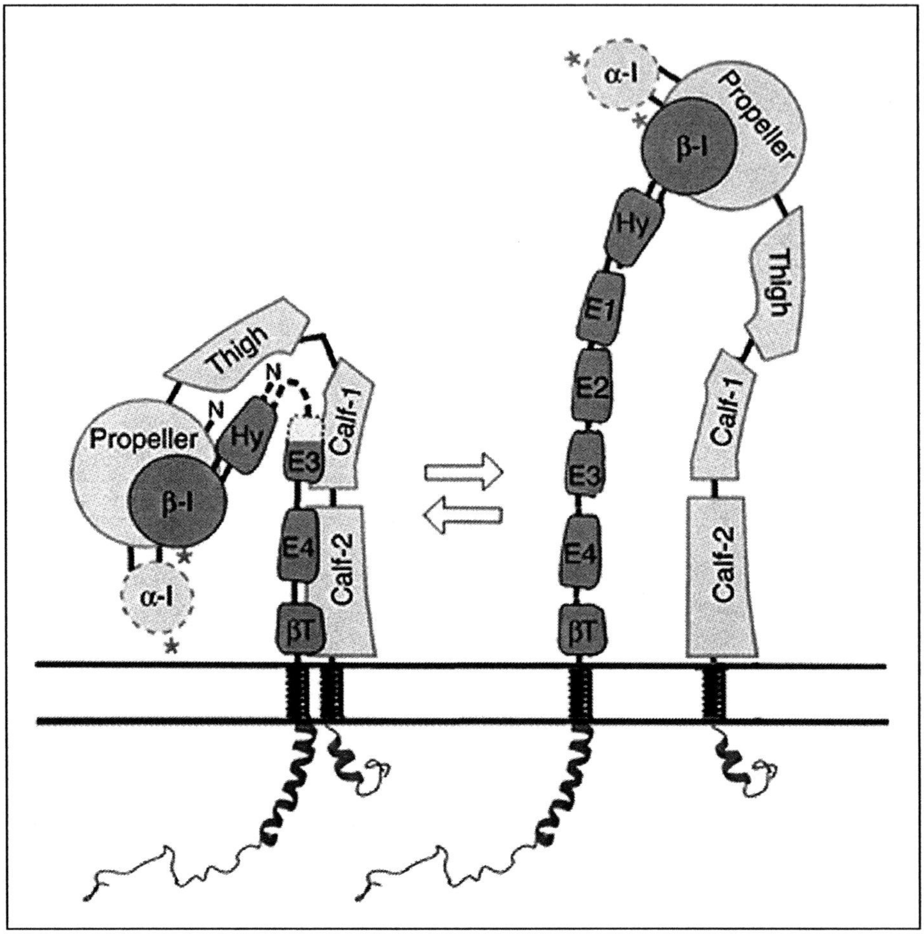

Figure 1. Cartoon of the αVβ3 structure derived by crystallography and EM studies.[1,7] At left, the inactive (resting) with low affinity stabilized by bonds between the head, legs and cytoplasmic tails. Binding extracellular ligand alters the head group structure, which perturbs interdomain interactions, leading to destabilization of the compact structure and a "switchblade" opening to the extended high affinity structure with cytoplasmic tails separated. The system is in equilibrium, so cytoplasmic proteins such as talin can activate integrin from the inside-out by sequestering one of the tails, triggering the same conformational change in the head.

"inside-out" signaling, cytosolic proteins bind and sequester one or both of the cytoplasmic tails, triggering conformational changes in the head that lead to a high affinity "active" integrin,[4,5] in which the integrin "stands up" (see below).

The α-I Domain

The first crystal structure of an α-I domain revealed a compact domain comprising a central mostly parallel β-sheet surrounded on both sides by amphipathic α-helices[14] (Fig. 2). Subsequent crystal structures of recombinant αL, α1 and α2 I domains display the same three-dimensional fold, as expected given their reasonable sequence similarity.[15-17] The MIDAS motif lies at the C-terminal end of the central β-sheet, with three loops contributing sidechains

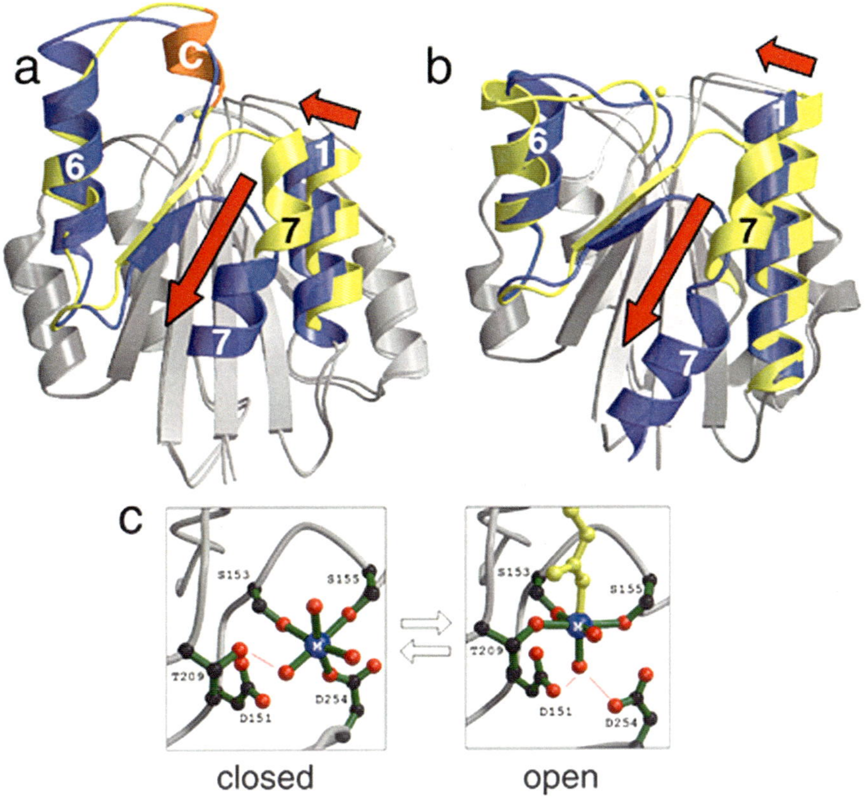

Figure 2. Two conformations of the α-I domain.[2,27] a. conformational changes in the α2-I domain on binding collagen. b. conformational changes in the αM-I domain on binding a ligand mimetic. Note the similarity in the changes (also seen in αL[3]), suggesting that all α-I domains will undergo the same switch. c. The MIDAS motif. Three loops from the α-I domain coordinate the metal ion (Mg^{2+}). At left, the "closed" conformation observed in the absence of ligand; at right, the "open" conformation seen when ligand is bound. These changes are mechanically linked to the tertiary changes in the domain (a. and b.).

that coordinate the metal ion (Fig. 2c). The metal-coordinating MIDAS residues are invariant among α-I domains, and mutagenesis of any of these residues abrogates ligand binding. Surface-exposed sidechains surrounding the MIDAS motif are more variable; they provide additional ligand contact points and hence ligand specificity.[8,18-26]

The structures of two ligand-bound α-I domains have been determined. The first structure is of the α2-I domain bound to a collagen-like triple helix.[2] More recently, the structure of the αL-I domain in complex with a fragment of ICAM-1 has been determined.[3] These structures, along with an earlier structure of the αM I domain bound to a "ligand-mimetic" crystal contact,[27] demonstrate that ligand binding triggers a profound conformational switch in the α-I domain that underlies affinity regulation and signal transduction (see below). The conformational switch is essentially identical in these three examples, indicating that all α-I domains will undergo the same switch.

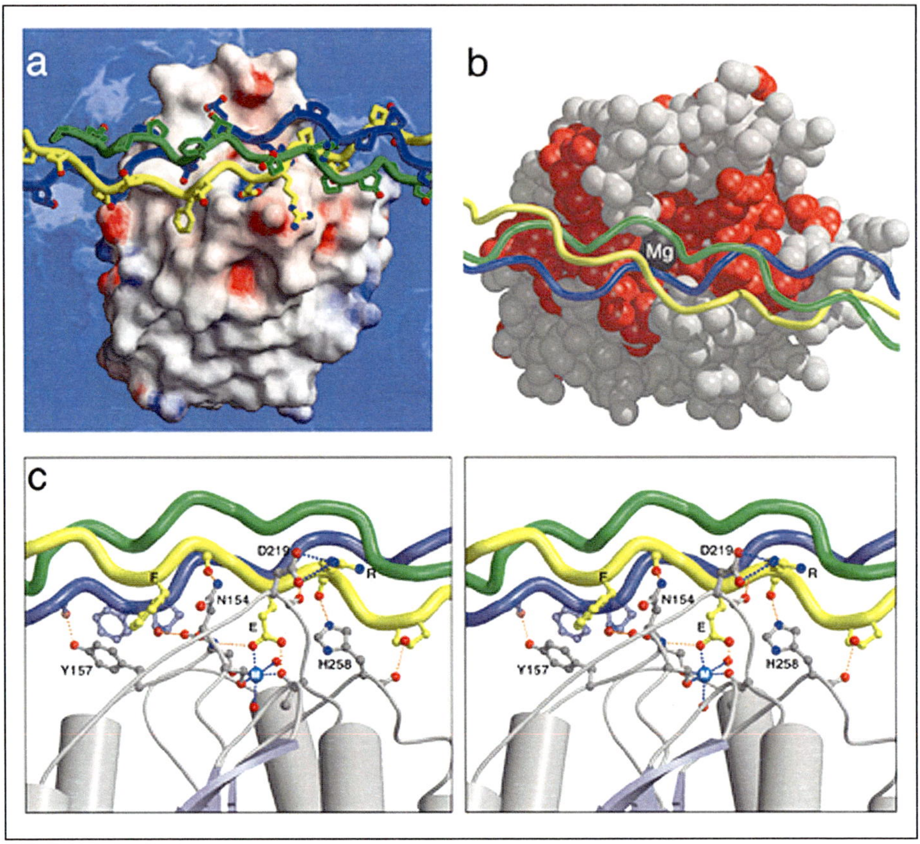

Figure 3. Collagen binding to the α2-I domain.[2] a. Surface model of the α2-I domain colored by surface charge (red=negative, blue=positive) with a triple helical fragment of collagen bound. b. Space-filling model of the complex (rotated about a horizontal axis compared with a.), showing residues (in red) that are invariant in the collagen-binding integrins, α1β1, α2β1 and α10β1. The strong conservation of the binding surface suggests that these integrins will engage collagen in the same fashion. c. Stereo close-up image of the α2-I:collagen complex.

Structural Determinants of Collagen Binding

Recombinant α2-I domain was crystallized as a complex with a homotrimer of a 21-mer peptide containing a critical GFOGER motif.[2,28] The peptide closely resembles the structure of uncomplexed collagen-like peptides previously determined,[29] and has the properties of a folded protein domain (i.e., a stable secondary and tertiary structure). Three loops on the upper surface of the α-I domain that comprise the MIDAS motif also engage the collagen, with a collagen glutamate completing the coordination sphere of the metal (Fig. 3). The critical role of both the MIDAS and surrounding residues have been confirmed by mutagenesis.[25,26]

The buried surface area on complex formation - 1230 Å^2 - is at the lower limit of known protein-protein interfaces (the value is almost identical for the αL-I:ICAM complex[3]), especially given the fact that some of the binding energy must be expended in switching the conformation of the I domain from closed to open (see below). The quite reasonable affinity of the interaction (Kd = 35-90 nM[30]) probably reflects the unusually strong bonds formed by the glutamate-metal-I domain bridge. This bridge is indeed critical, since the conservative substitution of collagen Glu to Asp in the GFOGER motif eliminates binding,[31] presumably because the aspartic acid is too short to reach down to bind to the partly buried metal ion.

The MIDAS motif and much of the collagen-binding surface are strictly invariant amongst the collagen-binding I domain integrins (α1, α2, α10 and α11), suggesting that these integrins will all engage collagen in a similar fashion, with a strict requirement for glutamate in the collagen motif. The periphery of the binding surface is more variable, however (Fig. 3b), which would explain their collagen type preferences (see Chapter 3 by Zutter & Santoro).

Conformational Changes in the α-I Domain

Ligand binding alters the conformation of α-I domains in a dramatic way. In all three cases studied (α2, αM and αL), there is a switch in Mg^{2+} coordination in which a direct bond to a MIDAS threonine is gained while a direct bond to an aspartic acid is lost (Fig. 2c). These subtle changes in metal coordination are linked to extensive secondary and tertiary changes that create a complementary surface for binding collagen and generate a 10 Å downward movement of the C-terminal helix. The movement of the helix propagates the change in the MIDAS surface to the lower surface of the domain. The shift of the C-terminal helix is highly significant in the context of the whole integrin, since the helix is packed against the propeller and β-I domain (see below).

The close similarity between the structural changes seen in the α2-I, αM-I and αL-I domains suggests that there are just two principal conformations for I domains, "open" and "closed" (Fig. 2). The "open" conformation is seen in the presence of ligand or ligand mimetic, while the "closed" conformation is seen in the absence of ligand. It therefore appears to be the formation of a strong ligand-metal bond, requiring a change in metal coordination, that triggers the conformational switch. Recently, Springer's group has engineered disulfide-linked αL-I domains with both intermediate affinity and packing of the C-terminal helix[3]. However, in these structures the MIDAS motif exists in only two conformations, and it remains to be seen whether the intermediate conformation has biological relevance or is an artifact of the engineered disulfide. It should be noted that it is not necessary to invoke an intermediate tertiary conformation in order to explain an intermediate affinity. In principle, a shift in the position of the equilibrium between two states is sufficient.[32]

Various studies have now been published in support of the hypothesis that the open and closed conformations of the α-I domain equate with high and low affinity states. Thus, mutants of the αM-I domain that are predicted on structural grounds to destabilize the closed conformation and favor the open conformation increase the affinity for the ligand C3bi.[23] The epitope for an antibody that binds only to the high affinity form of the αMβ2 integrin maps to a region that undergoes extensive conformational changes between the closed and open forms.[33] Disulfide engineering studies on recombinant I domains and full-length integrins, which lock the domain either into the open or closed state, also support the hypothesis.[34-37] So does the structure of the αL-I domain in complex with the inhibitor lovastatin:[38] it reveals allosteric inhibition by binding between the β-sheet and the C-terminal helix, preventing the C-terminal helix shift.

The β-I Domain

The existence of a β-I domain was initially predicted based on the conserved and critical MIDAS-like sequence, DxSxS, and hydropathy plot comparisons with the α-I domain,[14] and later from more sophisticated sequence analysis.[39] The structure of the β3-I domain, contained within the αVβ3 crystal structure, confirms that the basic fold is very similar to the α-I domain.[1] The MIDAS motif is similar to that in the α-I domain, except that the MIDAS threonine is replaced by glutamate. This difference likely explains the different cation specificities – in α-I, the smaller Mg^{2+} ion favors ligands lacking a formal charge; while for β-I, the larger Ca^{2+}

ion favors charged ligands. The structure of αVβ3 in complex with an RGD-style peptide shows that the Asp sidechain completes the coordination sphere of the MIDAS metalion,[40] as expected. There are two elaborate loops emanating from the upper surface of the β-I domain which contact the α-subunit propeller. One of these emanates from MIDAS Loop 3, thus providing a potential link between ligand binding and the quaternary structure.

The conformation of the β-I domain in the crystal structure is much more similar to the closed conformation of the α-I domain, although Xiong et al proposed the opposite.[1] Furthermore, the conformational changes induced by soaking the RGD-style peptide into the crystals[40] are consistent with those expected for a liganded I domain within the context of a closed quaternary structure; that is, the tertiary changes are in the direction of those observed in the α-I domain, but appear to be frustrated by the closed quaternary structure.[41] A model has been proposed in which conformational changes comparable to those observed in the α-I domain occur in the β-I domain of the active integrin, requiring a large-scale reorganization of the headpiece (Fig. 4).[42] The structure of the headpiece in complex with an authentic ligand in the active conformation is required to confirm this model.

In integrins that lack an α-I domain, the β-I domain and α-subunit propeller are the major recognition elements.[43] However, in integrins that contain an α-I domain, there is no evidence that the β-I and α-propeller play direct roles in ligand recognition. Instead, they play important regulatory roles. For example, mutation of the β-MIDAS motif leads to loss of iC3b binding.[44] This was initially interpreted as evidence for a direct role in ligand binding, but it now seems clear that the mutation works allosterically, by preventing conformational changes in the α-I domain[27] (see below).

Quaternary Changes in Integrins Lacking an α-Domain

Takagi et al[7] have recently shown that the inactive (resting) form of the integrin αVβ3, observed in physiological concentrations of Ca^{2+} and Mg^{2+}, is largely bent, and closely resembles the crystal structure, in which the C-termini of both chains are closely apposed. Other integrins tested had a lower propensity to adopt the bent conformation; however, the experiments were performed with extracellular heterodimers truncated near the plasma membrane, so that they lack the transmembrane helices and cytoplasmic tails that are known to contribute critically to the stability of the inactive conformation. By engineering a disulfide link between the α-subunit propeller and the EGF4 domain of the β-subunit (which are 4 Å apart in the bent (crystal) structure, but would be very far apart in the "standing-up" conformation), Takagi et al[7] also showed that integrin expressed on the cell surface was in a low affinity state and could only be activated under reducing conditions. These data seem convincing; it remains to be seen whether α-I domain integrins adopt the same bent conformation.

On integrin activation in vitro, either by using Mn^{2+} or a ligand-mimetic RGD peptide, Takagi et al[7] showed that the integrin "stands up", breaking contacts between the β-hybrid domain and the membrane-proximal domains of both α- and β-subunits. This "switchblade" motion is linked to a separation of the legs (Fig. 1). The large quaternary motion is associated with a more subtle change in the angle between the β-I domain and hybrid domain, consistent with the idea that the C-terminal helix of β-I undergoes a shift comparable to that in α-I. That is, the data support the prediction that the trigger for the integrin switch is the same as in the α-I domain integrins - a subtle change in metal coordination at the MIDAS motif.[2]

As noted above, these experiments were performed with truncated integrins and small peptide ligands. Although the nature of the trigger in the integrin head seems secure, it remains to be seen whether the quaternary consequences of the trigger are the same in the intact integrin. Other lines of evidence support a separation of the head and a more subtle separation of the tail.[6] Indeed, it may require too much energy to push a transmembrane helix through the

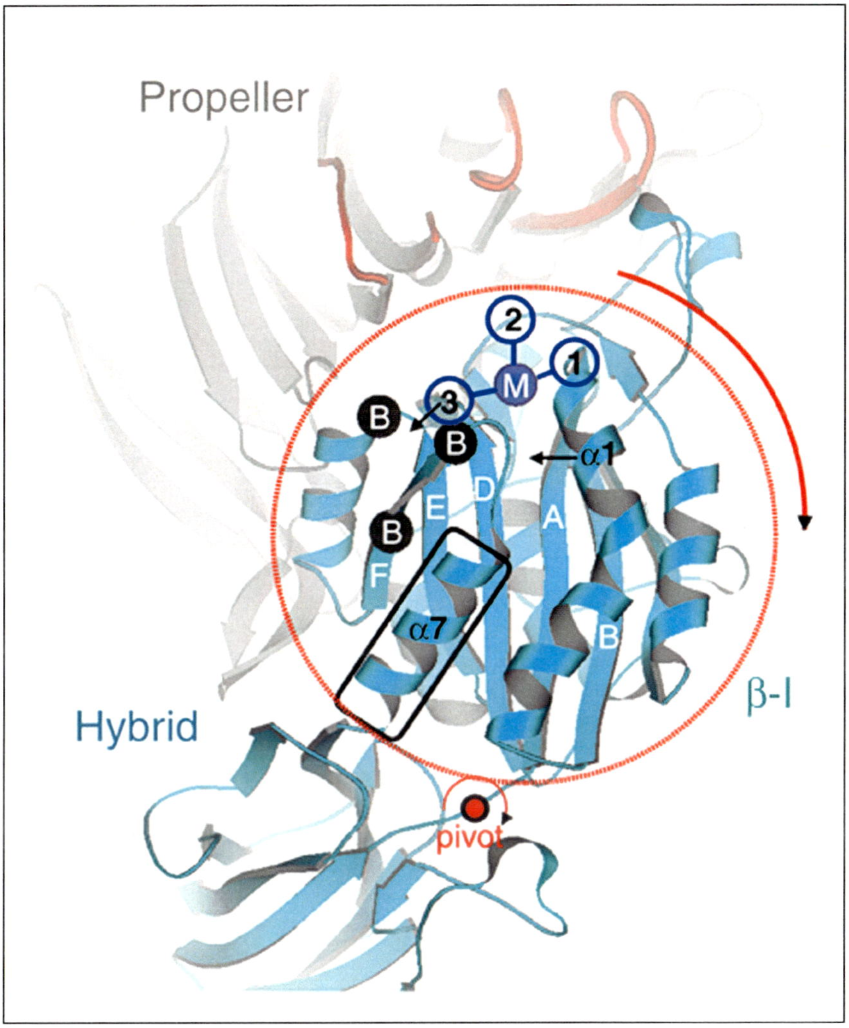

Figure 4. Model of the initial tertiary and quaternary changes triggered by ligand binding in integrins that lack an α-I domain. Ligand binding to the MIDAS motif of the β-I domain (M) triggers conformational changes in the three MIDAS loops (numbered 1–3) that cause a shift of Helix 1 and MIDAS Loop 3 towards the βF-7 loop (black arrows) and a loosening of the contacts between the β-I domain and the propeller. These movements cause steric clashes that are relieved by a rotation of the β-I domain around the indicated pivot point (red circle) at the hybrid–β-I domain boundary, separating the β-I from the propeller. Helix 7 remains in position, thus freeing itself from its contacts with MIDAS Loop 3 and Helix 1. Three mutations (B) shown previously to block activation[60] define a path between the movement of MIDAS loop 3 and helix 7; an activation epitope maps to the base of 1 (labeled H) at the β-I–hybrid interface . Red highlighted loops on the propeller map the location of chimeras that switch ligand recognition specificities between integrin αV and α5 subunits.

distances suggested from the truncated integrins. Simple modeling studies of fibronectin bound to α5β1, which involve an RGD motif and a "synergy" motif separated by 35 Å, suggest that the propeller and β-I domain must separate significantly in the active conformation.

Quaternary Changes in Integrins Containing an α-I Domain

How does the quaternary organization of the integrin regulate the affinity of the α-I domain? Currently, there is no crystal structure available for an integrin headpiece containing an α-I domain. Neither is there a crystal or EM structure for a headpiece with a physiological ligand bound in the "active" conformation. Until they become available, models of integrin allostery must remain speculative. Nevertheless, it seems likely that α-I domain integrins will undergo similar conformational switches to those that lack α-I domains. We know that regulation occurs allosterically (rather than by steric masking of the binding site), since the α-I domain is a major antibody epitope. It is straightforward to build a model of the α-I domain head group based on the crystal structure of αVβ3 (Fig. 5). The model predicts that the α-I domain will contact the α-subunit propeller and the β-I domain.

A priori, there are two possibilities of allosteric regulation: either the quaternary organization of the inactive (resting) integrin serves to reduce the inherent affinity of the isolated α-I domain by stabilizing its closed tertiary conformation (Model A); or the quaternary organization of the active integrin serves to stabilize the open conformation of the I domain (Model B). The two models are not mutually exclusive, and the true picture may lie between these extremes.

For αLβ2, some data favor Model B. Thus, the affinity of the disulfide-locked closed conformation is similar to that of the wild-type recombinant domain, roughly 10^4–fold lower than for the locked open domain.[35] The affinity of the disulfide-locked conformation of αL is similar to that of the activated integrin (~200 nM), suggesting that bonds in the active integrin play a similar role to the disulfide lock. It should be noted that not all α-I domains need behave in the same fashion. Thus, although the affinity difference between the open and closed forms of αM is probably similar, ~10^4, the equilibrium between the unliganded open and closed forms of the recombinant protein is much less pronounced (~10^1 in favor of the closed conformer), so that the major role of the integrin in that case may be to stabilize the closed conformation (Model A).

Mutagenesis of the lower surfaces of the I domain, which in the simple model (Figure) contact the α-subunit propeller and the β-I domain, have varying effects on integrin affinity. Many mutations are "gain-of-function", suggesting that these mutations disrupt intersubunit bonds that hold the α-I domain in the closed conformation, consistent with Model A.[45] However, a few mutations, particularly to the linker region between the C-terminal helix of α-I and the propeller, lead to loss of function.[46] This region includes a conserved glutamic acid, and Alonso et al[47] have proposed that the glutamic acid completes the coordination sphere of the α-I domain MIDAS motif in the active integrin; that is, the active quaternary conformation serves to stabilize the open α-I domain (Model B). This hypothesis is strikingly consistent with a detailed analysis of metal dependence of ligand binding,[48] which reveals two metal binding sites (1 and 2) that promote ligand binding to LFA-1. Site 1 is specific for Ca^{2+} (Kd = 15 μM). Occupancy of Site 1 enhances the affinity of Mg^{2+} for Site 2 (from 160 μM to 12 μM)). Ca^{2+} can also bind to Site 2 but with low affinity (1.5 mM), competitively displacing Mg^{2+} and decreasing the binding of soluble ligand. Site B can be equated with the α-MIDAS (Mg^{2+} specific), and site A with the β-MIDAS (Ca^{2+} specific).

Tail Interactions

There are abundant biochemical and genetic data supporting the notion that interactions between integrin α and β cytoplasmic tails hold the resting integrin in a low affinity conformation.[11,49,50] For example, a classic study by Ginsberg and colleagues showed that a salt bridge between the αIIb Arg995 and β3 Asp723 is necessary and sufficient to hold the integrin in its resting state.[11]

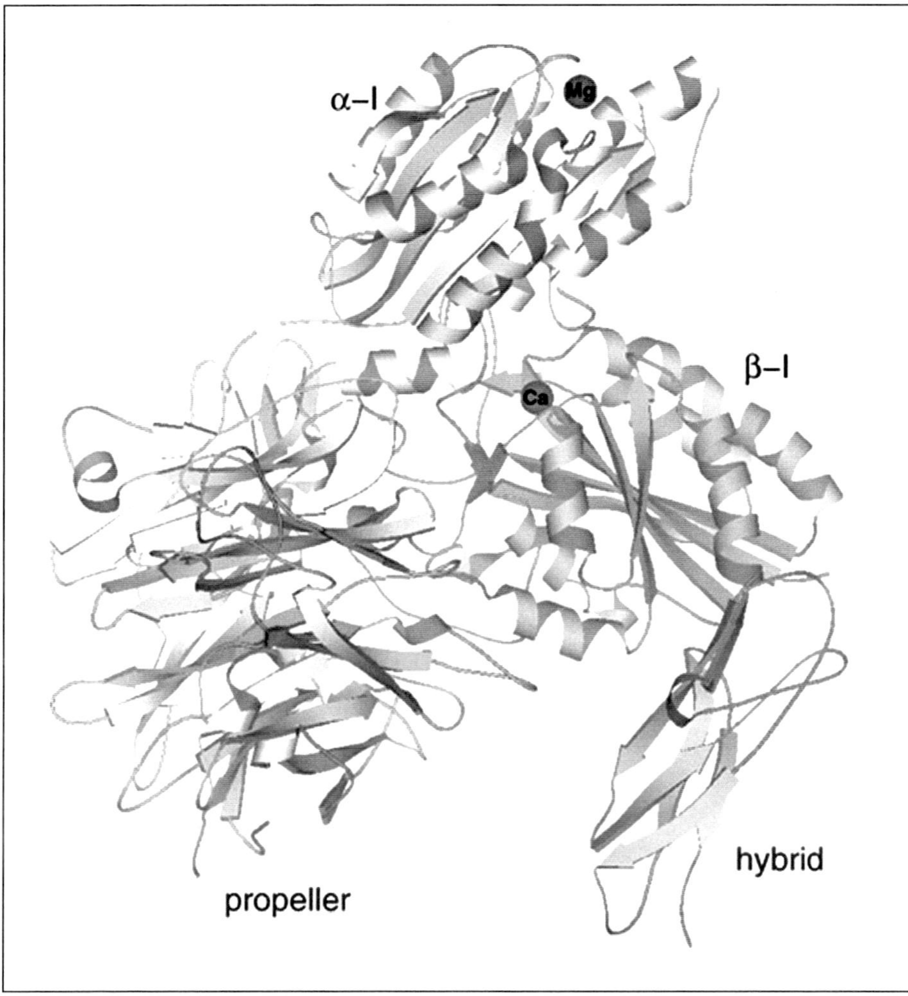

Figure 5. Hypothetical model of an α-I domain integrin head. The model is based on the crystal structures of αVβ3[1] and the α2 I domain.[17] Note that a Ca^{2+} ionat the β-MIDAS motif is in a position to coordinate a glutamic acid from the α-I domain, as proposed.[47]

A clear atomic resolution picture of the integrin αβ tail interaction has finally emerged,[5] along with models of inside-out activation that largely confirm models previously proposed from mutational and functional studies.[4,5] The atomic model of the αIIbβ3 tail (Fig. 6a), derived by NMR spectroscopy,[5] provides a good model for all integrin tails (except α6β4). The weak interaction of the recombinant peptide (~6 μM) made the studies difficult. This may explain the failure of earlier groups to detect the interaction; furthermore, it is possible that additional interactions occur in the context of the membrane-bound holoreceptor. Notwithstanding, clear interactions between membrane-proximal helical elements of the α and β- chains are evident, and are consistent with the key salt bridge predicted by Ginsberg and colleagues.[11]

The abundant cytoskeletal protein talin has been shown to bind isolated integrin β-tails, and the 40 kDa talin head domain activates platelets when overexpressed in cells.[51] Vinogradova

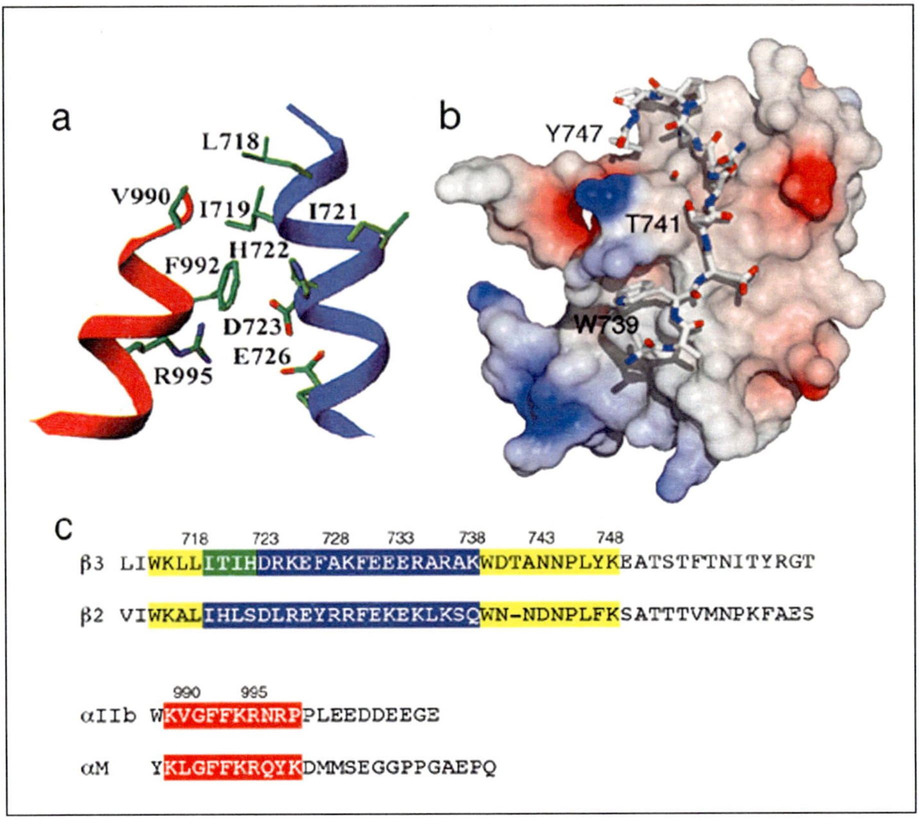

Figure 6. Integrin tail interactions. a. Atomic interactions between the αIIb and β3 tails as revealed by NMR.[5] b. Crystal structure of a fragment of the β3 tail in complex with the F3 domain of talin.[4] c. Sequences of α and β tails. Highlighted in blue and red are the interacting regions of the αβ tail complex. Highlighted in yellow are the regions shown to interact with talin by crystallography and NMR.[4,5]

et al also showed by NMR that the talin head disrupts αβ association. While the result was predictable, given the high affinity of Talin-H for the isolated β-tail in vitro (100 nM), nevertheless, this work is very important for the field, because it represents the first formal demonstration of an activating pathway that had long been predicted from biochemical and mutagenesis studies. In the absence of these data, competing models featuring homomeric association had been proposed.[52] While it remains possible that homomeric tail association may be related to integrin clustering and avidity changes, the fundamental hypothesis of regulated αβ association expounded 10 years ago has finally been vindicated.

The crystal structure of a talin head fragment in complex with part of the integrin β3 tail has recently been determined (Fig. 6b).[4] The talin head contains a three-lobed FERM (Band 4.1, Ezrin, Radixin Moesin) domain. The third lobe, F3, is sufficient for β3 tail binding in vitro and platelet activation.[53] F3 comprises a so-called PhosphoTyrosine-Binding (PTB) domain (despite its name, it should be noted that many members of this family bind unphosphorylated proteins). It had previously been shown that integrin β-tails contain a conserved and critical NPxY motif, similar to the recognition motifs of "classical" PTB domains. Thus, point mutations in this motif abrogate talin binding[54] and activation.[55] The crystal structure of the complex shows that talin engages integrin principally via a PTB-peptide interaction involving the F3 subdomain of talin and a central portion of the integrin tail, near to the end of the major helical segment observed in the αβ complex. Key interactions are the conserved NPLY motif, which adopts a reverse turn conformation, and an aromatic residue at position –8 (with respect to the tyrosine) at the beginning of a short β-strand that augments the β-sheet of talin. Given the conserved sequences of β-tails (Fig. 6c), the structure provides general models of integrin-talin engagment. There is some flexibility in the position of the aromatic residue, since in β2 for example, the aromatic is at position –7 with respect to the tyrosine, which could be accommodated by a less pronounced bulge between the β-strand and reverse turn segments.

The region of β3 defined in the crystal structure is immediately next to, although not part of, the region involved in direct interactions with the α-tail as observed by NMR.[5] However, there is also evidence that talin binding involves membrane-proximal parts of the β tail that do bind α.[5,56] One possibility that reconciles these data is a two-step pathway of activation, in which the talin head first recognizes the "available" high-affinity binding site in the membrane-distal region of the resting integrin, with subsequent binding to a second, lower affinity, membrane-proximal site involved in αβ association that triggers separation of the tails.

There are a growing number of proteins that bind integrin tails.[57] Some of these, such as cytohesin-1, which binds the β2 tail, also activate the integrin when overexpressed in cell.[58] No structure of the complex is available yet, but mutagenesis studies show that cytohesin-1 requires a membrane proximal segment of β2, at the N-terminal end of the αβ contacts, suggesting that activation proceeds by a pathway similar to that of talin via αβ tail disruption.

The model of αβ tail dissocation upon integrin activation predicts that the α-tail also becomes available to bind effector molecules involved in signal transduction. One such interaction is between the α4 tail and paxillin, which regulates cell migration.[59] Another is suggested from studies of α2β1-mediated platelet signaling (see Chapter 4 by Zutter & Santoro). Currently, no complex structures are available.

Summary and Conclusions

Recent structural and structure-function studies have revealed many of the major paradigms of integrin allostery underlying affinity regulation and bi-directional signal transduction. Notably lacking is the structure of an integrin head-piece bound to a physiological ligand that would reveal the "active" conformation. Recent EM studies that point to three distinct conformational states of the integrin will be an area of intense study. Do these represent "resting" "inside-out-activated" and fully liganded conformations? Further studies of the αβ tails and their complexes with intracellular binding partners are also urgently required. We still do not know how conformational changes in the headpiece propagate through the plasma membrane, and cryoelectron microscopy of intact integrins in different physiological ligation states is likely to be the method of choice here. Finally, this chapter has addressed the structural basis of affinity changes in integrins. Lateral association (clustering) of integrins in the plasma membrane at sites of ECM contact also plays a major role in integrin signaling, and we currently know next to nothing about its structural basis.

References

1. Xiong J-P, Stehle T, Diefenbach B et al. Crystal structure of the extracellular segment of integrin αVβ3. Science 2001; 294:339-345.
2. Emsley J, Knight CG, Farndale RW et al. Structural basis of collagen recognition by integrin α2β1. Cell 2000; 101:47-56.
3. Shimaoka M, Tsan Xiao T, Liu J-H et al. Structures of the αL I Domain and its Complex with ICAM-1 Reveal A Shape-Shifting Pathway for Integrin Regulation. Cell 2003; In press.

4. Garcia-Alvarez B, de Pereda JM, Calderwood DA et al. Structural Determinants of Integrin Recognition by Talin. Mol Cell 2003; In press.
5. Vinogradova O, Velyvis A, Velyviene A et al. A structural mechanism of integrin αIIbβ3 "inside-out" activation as regulated at its cytoplasmic face. Cell 2002; 110:587-597.
6. Hantgan RR, Paumi C, Rocco M et al. Effects of ligand-mimetic peptides Arg-Gly-Asp-X (X = Phe, Trp, Ser) on αIIbβ3 integrin conformation and oligomerization. Biochem 1999; 38:14461-14474.
7. Takagi J, Petre BM, Walz T et al. Global conformational rearrangements in integrin extracellular domains in outside-in and outside-in signaling. Cell 2002; 110:599-611.
8. Michishita M, Videm V, Arnaout MA. A novel divalent cation-binding site in the A domain of the β2 integrin CR3 (CD11b/CD18) is essential for ligand binding. Cell 1993; 72:857-867.
9. Kamata T, Takada Y. Direct binding of collagen to the I-domain of integrin α2β1 (VLA-2, CD49b/CD29) in a divalent cation-independent manner. J Biol Chem 1994; 269:26006-26010.
10. Tuckwell DS, Calderwood DA, Green LJ et al. Integrin α2 I-domain is a binding site for collagens. J Cell Sci 1995; 108:1629-1637.
11. Hughes P, Diaz-Gonzalez F, Leong L et al. Breaking the integrin hinge. J Biol Chem 1996; 271:6571-6574.
12. Takagi J, Erickson HP, Springer TA. C-terminal opening mimics 'inside-out' activation of integrin alpha5beta1. Nat Struct Biol 2001; 8:412-416.
13. Schwartz MA, Ginsberg MH. Networks and crosstalk: integrin signalling spreads. Nature Cell Biol 2002; 4:E65-E68.
14. Lee J-O, Rieu P, Arnaout MA et al. Crystal structure of the A-domain from the α subunit of integrin CR3 (CD11b/CD18). Cell 1995; 80:631-635.
15. Qu A, Leahy DJ. Crystal structure of the I-domain from the CD11a/CD18 (LFA-1, αLβ2) integrin. Proc Natl Acad Sci USA 1995; 92:10277-10281.
16. Nolte M, Pepinsky RB, Venyaminov SY et al. Crystal structure of the alpha 1 beta 1 integrin I-domain: insights into integrin I-domain function. FEBS Letters 1999; 452:379-385.
17. Emsley J, King S, Bergelson J et al. Crystal structure of the I domain from integrin α2β1. J Biol Chem 1997; 272:28512-28517.
18. Kamata T, Puzon W, Takada Y. Identification of putative ligand binding sites within I domain of integrin α2β1 (VLA-2, CD49b/CD29). J Biol Chem 1994; 269:9659-9663.
19. Kern A, Briesewitz R, Bank I et al. The role of the I domain in ligand binding of the human integrin α1β1. J Biol Chem 1994; 269:22811-22816.
20. Edwards CP, Champe M, Gonzales T et al. Identification of amino acids in the CD11a I-domain important for binding of the leukocyte function-associated antigen-1 (LFA-1) to intercellular adhesion molecule-1 (ICAM-1). J Biol Chem 1995; 270:12635-12640.
21. Huang C, Springer TA. A binding interface on the I domain of lymphocyte function associated antigen-1 (LFA-1) required for specific interaction with intercellular adhesion molecule 1 (ICAM-1). J Biol Chem 1995; 270:19008-19016.
22. Zhang L, Plow EF. Identification and reconstruction of the binding site with in αMβ2 for a specific and high affinity ligand, NIF. J Biol Chem 1997; 272:17558-17564.
23. Li R, Rieu P, Griffith DL et al. Two functional states of the CD11b A-domain: correlations with key features of two Mn2+-complexed crystal structures. J Cell Biol 1998; 143:1523-1534.
24. Zhang L, Plow EF. Amino acid sequences within the a subunit of integrin αMβ2 (Mac-1) critical for specific recognition of C3bi. Biochemistry 1999; 38:8064-8071.
25. Kamata T, Liddington RC, Takada Y. Interaction between collagen and the alpha(2) I-domain of integrin α2β1 - Critical role of conserved residues in the metal ion-dependent adhesion site (MIDAS) region. J Biol Chem 1999; 274:32108-32111.
26. Smith C, Estavillo D, Emsley J et al. Mapping the collagen binding site in the I domain of the Glycoprotein Ia/IIa (integrin α2β1). J Biol Chem 2000; 275:4205-4209.
27. Lee J-O, Bankston LA, Arnaout MA et al. Two conformations of the integrin A-domain (I-domain): a pathway for activation? Structure 1995; 3:1333-1340.
28. Knight CG, Morton LF, Onley D et al. Identification in collagen type I of an integrin alpha(2)beta(1)-binding site containing an essential GER sequence. J Biol Chem 1998; 273:33287-33294.

29. Kramer RZ, Bella J, Mayville P et al. Sequence dependent conformational variations of collagen triple-helical structure. Nature Struct Biol 1999; 6:454-457.
30. Jung SM, Moroi M. Platelets interact with soluble and insoluble collagens through characteristically different reactions. J Biol Chem 1998; 273:14827-14837.
31. Knight CG, Morton LF, Peachey AR et al. The collagen-binding A-domains of integrins α1β1 and α2β1 recognize the same specific amino acid sequence, GFOGER, in native (triple-helical) collagens. J Biol Chem 2000; 275:35-40.
32. Loftus JC, Liddington RC. New insights into integrin-ligand interaction. J Clin Invest 1997; 99:2302-2306.
33. Oxvig C, Lu C, Springer TA. Conformational changes in tertiary structure near the ligand binding site of an integrin I domain. Proc Natl Acad Sci USA 1999; 96:2215-2220.
34. Shimaoka M, Shifman JM, Jing H et al. Computational design of an integrin I domain stabilized in the open high affinity conformation. Nature Struct Biol 2000; 7:674-678.
35. Shimaoka M, Lu C, Palframan RT et al. Reversibly locking a protein fold in an active conformation with a disulfide bond: integrin alphaL I domains with high affinity and antagonist activity in vivo. Proc Natl Acad Sci USA 2001; 98:6009-6014.
36. Shimaoka M, Lu C, Salas A et al. Stabilizing the integrin αM inserted domain in alternative conformations with a range of engineered disulfide bonds. Proc Natl Acad Sci USA 2002; 10:1073/pnas.252633099.
37. McCleverty C, Liddington RC. Structure-function studies of allostery in the integrin αM I domain. Biochem J 2003; In press.
38. Kallen J, Welzenbach K, Ramage P et al. Structural basis for LFA-1 inhibition upon lovastatin binding to the CD11a I-domain. J Mol Biol 1999; 292:1-9.
39. Tuckwell DS, Humphries MJ. A structure prediction for the ligand-binding region of the integrin beta subunit: evidence for the presence of a von Willebrand factor A domain. FEBS Lett 1997; 400:297-303.
40. Xiong JP, Stehle T, Zhang R et al. Crystal structure of the extracellular segment of integrin αVβ3 in complex with an Arg-Gly-Asp ligand. Science 2002; 296:151-155.
41. Liddington RC. Will the real integrin please stand up? Structure. 2002; 10:605-607.
42. Liddington RC, Ginsberg MH. Integrin activation takes shape. J Cell Biol 2002; 158:833-839.
43. Mould AP, Askari JA, Aota S et al. Defining the topology of integrin α5β1-fibronectin interactions using inhibitory anti-α5 and anti-β1 monoclonal antibodies - Evidence that the synergy sequence of fibronectin is recognized by the amino-terminal repeats of the α5 subunit. J Biol Chem 1997; 272:17283-17292.
44. Bajt ML, Goodman T, McGuire SL. Beta 2 (CD18) mutations abolish ligand recognition by I domain integrins LFA-1 (alpha L beta 2, CD11a/CD18) and MAC-1 (alpha M beta 2, CD11b/CD18). J Biol Chem 1995; 270:94-98.
45. Zhang L, Plow EF. A discrete site modulates activation of I domains. J Biol Chem 1996; 271:29953-29957.
46. Lupher MLJ, Harris EA, Beals CR et al. Cellular activation of leukocyte function-associated antigen-1 and its affinity are regulated at the I domain allosteric site. J Immunol 2001, 167.1431-1439.
47. Alonso JL, Essafi M, Xiong JP et al. Does the Integrin alphaA Domain Act as a Ligand for its betaA Domain? Curr. Biol 2002; 12:R340-R342.
48. Labadia ME, Jeanfavre DD, Caviness GO et al. Molecular recognition of the interaction between leukocyte function-associated antigen-1 and soluble ICAM-1 by divalent metal cations. J Immunology 1998; 161:836-842.
49. Williams MJ, Hughes PE, O'Toole TE et al. The inner world of cell adhesion: integrin cytoplasmic domains. Trends Cell Biol 1994; 4:109-112.
50. Ginsberg MH, Yaspan B, Forsyth J et al. A membrane-distal segment of the integrin αIIb cytoplasmic domain regulates integrin activation. J Biol Chem 2001; 276:22514-22521.
51. Calderwood DA, Zent R, Grant R et al. The talin head domain binds to integrin beta subunit cytoplasmic tails and regulates integrin activation. J Biol Chem 1999; 274:28071-28074.
52. Li R, Babu CR, Lear JD et al. Oligomerization of the integrin αIIbβ3: Roles of the transmembrane and cytoplasmic domains. Proc Natl Acad Sci USA 2001; 98:12462-12467.

53. Calderwood DA, Yan B, de Pereda JM et al. The phosphotyrosine binding (PTB)-like domain of talin activates integrins. J Biol Chem 2002; 277:21749-21758.
54. Pfaff M, Liu S, Erle DJ et al. Integrin b cytoplasmic domains differentially bind to cytoskeletal proteins. J Biol Chem 1998; 273:6104-6109.
55. O'Toole TE, Ylanne J, Culley BM. Regulation of integrin affinity states through an NPXY motif in the beta subunit cytoplasmic domain. J Biol Chem 1995; 270:8553-8558.
56. Patil S, Jedsadayanmata A, Wencel-Drake JD et al. Identification of a talin-binding site in the integrin beta(3) subunit distinct from the NPLY regulatory motif of post-ligand binding functions. The talin n-terminal head domain interacts with the membrane-proximal region of the beta(3) cytoplasmic tail. J Biol Chem 1999; 274:28575-28583.
57. Liu S, Calderwood DA, Ginsberg MH. Integrin cytoplasmic domain-binding proteins. J Cell Sci 2000; 113:3563-3571.
58. Geiger C, Nagel W, Boehm T et al. Cytohesin-1 regulates beta-2 integrin-mediated adhesion through both ARF-GEF function and interaction with LFA-1. EMBO J 2000; 19:2525-2536.
59. Liu S, Thomas SM, Woodside DG et al. Binding of paxillin to α4 integrins modifies integrin-dependent biological responses. Nature 1999; 402:676-681.
60. Baker EK, Tozer EC, Pfaff M et al. A genetic analysis of integrin function: Glanzmann thrombasthenia in vitro. Proc Natl Acad Sci USA 1997; 94:1973-1978.

CHAPTER 9

Integrin Recognition Motifs in Collagens

Richard W. Farndale

The vertebrate integrins are a family of twenty four heterodimeric transmembrane signalling and adhesion molecules, found in very many cells types where their expression and affinity is carefully regulated, reviewed in Chapter 1 and.[1,2] Integrins mediate either cell–cell interactions, often with cellular adhesion molecules (CAMs) as counter-receptors,[3] or cell–matrix interactions.[4] The two matrix molecules best-known in the latter context are fibronectin and collagen. Fibronectin is a ligand for both β1 and β3 integrins, with α5β1 as its most widespread receptor, whereas collagen is considered to interact directly only with a subset of the β1 integrins which is characterized by the presence within the α–subunit of an inserted, or I, domain. For a recent review, see.[5] The α-I domains adopt the dinucleotide-binding, or Rossman, fold, and are compared in.[6] This structure, the prototype for which is the von Willebrand factor A–domain, is found in intracellular signalling species such as G protein α–subunits and in extracellular adhesive proteins including the terminal extensions of the non-fibrillar collagens.[7] Under physiological conditions, the integrin α-I domains, though not necessarily other A-domains,[8] constitutively bind a divalent cation, Mg^{2+}, in their metal ion dependent adhesion site (MIDAS) which is the focus of their interaction with collagens.[9] Of the four collagen-binding integrins, α1β1 and α2β1 have been studied for over a decade[10-25] whilst the properties of both α10β1, which was purified using its capacity to bind collagen II,[26-28] and α11β1[29-31] are just beginning to receive attention. Data is emerging to suggest differential ability of α10β1 and α11β1, like that of α1β1 and α2β1, to interact with specific collagen types.[32,33]

Other integrins which do not contain α-I domains, the platelet-specific αIIbβ3[34] and α3β1[35], have also been proposed to be collagen receptors, but their capacity to bind collagen perhaps resides in either an indirect interaction or in their ability to up-regulate the activity of other integrins such as α2β1.[36] It is of interest that α3β1 recognizes linear peptides derived from collagen, though not as well as triple-helical structures,[35] and that D-amino acid polypeptides serve the same function.[37] This suggests that the interaction of α3β1 is not specific for collagens in their native conformation, in line with which, denatured collagens also interact with α3β1.[38] This topic is discussed further in Section 2. This Chapter will focus on the properties of collagen which enable it to interact with the four α-I domain-containing β1 integrins.

Collagen Structure

The collagens are defined by their repeated GXX' sequences which confer upon the collagen α chains the capacity to assemble as trimers. These GXX' sequences allow each polypeptide chain to adopt a left-handed helical conformation, the polyproline II structure, and three such helices to assemble as a right-handed superhelix. Such close packing is only possible because glycine lacks the bulky sidechains of other amino acids, which

I Domains in Integrins, edited by Donald Gullberg. ©2003 Eurekah.com and Kluwer Academic / Plenum Publishers.

Table 1. Collagen types classified according to their structural properties

Group	Types	Homo-Trimeric	Hetero-Trimeric	Integrin Binding?	Other Modules
Fibrillar	I, II, III, V, XI	II, III, V	I, V, XI,	I, II, III, XI V with XI	
FACIT	IX, XII, XIV, XVI, XIX, XXI,	XII, XIV, XVI, XIX	IX		A,Fn(XII,XIV)
Hexagonal	IV, VIII, X(?)	X	IV, VIII	IV	
MACIT	XIII, XVII, XXV	XIII, XVII		XIII[a], XVII[a]	
Multiplexin	XV, XVIII	XV, XVIII[b]		XVIII[a]	
Other	VI, VII	VI, VII	VI	VI	A,Fn(VI,VII)
Not known	XX, XXII, XXIII, XXIV	?	?		

a. indicates collagens which lack GER motifs but are known to bind integrins.
b. splice variants of the same gene occur
Fibrillar collagens are described in the body of the text. FACITs (Fibril-Associated Collagens with Interrupted Triple-helices) are found on the surface of fibrillar collagens where they may regulate fibril assembly and interactions with other matrix components. The Hexagonal group is best-characterized by collagen IV, where four triple-helical monomers assemble in anti-parallel array, and their long N-terminal extensions interact with those of other tetramers, such that a chicken wire-like planar network is formed, the central structure of the basement membrane (BM). Collagens VIII and X may form similar components of specialized BMs. MACITs (Membrane-Associated Collagens with Interrupted Triple-helices) are expressed on the cell surface, sharing features with the monocyte scavenger receptor, MARCO, and the C1q complex. Multiplexins (Multiple Triple helix domains with Interruptions) are diversely expressed, at high level in the internal organs. Type VI collagen falls into no obvious grouping, but forms a 3-dimensional lattice in interstitial tissue. Von Willebrand factor A domains (A) and fibronectin III domains (Fn) add further defined binding properties to some collagens.

could not be accommodated inside the triple helix. Each left-handed helix has three residues per turn, so that glycine is always located close to the axis of the superhelix. As a consequence, both X and X' sidechains are exposed on the superhelix surface where they may be free to interact with other molecules.[39-41] The capacity to adopt the polyproline helical conformation arises from the distortion of the α chain backbone introduced by the iminoacids proline (P) and hydroxyproline (O) in the X and X' positions respectively. Twenty five human collagens have been identified to date,[42] the family expressing a surprising diversity in structure and function, with the presence of a GXX' collagen helix (Col domain) as either a continuous domain, or interrupted by non-helical sequence, being their common defining feature. These collagens fall into several identifiable groups (summarized in Table 1), with the fibrillar collagens, comprising types I, II, III, V and XI, being the prime example and the most abundant proteins within the vertebrate organism. Their fibrillar structure is achieved by the packing of the rod-like trimeric collagen monomers in the typical quarter-staggered array, aided by electrostatic interactions between the sidechains of adjacent triple helices. Detailed discussion of the structure of the different collagen types is beyond the scope of the present Chapter.

Collagens as a group undergo post-translational modification, notably the hydroxylation of proline and a proportion of those lysine residues which occur in the X' position. Subsequent crosslinking between triple helices occurs by condensation of hydroxylysine and lysine sidechains, adding to the stability of the fibril, whilst hydrogen bonding through water molecules within the hydration shell which surrounds the collagen monomer may stabilize both the fibril and the helix itself.[39] It is important to note that the collagen mRNA encodes precursors, procollagens, from which N- and C-terminal propeptides are cleaved proteolytically as the triple helix is secreted from the cell and assembles into the fibril. This process leaves short telopeptide extensions at both the N- and C-terminal ends of the triple-helical fibrillar collagen monomer.

In the non-fibrillar collagens, the triple-helices may be interrupted by non-helical domains, and the telopeptides may be quite long and support interactions between the collagen monomers, such that complex two- or three-dimensional networks can assemble. Collagens IV and VI exemplify such structures, which are important components of basement membrane and of connective tissues where flexibility of the collagen matrix may be more important than its tensile strength.

Both the helical and non-helical domains of collagens express the capacity to interact with integrins. Clearly-defined integrin-binding sequences occur within the triple-helical domains of the collagens, but other such sites located within the non-collagenous domains, such as within the C-terminal propeptide of collagen 1 which binds both α1β1 and α2β1,[43,44] have yet to be elucidated. Similarly endostatin, the C-terminal domain of collagen XVIII, interacts with α5β1 by a mechanism yet to be clarified.[45, 46]

Regions of Collagen That Bind Integrins

The possibility that collagens might interact with cells as well as with other matrix proteins acquired an experimental basis forty years ago, when collagen fibre preparations were found to have a rapid and dramatic activatory effect on platelet function.[47,48] Specific collagen receptors were not identified for a further twenty years, when a patient presenting with a bleeding disorder was found to lack VLA-2, the integrin α2β1, also known as platelet glycoprotein (Gp) Ia-IIa.[10] This discovery led to the assumption that α2β1 was crucial to the platelet activation process, to the point that no need for other platelet receptors was envisaged. However, digestion of purified collagen α chains using cyanogen bromide (CB) yielded long, defined collagen fragments which could be renatured into triple helical form.[49,50] These fragments did not behave uniformly. It became apparent that the collagen sequence contained a series of platelet-reactive sites, some of which supported adhesion, now considered to be mediated by α2β1, and some of which supported both platelet adhesion and activation, the latter property only apparent when the fragments were crosslinked chemically.[51] Yet other fragments showed no detectable platelet-reactivity at all despite being triple helical. Thus the two-site, two-step model was born, which envisaged firm adhesion as being mediated by integrin α2β1, and that platelet activation required a further receptor,[21,23,24,52] now identified as the immunoglobulin superfamily receptor, glycoprotein VI, reviewed in.[53,54] The ability of platelets to bind to collagen under the high shear rates which occur in the arterial circulation depends on an indirect interaction between collagen and glycoprotein Ib which is mediated by von Willebrand factor, but also requires the presence of integrin α2β1,[55] reviewed in.[54,56] The topic of platelet activation by collagen receives further discussion in Chapters 4 and 12.

Two fibrous collagens, I and III, have been studied in some detail, revealing that they contain rather few integrin recognition sites. The collagen I α1 chain yielded three sites when fragmented with CB and then renatured to form triple helices, in CB peptides 3, 7 and 8, whilst collagen III yielded two sites, in CB peptides 3 and 4.[52]

Two laboratories studied α1(I)CB3. The first group reported that integrin recognition resided in a tetrapeptide motif, DGEA, comprising residues 613-616 of the transcript sequence.[57] In linear form, this peptide inhibited platelet adhesion to collagen at about 5mM. Although subsequent studies from several laboratories use this and similar peptides to modify nucleated cell function, several groups have failed to reproduce the original data when applying DGEA to human platelets; in triple helical conformation, the sequence does not appear to bind α2β1 in resting platelets[58] or chondrocytes,[59] although denatured collagens may utilize the sequence to bind to other cells.[38,60-66] In our hands, it proved that neutralisation of the peptide which contains two acidic side chains (and is cleaved from the synthetic resin using trifluoracetic acid) caused its inhibitory activity to be lost. It is possible that the acidity of the peptide may have confounded the original report, especially given the high working concentration needed for its activity. Subsequent work, detailed below, identified quite unequivocally that another GE-containing sequence interacted with the I-domain MIDAS of the α2 integrin subunit. Despite these doubts, it remains plausible that under some conditions, DGEA might interact with the same site, especially when the integrin is in an activated state. Note that the sequence DGEA also occurs closer to the N-terminus of collagen I, in residues 232-235.

In a more comprehensive study, a series of peptides were synthesized in triple-helical form using the host-guest approach of Ramshaw,[67] in which sequences of interest are situated between GPP or GPO flanking sequences to ensure triple helical conformation. This topic is reviewed in.[68] In this way a 15 amino acid portion of bovine α1(I)CB3 was found to contain a site which bound α2β1,[58] and truncation and substitution within this peptide identified the sequence GFOGER as the pair of GXX' triplets essential for binding not only the platelet surface, but also that of HT1080 fibrosarcoma cells as well as purified α2β1 and the recombinant α2 I-domain. The same sequence bound α1β1,[69] and good inhibition of binding to collagens I and III was obtained using the peptide at about 10μM.

An important property of these peptides was established for their integrin reactivity, the need for presentation to the cell in triple-helical conformation. Peptides synthesized within non-helical hosts lacked the capacity to interact with α2β1. This requirement mirrors that of the CB peptides or of collagen itself, and triple-helical structure is a pre-requisite for the interaction with α2β1 as well as with other integrins. Thus far, α1β1, α2β1 and α11β1 each show the same requirement for triple-helical conformation in their collagenous peptide ligands,[33,58,69] and whilst α10β1 binds triple-helical collagens, an absolute dependence on this has not yet been shown.[32]

It is interesting to note in this context that each collagen type may contain several RGD sequences, and denatured collagen (gelatin) can interact with other integrins (such as α5β1 and αVβ3) through these motifs. In other proteins, such as the archetypic integrin ligand, fibronectin, the RGD sequence is presented as a flexible loop,[70] allowing it to interact with the integrin β-subunit,[71] whereas in native collagens, G is concealed within the helix as described above. Moreover, the curvature of the polyproline II helix is rather slight, so that collagen in its native conformation is unlikely to present a complementary surface to the RGD-binding site of integrins. Together, these effects may be presumed to account for the latency of the native collagen RGD sequences, and it seems unlikely that triple-helical collagen uses RGD as an integrin-recognition motif. Nonetheless, it seems very plausible that direct interaction between collagen and RGD-selective integrins may occur, either where RGD occurs within a telopeptide or a non-helical domain, or particularly, perhaps, after degradation and denaturation of the triple-helix. This may be the case in diseased tissue where collagenase or other metalloproteinase activity has clipped collagen so that the helix cannot be maintained. Such partial denaturation may account for the proposal that αIIbβ3 or α3β1 is a collagen receptor. Recently, a related sequence, KGD, was located in the Col domain of collagen XVII, which lacks GER sequences, which bound α5β1 and αvβ1 in a manner sensitive to blockade by RGD peptides, and which increased upon denaturation. The precise nature of this interaction remains to be established.[72]

It is also important to consider that other proteins may form indirect contact between cells and collagen. A clear example is that of von Willebrand factor, which binds through its A3 domain to sites in collagen as well as to the Glycoprotein Ibα subunit, a non-integrin leucine-rich repeat protein which is expressed in significant numbers on the platelet surface.[73-75] Extracellular proteins which interact with collagen and may also interact with the cell surface through integrins include laminins, which also bind to α6β1,[17,76,77] fibronectin, which may

Table 2. Recognized α2β1 sequences from the fibrillar collagens I, II and III

α1(I)	α2(I)	#	α1(II)	#	a1(III)	#
GLOGER	**GLOGER**	127	**GLOGER**	127	**GAOGER**	136
GPAGER	*GPOGSR*	328	**GAOGER**	328	*GPAGER*	
GFOGER	*GPOGES**	502	**GFOGER**	502	**GAOGER**	511
GMOGER	***GLOGER***	550	**GMOGER**	550	**GMOGER**	559
GAOGDR	*GSOGER*		**GAOGER**	619	*GSOGER*	
GASGER	*GPOGER*	811	*GASGDR*		**GLSGER**	820

Bolded text shows the recognized α2β1 sequences from the fibrillar collagens I, II and III. Corresponding sequences in the α2(I) chain are shown, and numbers represent position within the triple helical domain of these collagens. Sequences in italic either do not bind integrin* or are un-tested. Alignments courtesy of Dr D. Slatter, University of Cambridge.

also bind to α5β1[78-80] and decorin, which, like collagen, is also a ligand for α2β1.[81] The precise binding sites for such proteins within collagens are not yet known, although decorin is known to bind CB6, 7 and 8 of the collagen α1(I) chain, CB4 of the α2(I) and CB11 of α1(II).[82]

Collagen Sequences That Bind Integrins

Subsequent to the first report,[58] two further sequences were identified within collagen I which interacted, like GFOGER, with both α2β1 and α1β1. These were GLOGER, residues 127 to 132, and GASGER, residues 811 to 816 of the α1(I) triple-helix.[83] These most likely account for the integrin reactivity of the cyanogen bromide peptides α1(I)CB7 and 8. Whilst GLOGER expresses fair affinity for α2β1, GASGER is much less reactive. Collagens I, II and III share common evolutionary development. The position of a GER motif within collagen III in the same locus as that of GFOGER in collagen I led us to consider that GAOGER, residues 511-516 of the α1(III) triple-helix, might also be an integrin recognition motif, accounting for the reactivity of CB peptide α1(III)4. This sequence proved to be of modest reactivity in platelets, though bound well to other cells. It is striking that the spacing between this and a second GAOGER is identical with that of GLOGER and GFOGER within the collagen α1(I) chain. This analogy led us to inspect the site in collagen III which is equivalent to GASGER in the α1 chain of collagen I, and to the identification of GLSGER in collagen III, which appears to be a moderate-affinity ligand for α2β1 (P.R.-M. Siljander, unpublished observations). (It should be noted that an original report based on the use of covalently-linked triple-helical peptides detailing a different integrin-binding sequence in collagen III may have been in error as a consequence of the imperfectly characterized chemistry employed[84]).

Searching the sequences of the fibrillar collagens yields several candidate integrin-binding sites, one of which, GMOGER, is of particular interest since Methionine residues are the site of cleavage by cyanogen bromide. Thus, the capacity of this sequence to bind integrins cannot be investigated in CB peptides. It is the only sequence of close approximation to GFOGER which occurs in collagen V, and its activity should be investigated in this context. Thus a family of GER sequences has been identified for the conserved collagens I, II and III (Table 2). It is of interest that collagen II contains two additional GAOGER sequences which do not align with those in types I and III, starting from residues 328 and 619 within its triple-helical domain, and that the GFOGER sequence of collagen II comprises part of an immunogenic site associated with rheumatoid arthritis.[85] The α2 chain of the heterotrimeric collagen I provides additional flexibility in its capacity to bind integrin, through the sequence GLOGER, which complements both GLOGER and GMOGER at two sites in the α1 chain. GPOGES in the α2 chain, which aligns with GFOGER in α1(I), does not itself bind integrin when presented as a homotrimeric peptide (Author's unpublished data). From the other fibrillar collagens, only the sequences GMOGER occur, twice in the α2(V) chain, and GFOGER, in both the α1(XI) and α2(XI) chains.[86]

Collagen IV is also able to bind α2β1, although it expresses a preference for α1β1, detailed below. Two GER sequences from collagen IV are of interest, GFOGER which occurs in both the α1(IV) and α5(IV) chains, and GAOGER, present in the α4(IV) chain.

Other collagens may also express some of these sequences: GAOGER occurs in α1(VI) and α3(VI), whereas collagen VII α1 chain contains two copies of GAOGER as well as both GFOGER and GLOGER, suggesting that it binds integrins strongly. GAOGER and GMOGER each occur in collagen IX α3 chain, but no homologous sequences have been located in other collagens. The presence of integrin binding sites in the FACIT collagens is not unexpected, since this will enhance the integrin-binding properties of the fibrillar collagens. However, the poorly-defined integrin-binding properties of the MACIT collagens might imply that they have the capacity to mediate cell-cell interactions.

Why do these various GER sequences express different affinities for integrins? The crystal structure of the GFOGER:α2 I-domain complex provides insights into this question.[9] First, the phenylalanine ring is snugly located in a hydrophobic pocket on the surface of the I-domain. Substitution of various long hydrophobic residues does not abolish the capacity of GXOGER peptides to bind, but none of those tested so far exceeds the affinity of GFOGER in binding to α2β1. The alanine of GAOGER and GASGER is a much shorter residue than F, L and M, suggesting that it does not extend far enough to fit so well onto the surface of the integrin. Preliminary observations suggest that the same considerations may apply to α11.

The activation state of the integrin may dictate differential ability to bind specific peptides. Comparison of the crystal structures of the free α2 I domain with that in complex with the GFOGER-containing peptide ligand reveals an Active and a Resting conformation.[6,9] In platelets, stimulation with different ligands leads to different levels of affinity of α2β1.[87-89] If it is assumed that the α2 I-domain can exist in only two states, these observations are difficult to explain. It is now likely, given the recent crystal structures of αvβ3,[90,91] that the β1 subunit exists in close apposition to the α I-domain, conceivably giving rise to further conformational change not seen in the free α I-domain structures, reviewed in.[92-94] What can be inferred from the recent information is that conformational changes can occur in the stem (genu) of both integrin subunits, as well in both α- and β-I domains.[95] Thus eight permutations of conformations might be anticipated within αI-containing integrins, potentially giving rise to more than just two states of activation.[96] Conceivably, such conformational diversity gives rise to the differences in affinity of the same integrin expressed in different cell types for these different peptide ligands, although it will be important to differentiate kinetically between changes in affinity and in the number of activated integrins available.

Next the question of the Arg within GER sequences should be considered. Evidence from this laboratory suggests that K will not fully substitute for R in the GFOGER sequence when binding of triple-helical peptides to the human platelet integrin α2β1 is considered. GFOGEK will also support some binding in other cell types, such as HT1080 cells. This suggests that GEK residues have some affinity for human α2β1. Similar findings are reported using cells expressing integrin α11β1[31], although little further exploration of other GEK sequences has been undertaken.

A further question, yet to be resolved for the binding of collagen to the β1 integrins, is whether the β1 subunit is directly involved in integrin ligation. The paradigm for other complex ligands is that two suitably-spaced sequences might bind to specific sites on the α and β subunits. It is now clearly acknowledged that the universal integrin binding sequence, RGD, which occurs in both fibronectin and fibrinogen, can bind to the β1 and β3 subunits, whilst the sequence KQAGDV within the fibrinogen γ-chain is considered to bind to the αIIb subunit[97] and the synergy sequence, PHSRNL, in the Fn(III) repeat adjacent to that which contains the RGD sequence of fibronectin, may stabilize the interaction of fibronectin with α5β1.[98] These precedents suggest the possibility that sequences within collagen adjacent to the GER motifs identified above may be competent to bind the integrin β1 subunit. So far only limited evidence has been adduced to suggest that this is the case.[99]

In fibronectin, the RGD sequence is presented on an extended flexible loop which protrudes from the tenth Fn(III) repeat, suggesting that a ligand needs to extend quite deeply into the β-I MIDAS in order to interact with the β-subunit.[70] However, no such possibility occurs within the highly-constrained rod-like collagen structure. Hence sequences other than RGD should be considered as potential β-I binding motifs.

Several non-collagenous proteins also bind α2β1, although none has been identified which contains a GER sequence. Lysylhydroxylase, a cytoplasmic protein involved in the post-translational processing of collagen, is the only non-collagen mammalian protein containing GFPGER, a sequence which as a triple-helical homotrimeric peptide binds α2β1. Whether this GFPGER has any biological significance in the present context is not known.

Recognition of Integrins by Multiple Collagen α-Chains

The capacity of GFOGER to bind to α2β1 can substantially be accounted for by interaction between their middle strand and the α2-I MIDAS region. There is also some interaction between the leading strand and the I domain via the F residue, whereas the trailing strand makes no contact; these conclusions are based on analysis of the α2-I:GFOGER complex.[9]

For other integrins, there may not be such a strong dependence upon a single strand of the triple helix. A locus has been identified within type IV collagen which binds α1β1 where residues in at least two strands are considered important, α1(IV)D461 and α2(IV)R461.[15] Since this site occurs in the centre of an interrupted Col domain, the registration between the strands, staggered by a single residue, cannot be assumed to be the same throughout the sequence of collagen IV. Fluorescence resonance energy transfer has been used to indicate that the correct registration in the region of the integrin α1β1-binding site is α1',α1,α2, leading to the proposal that the binding motif spirals around the helix in the order D'DR, rather than D'RD or RDD'.[100] This remains to be proven by synthesis of appropriately constrained peptides. Evidence from this laboratory suggests that despite the capacity of GFOGER-like peptides to bind α1β1, they do not readily displace α1β1 from collagen IV, suggesting that they are of lower affinity than the D'DR sequence.[69]

Recently, the capacity of collagen XIII, which again lacks suitable GER sequences, to interact with α1β1 but not α2β1 was demonstrated.[101] This may conceivably reflect another incidence of an integrin recognition motif dispersed between the α-chains of the collagen helix.

This perspective suggests the possibility that other collagen binding integrins may recognize dispersed motifs, a topic which is difficult to address for heterotrimeric collagens in view of the complexity of the synthetic chemistry required. It must remain a topic for future research.

Acknowledgment

Michael Barnes, whose work laid the foundations for research on the recognition of collagen by cellular receptors, died on 4th May 2003.

References

1. Schwartz MA, Ginsberg MH. Networks and crosstalk: integrin signalling spreads. Nature Cell Biol 2002; 4:E65-68.
2. Hynes RO. Integrins:bidirectional allosteric signaling machines. Cell 2002; 110:673-687.
3. Casasnovas JM, Springer TA, Liu J-H et al. Crystal structure of ICAM-2 reveals a distinctive integrin recognition surface. Nature 1997; 387:312-315.
4. Humphries MJ. Integrin structure. Biochem Soc Trans 2000; 28:311-340.
5. Gullberg DE, Lundgren-Åkerlund E. Collagen-binding I domain integrins—what do they do? Prog Histochem Cytochem 2002; 37:3-54.
6. Emsley J, King SL, Bergelson JM et al. Crystal structure of the I-domain from integrin α2β1. J Biol Chem 1997; 272:28512-28517.
7. Colombatti A, Bonaldo P, Doliana R. Type A modules: Interacting domains found in several non-fibrillar collagens and in other extracellular matrix proteins. Matrix 1993; 13:297-306.
8. Bienkowska J, Cruz M, Atiemo A et al. The von Willebrand factor A3 domain does not contain a metal ion-dependent adhesion site motif J Biol Chem 1997; 272:25162-25167.
9. Emsley J, Knight CG, Farndale RW et al. Structural basis of collagen recognition by α2β1. Cell 2000; 101:47-56.
10. Nieuwenhuis HK, Akkerman JWN, Houdijk WPM et al. Human blood platelets showing no response to collagen fail to express surface glycoprotein Ia. Nature 1985; 318:470-472.
11. Nieuwenhuis HK, Sakariassen K, Houdijk WPM et al. Deficiency of platelet membrane glycoprotein Ia associated with a decreased platelet adhesion to subendothelium: a defect in platelet spreading. Blood 1986; 68:692-695.
12. Vandenberg P, Kern A, Ries A et al. Characterization of a type IV collagen major cell binding site with affinity to the alpha 1 beta 1 and the alpha 2 beta 1 integrins. J Cell Biol 1991; 113:1475-1483.
13. Gullberg D, Terracio L, Borg TK et al. Identification of integrin-like matrix receptors with affinity for interstitial collagens. J Biol Chem 1989; 264:12686-12694.
14. Kern A, Eble J, Golbik R et al. Interaction of type IV collagen with the isolated integrins alpha 1 beta 1 and alpha 2 beta 1. Eur J Biochem 1993; 215:151-159.
15. Eble JA, Golbik R, Mann K et al. The alpha 1 beta 1 integrin recognition site of the basement membrane collagen molecule [alpha 1(IV)]2 alpha 2(IV). EMBO J 1993; 12:4795-4802.
16. Gullberg D, Gehlsen KR, Turner DC et al. Analysis of α1β1, α2β1 and α3β1 integrins in cell-collagen interactions: identification of conformation dependent α1β1 binding sites in collagen type 1. EMBO J 1992; 11:3865-3873.
17. Calderwood DA, Tuckwell DS, Eble J et al. The integrin alpha1 A-domain is a ligand binding site for collagens and laminin. J Biol Chem 1997; 272:12311-12317.
18. Santoro SA. Identification of a 160000 dalton platelet membrane protein that mediates the initial divalent cation-dependent adhesion of platelets to collagen. Cell 1986; 46:913-920.
19. Zutter MM, Santoro SA. Widespread histologic distribution of the alpha 2 beta 1 integrin cell-surface collagen receptor. Am J Pathol 1990; 137:113-120.
20. Staatz WD, Walsh JJ, Pexton T et al. The alpha 2 beta 1 integrin cell surface collagen receptor binds to the alpha 1 (I)-CB3 peptide of collagen. J Biol Chem 1990; 265:4778-4781.
21. Santoro SA, Walsh JJ, Staatz WD et al. Distinct determinants on collagen support alpha 2 beta 1 integrin-mediated platelet adhesion and platelet activation. Cell Regul 1991; 2:905-913.
22. Staatz WD, Rajpara SM, Wayner EA et al. The membrane glycoprotein Ia-IIa (VLA_2) complex mediates the Mg^{2+}-dependent adhesion of platelets to collagen. J Cell Biol 1989; 108:1917-1924.
23. Morton LF, Zijenah LS, McCulloch IY et al. Integrin-dependent platelet recognition sites in collagen: identification of a short platelet-reactive sequence in the type III-derived fragment alpha 1(III) CB3. Biochem Soc Trans 1991; 19:439S.
24. Morton LF, Peachey AR, Zijenah LS et al. Conformation-dependent platelet adhesion to collagen involving integrin α2β1-mediated and other mechanisms: multiple α2β1-recognition sites in collagen type I. Biochem J 1994; 299:791-797.
25. Knutson JR, Iida J, Fields GB et al. CD44/chondroitin sulfate proteoglycan and alpha 2 beta 1 integrin mediate human melanoma cell migration on type IV collagen and invasion of basement membranes. Mol Biol Cell 1996; 7:383-396.

26. Camper L, Hellman U, Lundgren-Akerlund E. Isolation, cloning, and sequence analysis of the integrin subunit alpha10, a beta1-associated collagen binding integrin expressed on chondrocytes. J Biol Chem 1998; 273:20383-20389.
27. Camper L, Holmvall K, Wangnerud C et al. Distribution of the collagen-binding integrin alpha10beta1 during mouse development. Cell Tissue Res 2001; 306:107-116.
28. Bengtsson T, Camper L, Schneller M et al. Characterization of the mouse integrin subunit alpha10 gene and comparison with its human homologue. Genomic structure, chromosomal localization and identification of splice variants. Matrix Biol 2001; 20:565-576.
29. Velling T, Kusche-Gullberg M, Sejersen T et al. cDNA cloning and chromosomal localization of human alpha(11) integrin - A collagen-binding, I domain-containing, beta(1-)associated integrin alpha-chain present in muscle tissues. J Biol Chem 1999; 274:25735-25742.
30. Tiger CF, Fougerousse F, Grundstrom G et al. alpha11beta1 integrin is a receptor for interstitial collagens involved in cell migration and collagen reorganization on mesenchymal nonmuscle cells. Dev Biol 2001; 237:116-129.
31. Zhang WM, Popova SN, Bergman C et al. Analysis of the human integrin alpha11 gene (ITGA11) and its promoter. Matrix Biol 2002; 21:513-523.
32. Tulla M, Pentikäinen OT, Viitasalo T et al. Selective binding of collagen subtypes by integrin alpha 1I, alpha 2I, and alpha 10I domains. J Biol Chem 2001; 276:48206-48212.
33. Zhang W-M, Käpylä J, Puranen JS et al. $\alpha11\beta1$ integrin recognizes the GFOGER sequence in interstitial collagens. J Biol Chem 2003; 278:7270-7277
34. Shadle PJ, Ginsberg MH, Plow EF et al. Platelet-collagen adhesion: inhibition by a monoclonal antibody that binds glycoprotein IIb. J Cell Biol 1984; 99:2056-.
35. Miles AJ, Knutson JR, Skubitz AP et al. A peptide model of basement membrane collagen alpha 1 (IV) 531-543 binds the alpha 3 beta 1 integrin. J Biol Chem 1995; 270:29047-29050.
36. Riederer MA, Ginsberg MH, Steiner B. Blockade of platelet GPIIB-IIIA (Integrin alphaII(b)beta(3)) in flowing human blood leads to passivation of prothrombotic surfaces. Thromb Haemost 2002; 88:858-864.
37. Li C, McCarthy JB, Furcht LT et al. An all-D amino acid peptide model of alpha1(IV)531-543 from type IV collagen binds the alpha3beta1 integrin and mediates tumor cell adhesion, spreading, and motility. Biochemistry 1997; 36:15404-15410.
38. Yamamoto M, Yamato M, Aoyagi M et al. Identification of integrins involved in cell adhesion to native and denatured type I collagens and the phenotypic transition of rabbit arterial smooth muscle cells. Exp Cell Res 1995; 219:249-256.
39. Bella J, Eaton M, Brodsky B et al. Crystal and molecular structure of a collagen-like peptide at 1.9Å resolution. Science 1994; 266:75-81.
40. Prockop DJ, Kivirikko KI, Tuderman L et al. The biosynthesis of collagen and its disorders (Pt 2). New Engl J Med 1979; 301:77-85.
41. Persikov AV, Ramshaw JAM, Kirkpatrick A et al. Amino acid propensities for the collagen triple-helix. Biochemistry 2000; 39:14960-14967.
42. Hashimoto T, Wakabayashi T, Watanabe A et al. CLAC: a novel Alzheimer amyloid plaque component derived from a transmembrane precursor, CLAC-P/collagen type XXV. EMBO J 2002; 21:1524-1534.
43. Weston SA, Hulmes DJS, Mould AP et al. Identification of integrin $\alpha_2\beta_1$ as cell surface receptor for the carboxy-terminal propeptide of type I procollagen. J Biol Chem 1994; 269:20982-20986.
44. Davies D, Tuckwell DS, Calderwood DA et al. Molecular characterisation of integrin-procollagen C-propeptide interactions. Eur J Biochem 1997; 246:274-282.
45. Rehn M, Veikkola T, Kukk-Valdre E et al. Interaction of endostatin with integrins implicated in angiogenesis. Proc Natl Acad Sci USA 2001; 98:1024-1029.
46. Wickstrom SA, Alitalo K, Keski-Oja J. Endostatin associates with integrin alpha5beta1 and caveolin-1, and activates Src via a tyrosyl phosphatase-dependent pathway in human endothelial cells. Cancer Res 2002; 62:5580-5589.
47. Hugues J. Agglutination précoce des plaquettes au cours de la formation du clou hémostatique. Thromb Diath Haemorrh 1959; 3:177-186.
48. Bounameaux Y. L'accolement des plaquettes aux fibres sous-endothéliales. Thromb Diath Haemorrh 1961; 6:504-516.

49. Fitzsimmons CM, Cawston TE, Barnes MJ. The platelet reactivity of collagen Type I: evidence for multiple platelet-reactive sites in the Type I collagen molecule. Thromb Haemost 1986; 56:95-99.
50. Morton LF, Fitzsimmons CM, Rauterberg J et al. Platelet reactive sites in collagen. Biochem J 1987; 248:483-487.
51. Fitzsimmons CM, Barnes MJ. The platelet reactivity of the α2(I)-chain of type I collagen: Platelet aggregation induced by polymers of the molecule [α2(I)]3. Thromb Res 1985; 39:523-531.
52. Morton LF, Peachey AR, Barnes MJ. Platelet-reactive sites in collagens type I and type III. Evidence for separate adhesion and aggregatory sites. Biochem J 1989; 258:157-163.
53. Watson S, Berlanga O, Best D et al. Update on collagen receptor interactions in platelets: is the two-state model still valid? Platelets 2000; 11:252-258.
54. Clemetson KJ, Clemetson JM. Platelet collagen receptors. Thromb Haemost 2001; 86:189-197.
55. Saelman EUM, Nieuwenhuis HK, Hese KM et al. Platelet adhesion to collagen types I through VIII under conditions of stasis and flow is mediated by GPIa/IIa (α2β1-integrin). Blood 1994; 83:1244-1250.
56. Sixma JJ, van Zanten GH, Huizinga EG et al. Platelet adhesion to collagen: an update. Thromb Haemost 1997; 78:434-438.
57. Staatz WD, Fok KF, Zutter MM et al. Identification of a tetrapeptide recognition sequence for the α2β1-binding sites in collagen I. J Biol Chem 1991; 266:7363-7367.
58. Knight CG, Morton LF, Onley DJ et al. Identification in collagen type I of an integrin $\alpha_2\beta_1$-binding site containing an essential GER sequence. J Biol Chem 1998; 273:33287-33294.
59. Tuckwell DS, Ayad S, Grant ME et al. Conformation dependence of integrin-type II collagen binding. Inability of collagen peptides to support alpha 2 beta 1 binding, and mediation of adhesion to denatured collagen by a novel alpha 5 beta 1-fibronectin bridge. J Cell Sci 1994; 107(Pt 4):993-1005.
60. Mizuno M, Fujisawa R, Kuboki Y. Type I collagen-induced osteoblastic differentiation of bone-marrow cells mediated by collagen-alpha2beta1 integrin interaction. J Cell Physiol 2000; 184:207-213.
61. Mizuno M, Kuboki Y. Osteoblast-related gene expression of bone marrow cells during the osteoblastic differentiation induced by type I collagen. J Biochem (Tokyo) 2001; 129:133-138.
62. Sato M, Kojima N, Miura M et al. Induction of cellular processes containing collagenase and retinoid by integrin-binding to interstitial collagen in hepatic stellate cell culture. Cell Biol Int 1998; 22:115-125.
63. Takeuchi Y, Nakayama K, Matsumoto T. Differentiation and cell surface expression of transforming growth factor-beta receptors are regulated by interaction with matrix collagen in murine osteoblastic cells. J Biol Chem 1996; 271:3938-3944.
64. Fishman DA, Kearns A, Chilukuri K et al. Metastatic dissemination of human ovarian epithelial carcinoma is promoted by alpha2beta1-integrin-mediated interaction with type I collagen. Invasion Metastasis 1998; 18:15-26.
65. Johnson G, Jenkins M, McLean KM et al. Peptoid-containing collagen mimetics with cell binding activity. J Biomed Mater Res 2000; 51:612-624.
66. Xiao G, Wang D, Benson MD et al. Role of the alpha2-integrin in osteoblast-specific gene expression and activation of the Osf2 transcription factor. J Biol Chem 1998; 273:32988-32994.
67. Shah NK, Ramshaw JAM, Kirkpatrick A et al. A host-guest set of triple-helical peptides: stability of Gly-X-Y triplets containing common nonpolar residues. Biochemistry 1996; 35:10262-11026.
68. Barnes MJ, Knight CG, Farndale RW. The use of collagen-based model peptides to investigate platelet-reactive sequences in collagen. Biopolymers 1996; 40:383-397.
69. Knight CG, Morton LF, Peachey AR et al. The collagen-binding A-domains of integrin α1β1 and α2β1 recognize the same specific amino acid sequence, GFOGER, in native (triple-helical) collagens. J Biol Chem 2000; 275:35-40.
70. Main AL, Harvey TS, Baron M et al. The three-dimensional structure of the tenth type III module of fibronectin: an insight into RGD-mediated interactions Cell 1992; 71:671-678.
71. Irie A, Kamata T, Puzon-McLaughlin W et al. Critical Amino-Acid-Residues For Ligand-Binding Are Clustered in a Predicted Beta-Turn of the 3rd N-Terminal Repeat in the Integrin Alpha-4 and Alpha-5 Subunits. EMBO J 1995; 14:5550-5556.

72. Nykvist P, Tasanen K, Viitasalo T et al. The cell adhesion domain of type XVII collagen promotes integrin- mediated cell spreading by a novel mechanism. J Biol Chem 2001; 276:38673-38679.
73. d'Alessio P, Zwaginga JJ, de Boer HC et al. Platelet adhesion to collagen in subtypes of Type I von Willebrand's disease is dependent on platelet von Willebrand factor Thromb Haemost 1991; 64:227-231.
74. Huizinga EG, van der Plas RM, Kroon J et al. Crystal structure of the A3 domain of human von Willebrand factor: implications for collagen binding Structure 1997; 5:1147-1156.
75. Romijn RA, Bouma B, Wuyster W et al. Identification of the collagen-binding site of the von Willebrand factor A3-domain. J Biol Chem 2001; 276:9985-9991.
76. Sonnenberg A, Modderman PW, Hogervorst F. Laminin receptor on platelets is the integrin VLA-6. Nature 1988; 336:487-489.
77. Elices MJ, Hemler ME. The human integrin VLA-2 is a collagen receptor on some cells and a collagen/laminin receptor on others. Proc Natl Acad Sci USA 1989; 86:9906-9910.
78. Takada Y, Ylanne J, Mandelman D et al. A point mutation of integrin β_1 subunit blocks binding of $\alpha_5\beta_1$ to fibronectin and invasin but not recruitment to adhesion plaques. J Cell Biol 1992; 119:913-921.
79. Mould AP, Askari JA, Aota S-i et al. Defining the topology of integrin α5β1-fibronectin interactions using inhibitory anti-α5 and anti-β1 monoclonal antibodies. J Biol Chem 1997; 272:17283-17292.
80. Banères J-L, Roquet F, Green M et al. The cation-binding domain from the α-subunit of integrin α5β1 is a minimal domain for fibronectin recognition. J Biol Chem 1998; 273:24744-24753.
81. Guidetti G, Bertoni A, Viola M et al. The small proteoglycan decorin supports adhesion and activation of human platelets. Blood 2002; 100:1707-1714.
82. Tenni R, Viola M, Welser F et al. Interaction of decorin with CNBr peptides from collagens I and II. Eur J Biochem 2002; 269:1428-1437.
83. Xu Y, Gurusiddappa S, Rich RL et al. Multiple binding sites in collagen type I for the integrins α1β1 and α2β1. J Biol Chem 2000; 275:38981-38989.
84. Morton LF, Peachey AR, Knight CG et al. The platelet-reactivity of synthetic peptides based on the collagen III fragment, α1(III)CB4. Evidence for an integrin α2β1-recognition site involving residues 522-528 of the α1(III) collagen chain. J Biol Chem 1997; 272:11044-11048.
85. Kraetsch HG, Unger C, Wernhoff P et al. Cartilage-specific autoimmunity in rheumatoid arthritis: characterization of a triple helical B cell epitope in the integrin-binding-domain of collagen type II. Eur J Immunol 2001; 31:1666-1673.
86. Tuckwell DS, Reid KBM, Barnes MJ et al. The A-domain of integrin alpha 2 binds specifically to a range of collagens but is not a general receptor for the collagenous motif. Eur J Biochem 1996; 241:732-739.
87. Jung SM, Moroi M. Platelets interact with soluble and insoluble collagens through characteristically different reactions. J Biol Chem 1998; 273:14827-14837.
88. Jung SM, Moroi M. Signal-transducing mechanisms involved in activation of the platelet collagen receptor integrin $\alpha_2\beta_1$. J Biol Chem 2000; 275:8016-8026.
89. Moroi M, Onitsuka I, Imaizumi T et al. Involvement of activated integrin α2β1 in the firm adhesion of platelets onto a surface of immobilized collagen under flow conditions. Thromb Haemost 2000; 83:769-776.
90. Xiong J-P, Stehle T, Diefenbach B et al. Crystal structure of the extracellular segment of integrin αVβ3. Science 2001; 294:339-345.
91. Xiong JP, Stehle T, Zhang R et al. Crystal structure of the extracellular segment of integrin alpha Vbeta3 in complex with an Arg-Gly-Asp ligand. Science 2002; 296:151-155.
92. Liddington RC. Will the real integrin please stand up? Structure (Camb) 2002; 10:605-607.
93. Liddington RC, Ginsberg MH. Integrin activation takes shape. J Cell Biol 2002; 158:833-839.
94. Shimaoka M, Takagi J, Springer TA. Conformational regulation of integrin structure and function. Annu Rev Biophys Biomol Struct 2002; 31:485-516.
95. Mould AP, Askari JA, Barton S et al. Integrin activation involves a conformational change in the alpha 1 helix of the beta subunit A-domain. J Biol Chem 2002; 277:19800-19805.
96. Gottschalk KE, Gunther R, Kessler H. A three-state mechanism of integrin activation and signal transduction for integrin alpha(v)beta(3). ChemBioChem 2002; 3:470-473.

97. Holmback K, Danton MJS, Suh TT et al. Impaired platelet aggregation and sustained bleeding in mice lacking the fibrinogen motif bound by $\alpha_{IIb}\beta_3$. EMBO J 1996; 15:5760-5771.
98. Mould AP, Askari JA, Aota S et al. Defining the topology of integrin alpha 5 beta 1-fibronectin interactions using inhibitory anti-alpha 5 and anti-beta 1 monoclonal antibodies - Evidence that the synergy sequence of fibronectin is recognized by the amino-terminal repeats of the alpha 5 subunit. J Biol Chem 1997; 272:17283-17292.
99. Hayden J, Aleman J, Cruz MA. The beta1-subunit of the collagen-binding integrin alpha2beta1 contains an I-like domain that binds collagen directly. Thromb Haemost 2001; 86 Suppl:P1872.
100. Golbik R, Eble JA, Ries A et al. The spatial orientation of the essential amino acid residues Arginine and Aspartic acid within the alpha 1-beta-1 integrin recognition site of collagen 4 has been resolved using Fluorescence Resonance Energy Transfer. J Mol Biol 2000; 297:501-509.
101. Nykvist P, Tu H, Ivaska J et al. Distinct recognition of collagen subtypes by alpha(1)beta(1) and alpha(2)beta(1) integrins. Alpha(1)beta(1) mediates cell adhesion to type XIII collagen. J Biol Chem 2000; 275:8255-8261.

CHAPTER 10

Cellular Signaling by Collagen-Binding Integrins

Jyrki Heino

Collagens

Collagens are structural proteins of extracellular matrix that typically have triple helical domains of variable length.[1] Collagens form, for example, the large fibers of connective tissues and networks in basement membranes, while some collagens are transmembrane proteins. Metazoans from sea sponges[2] to mammals express collagens and in man, twenty two structurally and functionally different collagen subtypes have been published.[1,3,4,5,6] The subtypes have been named from type I to XXVII in the order they have been found. It is notable that the number of collagen genes is larger than the number of subtypes. For example, type I collagen is formed by two α1(I) proteins and one α2(I) protein and in the case of type IV collagen six genes coding for different α-subunits are known. The collagen family has been divided into several subgroups (Table 1). The fibril-forming collagens have a long, continuous triple helix that gives the molecule a rigid, rod-like structure. These collagens form large fibrils that give structural integrity and tensil-strength to the tissues. Type I collagen is the most abundant protein in the human body forming extracellular matrix in bone, tendons, dermis etc. Type II collagen has a similar function in cartilage. The network-forming collagens have interruptions in the triple helix. Type IV collagen belongs to this subfamily. It is an important structural component of basement membranes. Two collagen subtypes have unique functions and they alone form their own subgroups: beaded-filaments are formed by type VI collagen and anchoring fibers by type VII collagen. Fibril-associated collagens with interruptions in triple-helices (FACIT) form a large subgroup. Type IX collagen is a typical member of FACIT collagens. In cartilage it binds to surfaces of fibers formed by type II collagen. All FACIT collagens may not, however, be able to bind to fibers, and some of these collagens may have to be reclassified as FACIT-like collagens. Collagen types XIII and XVII were the first collagens shown to be transmembrane proteins. Type XVII is a structural component of hemidesmosomes, whereas type XIII is found, for example, in muscle, bone and skin. Multiplexins are collagens that are associated to basement membranes. Types XV and XVIII belong to this subfamily. The C-terminal cleavage product of type XVIII collagen has become known as angiogenesis-blocking endostatin.[7]

The I Domain-Containing Collagen-Binding Integrins

Cells and platelets have multiple ways to bind to different collagen subtypes. Many extracellular matrix or plasma glycoproteins, such as fibronectin and von Willebrand factor, can

I Domains in Integrins, edited by Donald Gullberg. ©2003 Eurekah.com and Kluwer Academic / Plenum Publishers.

Table I. The collagen family of extracellular matrix proteins

Fibril-forming collagens	
I	Forms large fibrils in tendons, bones, dermis, loose connective tissues, etc.
II	Replaces type I collagen in cartilage
III	Minor component in the same fibrils as type I collagen
V	Minor component in type I collagen fibrils
XI	In cartilage may from the core of type II collagen fibrils
Network forming collagens	
IV	Basement membranes
VIII	In Descement's membrane (basement membrane separating corneal endothelial cells from corneal stroma) and also in numerous other tissues
X	In hypertrophic cartilage
Beaded filaments -forming collagen	
VI	A component of stromal and pericellular matrix in various tissue, including skin, muscle, bone and cartilage
Anchoring fibrils forming collagen	
VII	Connects basement membranes to underlaying stroma
FACIT collagens and FACIT-like collagens	
XI	In cartilage on fibril formed by type II collagen.
XII	In dense connective tissues rich in type I collagen (e.g. tendons, ligaments)
XIV	In a number of tissues rich in type I collagen.
XVI	Broad distribution
XIX	Widely expressed during embryogenesis. In adults e.g. brain, eye and testis.
XX, XXI, (XXV1)	Found very recently
Transmembrane collagens	
XIII	Expressed, for example, in epidermis, cartilage, bone and muscle.
XVIII	Hemidesmosomes (also known as bullous pemphigoid antigen 180, PB180)
XXV	Alzheimer amyloid plague component
Multiplexins	
XV, XVIII	Concentrated in basement membrane regions, especially in internal organs.

Based on references 3,4,5,6,65,66

bind to collagen at the same time they have motifs which are recognized by specific adhesion receptors. There are also cellular receptors which can bind directly to collagen; GPVI on platelets is critical for their response to collagen,[8] and annexin V on bone forming cells participates in the organization of mineralization-related matrix vesicles.[9] Discoidin domain receptors (DDR) are tyrosine kinases, which bind to collagens and regulate cell proliferation.[10] Among the integrins, the collagen receptors form a separate subclass which is structurally distinct from other integrin-type matrix receptors in the sense that the ligand is recognized by an I domain in the α subunit. Four integrin heterodimers belong to this category: $\alpha 1\beta 1$, $\alpha 2\beta 1$, $\alpha 10\beta 1$ and $\alpha 11\beta 1$.[11,12,13,14] It is unclear why cells need so many closely related receptors, especially when they can be expressed simultaneously in the same tissue, such as cartilage, and even on a single cell (Nissinen and Heino, unpublished results). The fact that the cytoplasmic domains of the collagen binding α subunits are so different[11,12,13,14] suggests that the signaling functions of these receptors may be different and, as described bellow, this seems to be the case. The receptors also seem to have some differences in the way they recognize various collagen subtypes, which may have implications for the signaling function of the collagen receptors. Table 2 is a summary of the published information about recognition of collagen subtypes by integrins.

Table II. The recognition of distinct collagen subtypes by collagen receptor integrins

Fibril-forming collagens
Binding of $\alpha 2\beta 1$ integrin or the corresponding $\alpha 2$-I domain has been shown to collagen types I, II, III, V and XI.[56,57,58]
Integrin $\alpha 1\beta 1$ or the corresponding $\alpha 1$-I domain as well as $\alpha 10$-I domain bind can bind to collagen types I, II, III and V.[57] However, the approximated affinity is lower than that of $\alpha 2\beta 1$.[57]
Integrin $\alpha 11\beta 1$ or the corresponding $\alpha 11$-I domain has been shown to bind to collagen, types I and II.[14,67]
Network forming collagens
Integrin $\alpha 1\beta 1$ and $\alpha 2\beta 1$ as well as $\alpha 1$-I, $\alpha 2$-I and $\alpha 10$-I domains bind to type IV collagen. Integrin $\alpha 1\beta 1$ seems to prefer type IV over fibril-forming collagens, opposite to $\alpha 2\beta 1$ integrin. The same pattern can be seen at α-I domain level. $\alpha 10$-I domain binds to type IV as well.[57,59] Platelet binding (weak) to type VIII collagen is mediated by $\alpha 2\beta 1$.[56] Binding of $\alpha 11$-I domian to type IV is much weaker that binding to type I.[67]
Beaded-filaments forming collagen
Integrin $\alpha 1\beta 1$ and $\alpha 1$-I, $\alpha 10$-I domains bind to type VI collagen. Binding by $\alpha 2\beta 1$ and $\alpha 2$-I domain is much weaker.[57]
Anchoring fibrils forming collagen
NC1 domain in type VII collagen is recognized by $\alpha 2\beta 1$ on human fibroblasts.[60] Platelet binding to type VII collagen is mediated by $\alpha 2\beta 1$.[56]
FACIT collagens and FACIT-like collagens
Very little published information available. CD44 might be a receptor for type XIV collagen.[61]
Transmembrane collagens
Type XIII collagen is recognized by $\alpha 1\beta 1$ integrin as well as by the corresponding $\alpha 1$-I domain. Binding by $\alpha 2\beta 1$ or $\alpha 2$-I domain is much weaker.[62]
The largest collagenous domain (COL 15) of type XVII collagen cannot be recognized by the collagen receptors. However, when denatured the multiple KGD motifs can be used by $\alpha 5\beta 1$ and αV-integrins.[63]
Multiplexins
Endostatin, the C-terminal cleavage product of type XVIII collagen is recognized by $\alpha 5\beta 1$ and αV-integrins.[64]

Signaling by Collagen Receptors Is Dependent on the Organization of the Collagenous Matrix and the Cytoplasmic Domain of the α Subunit

Cell adhesion to collagen is known to have an effect on many cellular functions, suggesting that the receptors mediating cell–collagen interaction can regulate cellular signaling pathways. Some adhesion-related signaling events are not specific to any single matrix protein or integrin. In two-dimensional monolayers the formation of focal adhesion sites is associated with the activation of many tyrosine kinases, including focal adhesion kinase (FAK).[15] Activation of FAK-related pathways can be seen on different matrix proteins, including collagen. Often cell adhesion to monomeric collagen generates similar cellular responses as adhesion to fibronectin or other matrix glycoproteins, whereas cell adhesion to fibrillar collagen may have very different consequences.[16] Functions of the collagen receptors can also be tested inside three-dimensional collagen gels.[17] Inside three-dimensional matrix the cellular adhesion sites are very different when compared to prominent focal adhesions typical to cells cultured in monolayers.[18] For example inside collagen, FAK may not be phosphorylated,[19] and the integrin-mediated signaling seems to be very different when compared to monolayers. Indeed, the differences in signaling functions of $\alpha 1\beta 1$ and $\alpha 2\beta 1$ are sometimes nouticeable only inside collagen gels.[18] Physical stress related to collagen gels may have an important role in the modification of integrin signaling. For example, fibroblasts seem to be growth arrested in freely-floating, contracting gels, whereas they can proliferate inside anchored gels.[20] Thus, function of the collagen receptors is dependent on the organization of the collagenous matrix. This fact may help in the interpretation of results from different experimental models. Importantly, it lends strong support to the possibility that adhesion to distinct collagen subtypes may lead to different cellular responses. In extracellular fibrils and networks formed by distinct collagens, the number of receptor binding sites, their affinity to different integrins and their distances from each other are probably different. Without good experimental models, this question has been difficult to address. Still it is probable that the multiplicity of collagen receptors is related to the large number of collagen subtypes and that in the integrin–collagen interaction, both the receptor and the ligand determine the cellular response. Finally, degraded collagen fragments may have unique effects on cellular functions.[21] It is not clear whether these effects are mediated by the collagen receptors. However, when experimental cells can produce active proteinases, the fragments may influence the results.

The mechanism of signaling through the collagen receptors has been studied by deletions, mutations and swaps of the $\alpha 2$ subunit cytoplasmic domain.[19,22,23,24] Deletion of the entire domain targets the integrin to the focal adhesion sites, even in the absence of collagen.[22] The essential role of the $\alpha 2$ cytoplasmic domain in other functions of $\alpha 2\beta 1$ integrin, such as contraction of collagen gels, has also been well characterized.[23] Replacement of the $\alpha 2$ cytoplasmic domain with one from the $\alpha 1$ integrin renders the integrin unable to signal normally.[19,24] This also supports the idea that different collagen receptors have distinct signaling functions. Furthermore, these observations suggest that the cytoplasmic domains of the collagen receptor α subunits may be directly connected to cellular signaling proteins, or that they modify the molecular interactions of the $\beta 1$ cytoplasmic domain.

Regulation of Cell Proliferation and Survival by the Collagen Receptors

Integrin $\alpha 1\beta 1$ is among the integrins that can activate Shc in a process requiring caveolin-1 and Fyn.[25,26] Shc activation subsequently leads to activation of Ras and the growth promoting mitogen activated protein kinases (MAPKs). Analysis of $\alpha 1$ integrin-deficient knock-out mice has supported the idea that $\alpha 1$ might promote cell proliferation, since the dermis of the animals seems to be hypocellular[26] and their bone marrow -derived mesenchymal stem cells proliferate more slowly than those of control animals. [27]

In mesenchymal cells, including smooth muscle cells and fibroblasts, fibrillar collagen seems to prevent proliferation.[20,28] Similar results have been reported with melanoma cells.[16] In fibroblasts growth arrest may require that the cells are inside floating and contracting collagen gels.[20] In smooth muscle and melanoma cells as well as in fibroblasts, growth arrest has been connected to $\alpha 2\beta 1$ function and the accumulation of cyclin/cdk inhibitor, $p27^{kip}$ in the cells.[16,20,28] However, the organization of the collagenous matrix maybe critical for the action of $\alpha 2\beta 1$. It has been speculated that clustering of $\alpha 2\beta 1$ by antibodies or nonfibrillar collagen may actually promote proliferation, while fibrillar collagen prevents $\alpha 2\beta 1$ clustering and therefore inhibits proliferation.[16] The mechanism of $p27^{kip}$ regulation by $\alpha 2\beta 1$ is unclear.

In murine mammary gland-derived epithelial cells, $\alpha 2\beta 1$ has been reported to increase proliferation, when tested in monolayer cultures on nonfibrillar collagen.[29] In these cells the intracellular domain of $\alpha 2$ subunit has been analyzed by targeted mutations, and two distinct sites have been identified which regulate p38 and ERK pathways,[30] connected to cell migration and proliferation, respectively.

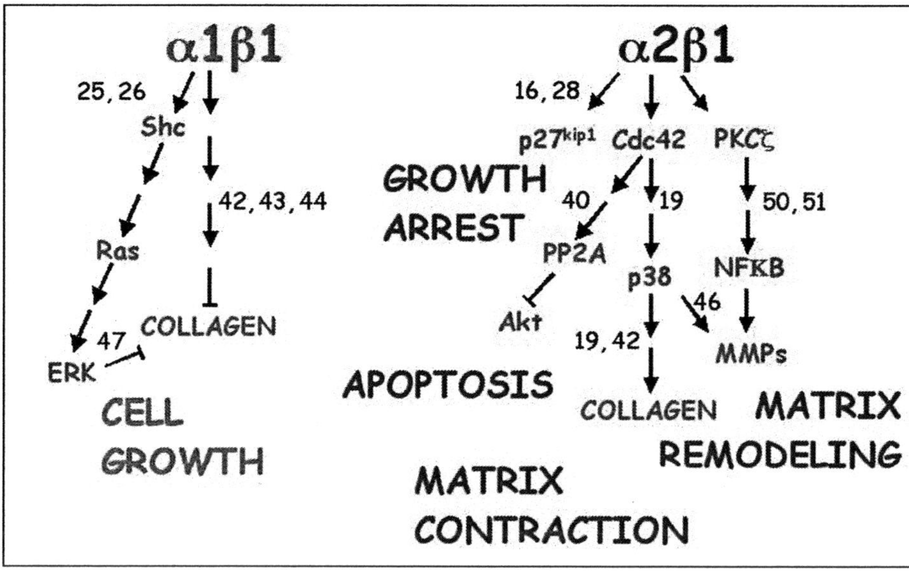

Figure. 1. Examples of cellular signaling pathways activated by α1β1 and α2β1 integrins. Note that the figure describes α2β1 signaling in mesenchymal cells inside a collagenous matrix. The response of other cell types may be different and, importantly, cell adhesion to monomeric collagen in monolayer cultures may activate distinct pathways. Numbers indicate the references cited.

Thus, it is not clear exactly how α2β1 modulates cell proliferation. Some of the effects may be cell type dependent, but the organization of collagen seems to be the most important determining factor for α2β1 function. In vivo, cells are naturally living in a three-dimensional world, but cell culture inside collagen gels may also differ from an intra-tissue environment. Therefore, it is difficult to speculate on the function of α2β1 in tissues, which makes the testing of integrin function in animal models even more important. The first reports on mice that are deficient in the α2 subunit described no obvious cell growth -related defects,[31,68] but more experiments are needed. Studies on human cancer have not helped to confirm the role of α2β1 in growth regulation. In melanoma, α2β1 expression has been connected to a highly aggressive cell phenotype,[32] but in breast cancer α2β1 is downregulated.[33]

Release of mechanical tension in a three-dimensional collagen gel model triggers apoptosis in fibroblasts.[34,35] In these conditions α2β1 mediates cell adhesion to collagen and is essential for contraction. Indeed, function blocking antibodies against α2β1 and α1β1 integrin can reduce the number of apoptotic cells, and α2-negative rhabdomyosarcoma cells undergo apoptosis only if they are cDNA-transfected to express the α2 subunit.[35] However, a large number of studies indicate that α2β1 can actually protect cells from apoptosis. This has been shown in human mammary epithelial cells,[36] Jurkat T cells,[37] A431 cells[38] and Madin-Darby canine kidney cells.[39] As in the case of α2β1-related regulation of cell proliferation also here the seemingly controversial results might reflect not only cell type -dependent differences, but also altered α2β1 function when the receptor is bound to monomeric instead of fibrillar collagen. Apoptosis may be partially regulated by the same signaling pathways as proliferation. In addition to the ERK pathway, Akt/Protein kinase B (PKB) seems to be involved. In fibroblasts and osteosarcoma cells, protein phosphatase 2A is activated in a process that requires the presence of α2 cytoplasmic domain and Cdc42 activity.[40] Activation of PP2A leads consequently to dephosphorylation of Akt/PKB,[40] a well-known promoter of cell survival. PP2A can also inhibit the cell cycle in several different ways, and α2β1 dependent activation of PP2A may be one mechanism leading to growth arrest. Dephosphorylation of Akt/PKB inside collagen gel can be prevented by anti-α2 antibodies.[40] Another study indicates that apoptosis of fibroblasts inside contracting collagen gels can be prevented by anti-β1 antibodies which also prevent Akt/PKB dephosphorylation.[41]

Regulation of Matrix Gene Expression by the Collagen Receptors

Several lines of evidence support the idea that α1β1 integrin is a negative regulator of collagen synthesis, especially in cells that are surrounded by a three-dimensional collagenous matrix. Early observations using osteosarcoma cells with a low expression level of α1β1 indicated that these cells do not down-regulate collagen sythesis inside collagen.[42] This was later confirmed by experiments performed with cells derived from α1 integrin deficient knock-out animals.[43] Furthermore, experiments utilizing functional anti-integrin antibodies have led to the same conclusion.[44] Similar regulatory mechanisms may also function in tissues since α1 knock-out mice have increased collagen synthesis rate in their dermis while a simultaneous increase in matrix metalloproteinase (MMP) expression prevents the accumulation of collagen.[43] Overexpression of α2β1 integrin in cells increases collagen synthesis, suggesting that α2β1 is a positive regulator of collagen gene expression.[19,42] Increased collagen synthesis can be prevented by selective inhibitors of the α isoform of p38 MAPK, which is known to be activated by α2β1 after contact which collagen.[19] The activation of the p38 pathway seems to be one of the most pronounced events observed after α2β1-collagen interaction since it has been observed to take place in several different cell lines and experiment models.[19,24,45,46] The mechanism of collagen gene suppression by α1β1 is not clear but its activation of ERK,[26] known to be a negative regulator of collagen synthesis[47] gives one possible mechanism. It is possible to speculate that the opposite effects of α1β1 and α2β1 on many cellular functions may partially be due to the alterations in the balance between p38 and ERK pathways. Other matrix genes activated by α2β1 integrin include osteocalsin and bone sialoprotein.[48] The activation of bone-related genes has been linked to α2β1-dependent activation of Osf2 transcription factor in osteoblasts.[48]

Cell contact to three-dimensional collagen activates the expression of collagenase-1 (matrix metalloproteinase-1, MMP-1).[17] This phenomenon has been linked to signaling through the α2β1 integrin.[42] Other MMPs regulated by either α1β1 or α2β1 include stromalysin-1 (MMP-3)[49] and collagenase 3 (MMP-13).[46] In skin fibroblasts inside collagen collagenase-1 (MMP-1) expression seems to be activated by a pathway involving protein kinase (PK) C-ζ and nuclear factor (NF) κB.[50,51] However, the above described p38 pathway may also participate in the process.[45] Interestingly, α2β1 can also regulate its own expression by a positive signaling loop involving PKC-ζ/NF-κB.[50] The p38 pathway seems to mediate the upregulation of MMP-13 by α2β1 integrin.[46]

Thus, α1β1 may simultaneously support the proliferation of fibroblasts and prevent matrix synthesis, while in the same cells α2β1 has opposite effects: inhibition of cell cycle, remodeling of extracellular matrix and apoptosis. It is an interesting possibility that, for example during wound healing, changes in the expression pattern of collagen receptors could reflect changes in the metabolism of a fibroblast.

Regulation of Collagen Binding by Integrin Inside-Out Signaling

Cellular signaling pathways may also regulate the ability of the integrins to bind to their ligands. This is termed "inside-out signaling". Activation of integrin function is a well-known phenomenon in inflammatory cells and platelets, but its molecular mechanisms are still uncertain. In platelets α2β1 may exist in inactive, active and high-affinity conformations.[52] ADP is known to activate α2β1 on platelets.[52] In addition to platelets, in many cell lines the transition from active to high-affinity state may be promoted by activators of PKC, such as TPA (12-O-tetradecanoylphorbol-1,3-acetate).[52,53,54] It is possible that integrin function is regulated by intracellular proteins binding to their cytoplasmic domains. Interestingly, a protein called calreticulin can bind to the cytoplasmic tail of α2 subunit and it may participate in the stabilization of the high-affinity state.[53] Integrins can also bind to other transmembrane proteins on cell surface. Integrin associated protein (IAP or CD47) is actually a receptor for trombospondin, an extracellular matrix protein.[55] A trombospondin-derived IAP-agonist peptide enhances α2β1-dependent functions, including cell migration. The intracellular part of IAP can regulate an inhibitory G-protein (Gi) and reduce the cellular levels of cAMP and ERK.[55]

Conclusions

The four collagen-binding I-domain integrins form their own subgroup among cell adhesion receptors. The signaling functions of α1β1 and α2β1 integrins have been analyzed in many experimental models, whereas studies on the recently found α10β1 and α11β1 heterodimers have only been initiated. Interestingly, collagen binding by α1β1 and α2β1 often generates opposite cellular responses. In the case of α1β1, cell culture experiments suggesting that α1β1 promotes cell proliferation and suppresses collagen synthesis have been supported by results from α1 deficient mice. The proposed signaling functions of α2β1 are still awaiting in vivo confirmation.

More than twenty collagen subtypes form fibrils and networks in extracellular matrix. The cellular response to collagenous matrix may be dependent on several factors. First, different collagen receptors regulate distinct signaling pathways. Second, collagen subtypes may be differentially recognized by the collagen receptors. Third, the organization of the collagenous matrix may be critical. More information about integrin function and integrin–collagen interaction is needed to address the structural requirements for integrin signaling.

Acknowledgement

The author wants to thank Dr. Wendy Lee Connors for critical reading of the manuscript.

References

1. Prockop DJ, Kivirikko KI. Collagens: molecular biology, diseases, and potentials for therapy. Annu Rev Biochem 1995; 64:403-434.
2. Aho S, Turakainen H, Onnela ML et al. Characterization of an intronless collagen gene family in the marine sponge Microciona prolifera. Proc Natl Acad Sci USA 1993; 90:7288-7292.
3. Ricard-Blum S, Dudler B, van der Rest M. Unconventional collagens. Oxford, UK: Oxford University Press, 2000.
4. Kreis T, Vale R. Extracellular matrix, anchor and adhesion proteins. Oxford, U.K: Oxford University Press, 1999.
5. Tuckwell D. Identification and analysis of collagen alpha 1(XXI), a novel member of the FACIT collagen family. Matrix Biol 2002; 21:63-66.
6. Koch M, Foley JE, Hahn R et al. Alpha 1(XX) collagen, a new member of the collagen subfamily, fibril-associated collagens with interrupted triple helices. J Biol Chem 2001; 276:23120-23126.
7. O'Reilly MS, Boehm T, Shing Y et al. Endostatin: an endogenous inhibitor of angiogenesis and tumor growth. Cell 1997; 88:277-285
8. Barnes MJ, Knight CG, Farndale RW. The collagen-platelet interaction. Curr Opin Hematol 1998; 5:314-20.
9. Pfaffle M, Ruggiero F, Hofmann H et al. Biosynthesis, secretion and extracellular localization of anchorin CII, a collagen-binding protein of the calpactin family. EMBO J 1988; 7:2335-2342.
10. Labrador JP, Azcoitia V, Tuckermann J et al. The collagen receptor DDR2 regulates proliferation and its elimination leads to dwarfism. EMBO Rep 2001; 2:446-452.

11. Briesewitz R, Epstein MR, Marcantonio EE. Expression of native and truncated forms of the human integrin alpha 1 subunit. J Biol Chem 1993; 268:2989-2996.
12. Takada Y, Hemler ME. The primary structure of the VLA-2/collagen receptor alpha 2 subunit (platelet GPIa): homology to other integrins and the presence of a possible collagen-binding domain. J Cell Biol 1989; 109:397-407.
13. Camper L, Hellman U, Lundgren-Akerlund E. Isolation, cloning, and sequence analysis of the integrin subunit alpha10, a beta1-associated collagen binding integrin expressed on chondrocytes. J Biol Chem 1998; 273:20383-20389.
14. Velling T, Kusche-Gullberg M, Sejersen T et al. cDNA cloning and chromosomal localization of human alpha(11) integrin. A collagen-binding, I domain-containing, beta(1)-associated integrin alpha-chain present in muscle tissues. J Biol Chem 1999; 274:25735-25742.
15. Parsons JT, Martin KH, Slack JK et al. Focal adhesion kinase: a regulator of focal adhesion dynamics and cell movement. Oncogene 2000; 19:5606-5613.
16. Henriet P, Zhong ZD, Brooks Pc et al. Contact with fibrillar collagen inhibits melanoma cell proliferation by up-regulating $p27^{KIP1}$. Proc Natl Acad Sci USA 2000; 97:10026-10031.
17. Grinnell F. Fibroblasts, myofibroblasts, and wound contraction. J Cell Biol 1994; 124:401-404.
18. Cukierman E, Pankov R, Stevens DR et al. Taking cell-matrix adhesions to the third dimension. Science 2001; 294:1708-1712.
19. Ivaska J, Reunanen H, Westermarck J et al. Integrin alpha2beta1 mediates isoform-specific activation of p38 and upregulation of collagen gene transcription by a mechanism involving the alpha2 cytoplasmic tail. J Cell Biol 1999; 147:401-416.
20. Fringer J, Grinnell F. Fibroblast quiescence in floating or released collagen matrices: contribution of the ERK signaling pathway and actin cytoskeletal organization. J Biol Chem 2001; 276:31047-31052.
21. Carragher NO, Levkau B, Ross R et al. Degraded collagen fragments promote rapid disassembly of smooth muscle focal adhesions that correlates with cleavage of pp125(FAK), paxillin, and talin. J Cell Biol 1999; 147:619-630.
22. Kawaguchi S, Bergelson JM, Finberg RW et al. Integrin alpha 2 cytoplasmic domain deletion effects: loss of adhesive activity parallels ligand-independent recruitment into focal adhesions. Mol Biol Cell 1994; 5:977-988.
23. Chan BM, Kassner PD, Schiro JA et al. Distinct cellular functions mediated by different VLA integrin alpha subunit cytoplasmic domains. Cell 1992; 68:1051-1060.
24. Klekotka PA, Santoro SA, Zutter MM. Alpha 2 integrin subunit cytoplasmic domain-dependent cellular migration requires p38 MAPK. J Biol Chem 2001; 276:9503-9511.
25. Wary KK, Mariotti A, Zurzolo C et al. A requirement for caveolin-1 and associated kinase Fyn in integrin signaling and anchorage-dependent cell growth. Cell 1998; 94:625-634.
26. Pozzi A, Wary KK, Giancotti FG et al. Integrin alpha1beta1 mediates a unique collagen-dependent proliferation pathway in vivo. J Cell Biol 1998; 142:587-594.
27. Ekholm E, Hankenson KD, Uusitalo H et al. Diminished callus size and cartilage synthesis in alpha1beta1 integrin deficient mice during bone fracture healing. Am J Pathol 2002; 160:1779-1785.
28. Koyama H, Raines EW, Bornfeldt KE et al. Fibrillar collagen inhibits arterial smooth muscle proliferation through regulation of Cdk2 inhibitors. Cell 1996; 87:1069-1078.
29. Zutter MM, Santoro SA, Wu JE et al. Collagen receptor control of epithelial morphogenesis and cell cycle progression. Am J Pathol 1999; 155:927-940.
30. Klekotka PA, Santoro SA, Wang H et al. Specific residues within the alpha 2 integrin subunit cytoplasmic domain regulate migration and cell cycle progression via distinct MAPK pathways. J Biol Chem 2001; 276:32353-32361.
31. Holtkotter O, Nieswandt B, Smyth N et al. Integrin alpha 2-deficient mice develop normally, are fertile, but display partially defective platelet interaction with collagen. J Biol Chem 2002; 277:10789-10794.
32. Klein CE, Dressel D, Steinmayer T et al. Integrin alpha 2 beta 1 is upregulated in fibroblasts and highly aggressive melanoma cells in three-dimensional collagen lattices and mediates the reorganization of collagen I fibrils. J Cell Biol 1991; 115:1427-1436.
33. Zutter MM, Santoro SA, Staatz WD et al. Reexpression of the alpha 2 beta 1 integrin abrogates the malignant phenotype of breast carcinoma cells. Proc Natl Acad Sci USA 1995; 92:7411-7415.

34. Grinnell F, Zhu M, Carlson MA et al. Release of mechanical tension triggers apoptosis of human fibroblasts in a model of regressing granulation tissue. Exp Cell Res 1999; 248:608-619.
35. Niland S, Cremer A, Fluck J et al. Contraction-dependent apoptosis of normal dermal fibroblasts. J Invest Dermatol 2001; 116:686-692.
36. Baeckstrom D, Lu PJ, Taylor-Papadimitriou J. Activation of the alpha2beta1 integrin prevents c-erbB2-induced scattering and apoptosis of human mammary epithelial cells in collagen. Oncogene 2000; 19:4592-603.
37. Aoudjit F, Vuori K. Engagement of the alpha2beta1 integrin inhibits Fas ligand expression and activation-induced cell death in T cells in a focal adhesion kinase-dependent manner. Blood 2000; 95:2044-2051.
38. Smida Rezgui S, Honore S, Rognoni JB et al. Up-regulation of alpha 2 beta 1 integrin cell-surface expression protects A431 cells from epidermal growth factor-induced apoptosis. Int J Cancer 2000; 87:360-367.
39. Saelman EU, Keely PJ, Santoro SA. Loss of MDCK cell alpha 2 beta 1 integrin expression results in reduced cyst formation, failure of hepatocyte growth factor/scatter factor-induced branching morphogenesis, and increased apoptosis. J Cell Sci 1995; 108:3531-3540.
40. Ivaska J, Nissinen L, Immonen N et al. Integrin alpha2beta1 promotes activation of protein phosphatase 2A and dephosphorylation of Akt and GSK3 beta. Mol Cell Biol 2002; 22:1352-1359.
41. Tian B, Lessan K, Kahm J et al. Beta1 integrin regulates fibroblast viability during collagen matrix contraction through a phosphatidylinositol 3-kinase AKT/protein kinase B signaling pathway. J Biol Chem 2002 (in press).
42. Riikonen T, Westermarck J, Koivisto L et al. Integrin alpha 2 beta 1 is a positive regulator of collagenase (MMP-1) and collagen alpha 1(I) gene expression. J Biol Chem 1995; 270:13548-13552.
43. Gardner H, Broberg A, Pozzi A et al. Absence of integrin alpha1beta1 in the mouse causes loss of feedback regulation of collagen synthesis in normal and wounded dermis. J Cell Sci 1999; 112:263-272.
44. Langholz O, Rockel D, Mauch C et al. Collagen and collagenase gene expression in three-dimensional collagen lattices are differentially regulated by alpha 1 beta 1 and alpha 2 beta 1 integrins. J Cell Biol 1995; 131:1903-1915.
45. Xu J, Clark RA, Parks WC. p38 mitogen-activated kinase is a bidirectional regulator of human fibroblast collagenase-1 induction by three-dimensional collagen lattices. Biochem J 2001; 355:437-447.
46. Ravanti L, Heino J, Lopez-Otin C et-al. Induction of collagenase-3 (MMP-13) expression in human skin fibroblasts by three-dimensional collagen is mediated by p38 mitogen-activated protein kinase. J Biol Chem 1999; 274:2446-2455.
47. Reunanen N, Foschi M, Han J et al. Activation of extracellular signal-regulated kinase 1/2 inhibits type I collagen expression by human skin fibroblasts. J Biol Chem 2000; 275:34634-34639.
48. Xiao G, Wang D, Benson MD et al. Role of the alpha2-integrin in osteoblast-specific gene expression and activation of the Osf2 transcription factor. J Biol Chem 1998; 273:32988-32994.
49. Lochter A, Navre M, Werb Z et al. Alpha1 and alpha2 integrins mediate invasive activity of mouse mammary carcinoma cells through regulation of stromelysin-1 expression. Mol Biol Cell 1999; 10:271-282.
50. Xu J, Clark RA. A three-dimensional collagen lattice induces protein kinase C-zeta activity: role in alpha2 integrin and collagenase mRNA expression. J Cell Biol 1997; 136:473-483.
51. Xu J, Zutter MM, Santoro SA et al. A three-dimensional collagen lattice activates NF-kappaB in human fibroblasts: role in integrin alpha2 gene expression and tissue remodeling. J Cell Biol 1998; 140:709-719.
52. Jung SM, Moroi M. Signal-transducing mechanisms involved in activation of the platelet collagen receptor integrin alpha(2)beta(1). J Biol Chem 2000; 275:8016-8026
53. Coppolino M, Leung-Hagesteijn C, Dedhar S et al. Inducible interaction of integrin alpha 2 beta 1 with calreticulin. Dependence on the activation state of the integrin. J Biol Chem 1995; 270:23132-23138.
54. Rosfjord EC, Maemura M, Johnson MD et al. Activation of protein kinase C by phorbol esters modulates alpha2beta1 integrin on MCF-7 breast cancer cells. Exp Cell Res 1999; 248:260-271.

55. Wang XQ, Lindberg FP, Frazier WA. Integrin-associated protein stimulates alpha2beta1-dependent chemotaxis via Gi-mediated inhibition of adenylate cyclase and extracellular-regulated kinases. J Cell Biol 1999; 147:389-400.
56. Saelman EU, Nieuwenhuis HK, Hese KM et al. Platelet adhesion to collagen types I through VIII under conditions of stasis and flow is mediated by GPIa/IIa (alpha 2 beta 1-integrin). Blood 1994; 83:1244-1250.
57. Tulla M, Pentikäinen O, Viitasalo T et al. Selective binding of collagen subtypes by integrin alpha1I, alpha2I, and alpha10I domains. J Biol Chem 2001; 276:48206-48212.
58. Tuckwell DS, Reid KB, Barnes MJ et al. The A-domain of integrin alpha 2 binds specifically to a range of collagens but is not a general receptor for the collagenous motif. Eur J Biochem 1996; 241:732-739.
59. Kern A, Eble J, Golbik R et al. Interaction of type IV collagen with the isolated integrins alpha 1 beta 1 and alpha 2 beta 1. Eur J Biochem 1993; 215:151-159.
60. Chen M, O'Toole EA, Li YY et al. Alpha 2 beta 1 integrin mediates dermal fibroblast attachment to type VII collagen via a 158-amino-acid segment of the NC1 domain. Exp Cell Res 1999; 249:231-239.
61. Ehnis T, Dieterich W, Bauer M et al. A chondroitin/dermatan sulfate form of CD44 is a receptor for collagen XIV (undulin). Exp Cell Res 1996; 229:388-397.
62. Nykvist P, Tu H, Ivaska J et al. Distinct recognition of collagen subtypes by alpha(1)beta(1) and alpha(2)beta(1) integrins. Alpha(1)beta(1) mediates cell adhesion to type XIII collagen. J Biol Chem 2000; 275:8255-8261.
63. Nykvist P, Tasanen K, Viitasalo T et al. The cell adhesion domain of type XVII collagen promotes integrin-mediated cell spreading by a novel mechanism. J Biol Chem 2001; 276:38673-38679.
64. Rehn M, Veikkola T, Kukk-Valdre E et al. Interaction of endostatin with integrins implicated in angiogenesis. Proc Natl Acad Sci USA 2001; 98:1024-1029.
65. Hashimoto T, Wakabayashi T, Watanabe A et al. CLAC:: a novel Alzheimer amyloind plaque component derived from a transmembrane precursor, CLAC-P/collagen type XXV. EMBO J 2002; 21:1524-1534.
66. Sato K, Yomogida K, Wada T et al. Type XXVI collagen, a new member of the collagen family, is specifically expressed in the testis and ovary. J Bio Chem 2002; 277:37678-37684.
67. Zhang WM, Käpylä J, Puranen JS et al. ALpha11beta1 integrin recognizes the GFOGER sequence in interstitial collagens. J Biol Chem 2002; in press.
68. Chen J, Diacovo TG, Grenache DG et al. The alpha(2) integrin subunit-deficient mouse: a multi-faceted phenotype including defects of branching morphogenesis and hemostasis. Am J Pathol 2002; 161:337-344.

CHAPTER 11

Collagen-Binding Integrins and Disease

Beate Eckes, Bernhard Nieswandt and Thomas Krieg

Among the β1 family of integrin receptors there are 4 receptors that mediate binding to different collagens. These are α1β1, α2β1, α10β1 and α11β1 (Fig. 1). Their specific structural and functional properties are covered in detail in different chapters of this book. Since collagens represent the most abundant constituent of the extracellular matrix (ECM) in nearly the entire vertebrate body, correct recognition of this environment and signal transduction elicited by integrin-collagen interaction appears vital for proper function. Common to all four receptors and prerequisite for collagen binding is the "inserted" or I-domain (also called von Willebrand A-domain), a structural motif of 180-200 amino acids in length present in the N-terminal half of the α-ectodomains (for recent reviews, see1,2).

Obviously, diseases associated with collagen-binding integrins could result from mutations in either receptor or ligand that affect binding, activation and/or signaling. Moreover, the tissue distribution of each individual receptor will determine to a large extent the potential damage of its dysfunction, while expression of two or more of these receptors on a given cell type will possibly alleviate the damage arising from one particular malfunctioning receptor. To understand the specific functions of the individual collagen receptors, and which functions can also be fulfilled by the others, knockout mouse models have been generated[3-5] or will be available in the future for functional studies.

This chapter will concentrate on the receptors α1β1 and α2β1, because so far no association of the recently described integrins α10β1 and α11β1 with pathological disorders has been reported and any assumptions, based on their restricted expression patterns, can only be speculative at this time.

In the following, the contribution of α1β1 and α2β1 to invasive tumor growth and metastasis, to tissue fibrosis, and the role of α2β1 in bleeding disorders/thrombosis will be highlighted. The involvement of collagen-binding integrin receptors in inflammatory disorders has been covered in a separate chapter.

Invasive Tumor Growth and Metastasis

An important feature of malignancy is the ability of tumor cells to invade adjacent tissues, to metastasize through blood and lymphatic vessels and to form metastatic foci at distant sites of the primary tumor in the body. Among other things, this process requires the cells to migrate. Mechanistically, migration is a highly complex multi-step sequence of events consisting of extending cellular protrusions, attaching to the substrate, cell contraction and subsequent detachment of the rear end of the cell body, and readhesion and spreading. Migration through interstitial tissue is facilitated by local proteolysis of matrix constituents, and most cells have developed an amazing armamentarium of matrix-degrading enzymes. Moreover, metastasis involves breakdown of basement membranes, which represent a special type of ECM, such as

I Domains in Integrins, edited by Donald Gullberg. ©2003 Eurekah.com
and Kluwer Academic / Plenum Publishers.

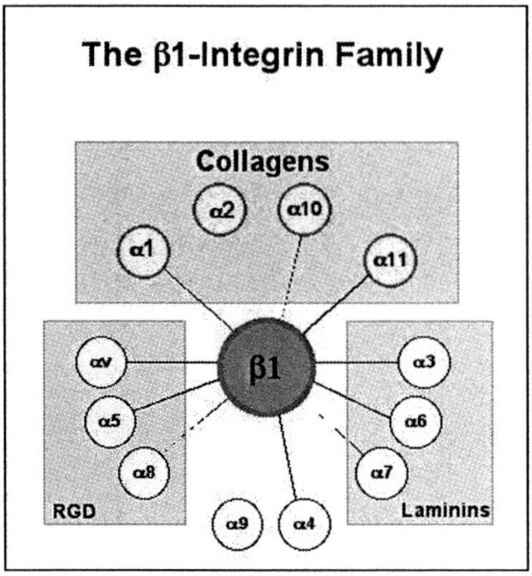

Figure 1. The Family of β1-Integrin Receptors. Interactions of cells with the ECM are mediated by a total of 12 integrin receptors sharing the β1 subunit. The α subunits of four of these contain an I-domain that confers binding to collagens. Of the other 8 α subunits, three dimerize with β1 to form the laminin-binding subfamily. Another subset binds to RGD-containing extracellular ligands. Altogether, these 12 receptors mediate attachment to more than 60 ligands present in the extracellular matrix.

those surrounding blood or lymphatic vessels. Once a tumor has escaped immune surveillance by the host, blood vessels need to be formed to supply nutrients and oxygen, lest the tumor necrotisizes. The complex process of angiogenesis involves ECM remodeling, migration and proliferation of endothelial cells as well as endothelial cell maturation into mature blood vessels.

Information summarized in the next paragraphs demonstrates that the interaction of cells with collagens plays an important role in all of these steps and is mediated by the activity of either α1β1 or α2β1 or both (Fig. 2).

Cell Migration

One major function of α1β1 and α2β1 is to mediate attachment of cells to the collagen fibrils constituting their three-dimensional environment. For a cell to relocate, loss of adhesive contact with the surroundings must therefore occur. Detachment from the environment and subsequent migration may become necessary or provide an advantage for the cell; examples are the migration of cells in order to close a tissue defect and during tumor cell metastasis. Integrins are important in mediating the loss of attachment to extracellular matrix (ECM), which results in apoptosis in a process termed "anoikis" (homeless), and resistence to anoikis is regarded as a hallmark of malignant conversion in epithelial cells.[6]

Reduced levels of α2β1 have frequently been correlated with induced malignancy, e.g., on adenocarcinoma cells.[7] The mechanisms that render these cells resistent to apoptosis are not well understood. Normal differentiated epithelial cells, e.g., of the mammary gland, express high levels of α2β1, and blockade by antibodies interferes with mammary gland development. Successive loss of receptor expression correlated with increasing loss of differentiation and conversion to an invasive and metastatic phenotype. Reexpression of α2β1 in such cells reverted the malignant phenotype, indicating that presence of α2β1 maintains the normal differentiated phenotype.[8] Interestingly however, mammary gland development is only mildly affected in α2-deficient mice,[5] suggesting that the organism can activate compensatory mechanisms.

In other cases, malignant conversion correlated with an increase in α2β1 levels. Examples were reported for malignant melanoma and rhabdomyosarcoma. Highly invasive melanoma cells characteristically express high levels of α2β1, adhere better to collagen and contract collagen lattices, while melanocytes or melanoma cells of low invasive potential express low α2β1 levels.[9] This particular feature lead to the identification of α2β1 as "melanoma progression

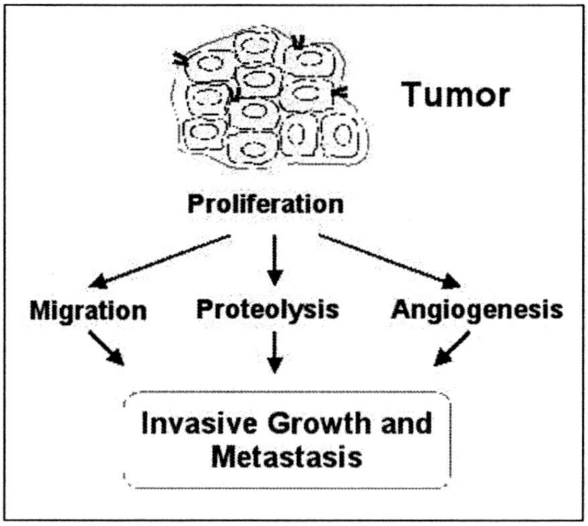

Figure 2. Invasive Tumor Growth and Metastasis. Invasive growth and metastasis of tumors depend on deregulated tumor cell proliferation. Tumor cell migration and invasion of extracellular matrices including basement membranes is facilitated by release of proteases (also see Table 1). In the process, endothelial cells are induced to proliferate and form new blood vessels that are required to supply the growing tumor with nutrients. In selected malignancies, integrins α1β1 and α2β1 (depicted schematically by dark bars on tumor cells) were shown to participate in all of these steps. Examples were reported for increased as well as decreased integrin levels.

antigen".[10] Later, migration of highly invasive melanoma cells through collagen lattices was studied using refined optical techniques.[11] These demonstrated that the tumor cells used α2β1 receptors for migration, to organize collagen fibrils by dynamic fiber traction including localized fiber disruption, and to form tube-like paths, used by the cells as paths of least resistence to migrate to distant sites just as they would through interstitial stroma to form metastases. Possibly these paths are created by localized proteolysis (see below).

Increased α2β1 expression also correlates with increased metastatic potential of rhabdomyosarcoma cells. Higher expression did not increase the growth properties of the tumor cells, i.e., tumor size was the same as in cells that did not express α2β1, but tumor foci increased significantly in number, and tumors were observed in a larger variety of tissues.[12] Whether increased tumor numbers were due to α2β1 mediating attachment to collagen IV in the subendothelium, thereby facilitating extravasation, has not been resolved.

In conclusion, depending upon the tissue and tumor type, both up- and downregulation of α2β1 may influence tumor initiation and progression.

Proteolysis

More recently, another property of α2β1 has emerged that facilitates migration of cells including tumor cells through collagenous substrates, namely the close association of α2β1 with matrix-degrading collagenases. Fibroblasts, but also other cell types, utilize α2β1 to migrate through collagen matrices resulting in collagen fibril reorganization and matrix contraction. In vitro, activation of α2β1, but not of α1β1 lead to induced synthesis of matrix metalloproteinase-1 (MMP-1).[13] In vivo, e.g., during skin wound healing, the activity of MMP-1 was shown to be required for migration of cells on a collagen substrate.[14,15] Migration could be abolished by mutant collagen resistant to MMP-1 cleavage, by antibodies to MMP-1 and also by antibodies against α2β1 (summarized by[16]). W. Parks et al then showed that the α2 integrin subunit physically interacts with proMMP-1, resulting in local activation and spatially restricted MMP-1 activity.[16] Thus, α2β1 locally concentrates enzyme and substrate and in this way is thought to facilitate collagen cleavage and subsequent cell migration.

Apart from MMP-1, collagen-mediated activation of α2β1 receptors also induces synthesis and activation of the membrane-bound MMP-14 (MT1-MMP) and activation of proMMP-2 (gelatinase A).[17] Although the authors demonstrated this link in fibroblasts, it could also be an intriguing mechanism used by invasive tumor cells, because MT1-MMP in turn activates proMMP2,[18-21] which is capable of degrading basement membrane structures. In fact, work by C. Mauch et al strongly suggests that highly invasive melanoma cells may employ such α2β1-associated mechanisms.[22,23]

MMP-13 is another collagen-degrading enzyme that differs from MMP-1 and MT1-MMP by an exceptionally wide substrate specificity. Induction and activation of MMP-13 by culture of fibroblasts within a collagen lattice was also correlated with α2β1, but interestingly, also α1β1 seems to play a role in this induction.[24] MMP-13 has been associated with a number of disease situations including destruction of rheumatoid synovium, cutaneous and intestinal ulcerations and a number of malignant tumors (summarized in[24]).

Such cell-proteinase association via integrins or other cell surface receptors is not unique to integrin α2β1, as it has also been reported for the proteinases/receptors MMP-2 / αvβ3,[25] MMP-9 / CD44[26] and MMP-7 / heparan sulfate proteoglycans[27] (Table 1). Thus, clearing the path for cells to migrate in collagen-rich stroma and during matrix remodeling appears to rely on receptor-mediated proteolysis in physiological as well as in malignant conditions.

Angiogenesis

Tumor growth and progression depends critically on blood vessels supplying nutrients and oxygen, and the role of RGD-binding integrins in this process is well documented. However, it seems plausible that new vessels also need to pervade collagen-rich ECM when sprouting from existing vasculature and reaching the tumor. Indeed, M. Detmar et al demonstrated an important role for both α1β1 and α2β1 in VEGF (vascular endothelial growth factor) -driven tumor angiogenesis.[28,29] VEGF, which also promotes αvβ3 expression on endothelial cells migrating through a provisional matrix containing vitronectin, fibronectin and fibrin, significantly induced α1β1 and α2β1 expresssion on dermal microvascular endothelial cells in vitro and in vivo. Blocking either one or both collagen receptors by antibodies markedly inhibited formation of new blood vessels and also reduced luminal vessel diameter in mouse skin, indicating an effect on cell maturation. In a tumor xenograft model (injection of squamous cell carcinoma cells), combined antagonism of α1β1 and α2β1 suppressed average vessel density and size, and significantly reduced tumor volume. Together, the two collagen integrins collaborated with VEGF in promoting chemotactic migration through a collagen-rich matrix in response to VEGF as well as endothelial cell proliferation.

These results agree with the observation that tumor vascularization was markedly reduced in α1-deficient mice.[30] This was shown to be due to increased levels of MMP-7 and MMP-9, which generate angiostatin from circulating plasminogen, which in turn acts as a potent inhibitor of endothelial cell proliferation.

Fibrosis

The hallmark of fibrotic tissue alterations is the excessive production of ECM constituents, mainly collagens, and deposition into the tissues.[31,32] Early stages of fibrosis are often associated with trauma or injury, resulting in inflammation. A number of reports suggests an important role for α1β1 and α2β1 in inflammatory reactions, which will be discussed in another chapter.

Table 1. Correlation of protease induction and/or activation with receptors

$\alpha1\beta1$	MMP-13	Ravanti et al. 1999
	MMP-1	Langholz et al. 1995
		Riikonen et al. 1995
$\alpha2\beta1$	MMP-2	Zigrino et al. 2001
	MMP-13	Ravanti et al. 1999
	MT1-MMP	Zigrino et al. 2001
$\alpha v\beta3$	MMP-2	Brooks et al. 1996
CD44	MMP-9	Yu et al. 2000
HSP	MMP-7	Yu et al. 2000

During the inflammatory phase of fibrotic diseases, cellular activation to produce excess matrix is due in part to the activity of cytokines such as TGF-β (transforming growth factor-β). However, disturbed recognition of and response to the ECM environment plays an important role in cellular activation[33] (Fig. 3). In fibrotic skin, e.g., in patients with systemic sclerosis, dermal fibroblasts are the major producers of excessive matrix;[34,35] in the liver, the activated cell type is the stellate or Ito cell; in kidney fibrosis, mesangial cells and podocytes contribute to the disease.

In all three organs, a correlation of the fibrotic phenotype with $\alpha1\beta1$ receptors has been observed. This agrees well with our results that show that contact of dermal fibroblasts with surrounding collagen normally results in strong downregulation of collagen levels,[36,37] which is mediated by $\alpha1\beta1$.[13] Integrin $\alpha2\beta1$ also contributes significantly to the net collagen amount in the tissue, however, it does so by inducing collagen-degrading MMP-1,[13,38] MMP-13[24] and MT1-MMP,[17] which all act on matrix remodeling. Analyzing fibroblasts cultured from the fibrotic skin of patients, which produce excessive collagen amounts, Ivarsson et al reported reduced levels of $\alpha1\beta1$ integrin receptors on the cell surface.[39] In a later study, we could not detect significantly altered expression levels of either $\alpha1\beta1$ or $\alpha2\beta1$.[40] Rather, we would like to propose that fibrosis results from defective integration of collagen-induced signals, mediated by intracellular proteins, which specifically associate with the cytoplasmic tail of integrin $\alpha1$ and result in altered signaling (Krieg, Eckes unpublished).

A further link between deregulated collagen and $\alpha1\beta1$ integrins was established using integrin $\alpha1$-deficient mice. These animals showed increased levels of collagen in experimental granulation tissue[41] in addition to reduced numbers of fibroblasts in the skin. The latter is due to the transduction of proliferative signals by $\alpha1\beta1$.[42] Fibrosis was not seen in these animals, presumably due to the concomitant induction of the interstitial collagenase by $\alpha2\beta1$.

In the kidney, progressive glomerulonephritis and subsequent tubulointerstitial fibrosis can result from mutations in collagen IV genes and lead to Alport syndrome.[43] Mice displaying this phenotype have been engineered, which die from renal fibrosis and failure. Breeding these mice with integrin $\alpha1$-deficient mice to eliminate $\alpha1\beta1$ resulted in marked attenuation of renal disease progression,[44] indicating that absence of $\alpha1\beta1$ receptors conferred a protective effect.

While the protease-inducing activity of $\alpha2\beta1$ in fibrosis has not finally been resolved, there is a clear correlation of $\alpha1\beta1$ with fibrosis in different tissues.

Thrombosis/Hemostasis

The role of integrin $\alpha2\beta1$ in platelet functions has interested hematologists for almost 20 years. Taking into account the recent results in mice arising from deficiency of $\beta1$ integrins and specifically of $\alpha2\beta1$, the role of these receptors in platelet biology may need to be reconsidered.

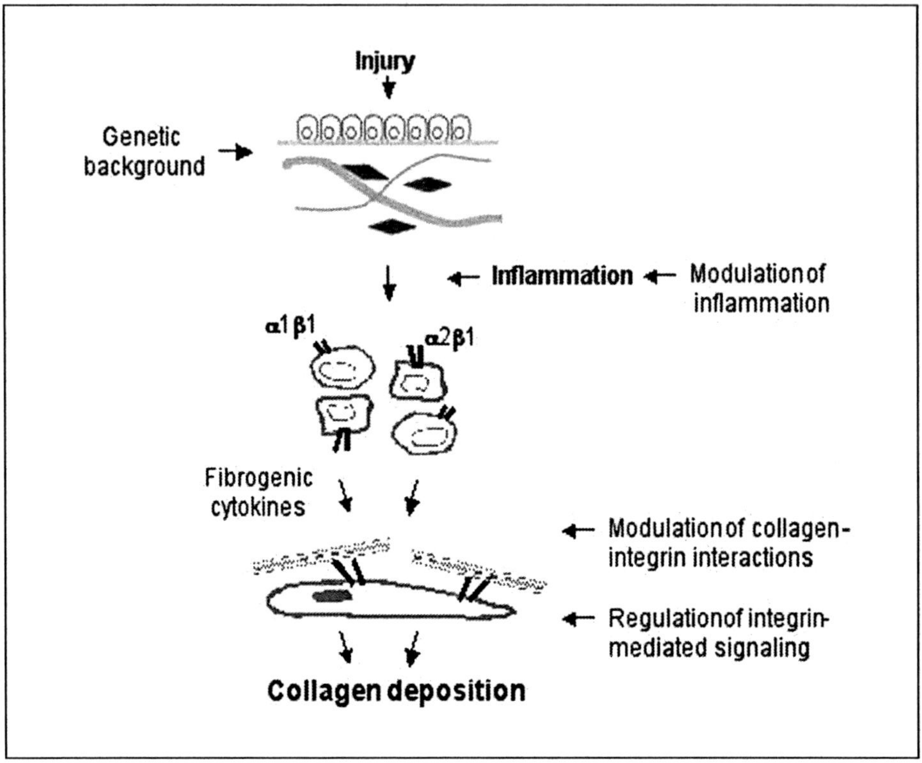

Figure 3. Fibrosis. Early events in the development of fibrosis often involve trauma or injury to the tissue, and genetic background can predispose individuals to be more susceptible. Injury triggers inflammatory responses with inflammatory cells being recruited, which release cytokines, e.g., TGF-β. Integrins α1β1 and α2β1 were both shown to play an important role in inflammation (see also the separate chapter on inflammation). Excessive collagen production results from the activation of fibroblasts (or stellate cells or mesangial cells etc.) by fibrogenic cytokines and from altered feedback regulation by the ECM. In particular, integrin α1β1 was reported to be involved in the defective recognition of ECM. Potential targets for therapeutic intervention are modulation of inflammation, of integrin-collagen interactions and of integrin-induced signaling.

Vessel wall injury triggers sudden platelet activation and platelet plug formation, followed by coagulant activity and the formation of fibrin-containing thrombi, which occlude the site of injury. These events are crucial to limit blood loss at sites of tissue trauma, but may also occlude diseased vessels leading to ischemia and infarction of vital organs.

The first step in the hemostatic cascade is platelet interaction with the exposed ECM at sites of injury. Among the macromolecular constituents of the ECM, collagen is considered to play a major role in this process as in vitro it not only supports platelet adhesion but also directly activates the cells and thereby enables their aggregation.[45] Platelet adhesion and aggregation on collagen is a complex and integrated process that involves many surface receptors including integrins, Ig-like receptors, G-protein coupled receptors and their various ligands and signaling pathways.

At high shear rates, such as found in arteries and arterioles, the rapid and reversible interaction between glycoprotein (GP) Ib-V-IX and von Willebrand factor (vWf), immobilized on collagen types I, III[46] and VI[47] is crucial for reducing the velocity of flowing platelets, thereby enabling their interaction with the ECM.[48,49] At this stage, collagen receptors mediate platelet activation and firm adhesion followed by second wave activation by platelet-released agonists, most importantly adenosine diphosphate (ADP) and thromboxane A2 (TxA2), leading to thrombus growth. Several platelet receptors have been proposed to be involved in these processes, particularly the integrin α2β1[50] and the Ig receptor GPVI.[51] Both are considered to be of clinical relevance. While integrin α2β1 serves mainly as an adhesion receptor, GPVI mediates platelet activation on collagen.

Bleeding Disorders

Integrin α2β1 (also known as platelet GPIa/IIa or lymphocyte VLA-2) was the first collagen receptor to be identified on platelets and was thought for a long time to play a crucial role in hemostasis. This hypothesis was largely based on the analysis of few patients lacking or expressing very low levels of α2β1 on their platelets. Two female α2β1-deficient patients were reported, whose platelets were largely unresponsive to collagen[52,53] and who suffered from posttraumatic bleeding and excessive menorrhagia. Interestingly, in both patients the defect spontaneously disappeared when they had passed menopause, indicating that their platelet phenotype was related to altered hormone activity. If this was indeed the reason, it appears very unlikely that such a defect exclusively affected α2β1 expression in platelets, but possibly also other receptors/signaling molecules. A third patient was described a few years later, whose platelets lacked α2β1 and were also unresponsive to collagen, and who suffered from a myeloproliferative disorder.[54] While the patient had a prolonged bleeding time, he suffered from profound thrombocytosis, the latter suggesting that platelet production was altered by undefined mechanisms, which most likely did not arise as a consequence of α2β1-deficiency.

Only recently, mice have been generated, which lack either the integrin β1 subunit (i.e., not expressing α2β1, α5β1, and α6β1)[55] or the α2 subunit on their platelets[4,5]. Strikingly, both mouse strains have normal platelet counts and bleeding times, demonstrating that α2β1 is not essential for normal platelet formation and hemostasis in mice. Furthermore, adhesion and activation of α2β1-deficient platelets on collagen is only moderately reduced as compared to wild type platelets. This strongly suggests that none of the above described patients had an isolated deficiency in α2β1. So far, no human with a defined genetic defect in the integrin α2 subunit has been described. Two possible explanations may account for this. First, such a defect would be embryonic lethal or, second, such humans exist but have not been identified due to a subtle phenotype. The first appears very unlikely because α2-deficient mice develop normally, are fertile, and display no obvious anatomical or hemostatic defects.[4,5]

Many integrins exist in low affinity states and require cellular activation to undergo a conformational change to efficiently bind their ligands, whereas other integrins have been shown to be constitutively in a high affinity state. For a long time, platelet α2β1 was thought to belong to the latter group of integrins. A previously accepted model of platelet-collagen interactions, the so-called "two-step, two-site" model incorporated the idea that platelets in a first step firmly bind to collagen via high affinity α2β1 allowing subsequent interaction of a second receptor leading to platelet activation.[56] This second, low affinity receptor was later proposed to be GPVI. However, the idea that α2β1 is active on unstimulated platelets was first challenged by Moroi and coworkers who demonstrated that the affinity of the integrin for soluble collagen increases upon cellular stimulation. This suggested that α2β1, like other platelet integrins, requires an agonist-induced conformational change (via inside-out signaling) to bind to collagen.[57,58] This hypothesis was then confirmed and extended by the analysis of mouse platelets lacking the activating collagen receptor GPVI. Their platelets, despite normal expression of α2β1, do not adhere to collagen under static or flow conditions.[55] However, when integrins on those platelets were activated by other agonists the cells firmly bound to collagen through α2β1. Together, these findings demonstrate that α2β1 is unable to bind to collagen without prior activation mediated by other receptors and can, therefore, not be involved in the very first steps of platelet attachment at sites of vascular injury. Only upon platelet activation by collagen-GPVI interactions or other pathways, integrin $\alpha2\beta1$ binds to its ligand and in concert with other receptors contributes to thrombus formation. Thus, despite its clear role in platelet-collagen interactions, $\alpha2\beta1$ is not essential for hemostasis. This strongly suggests that isolated defects in this integrin may not cause severe bleeding.

Arterial Thrombosis

There is a very wide range of $\alpha2\beta1$ expression levels in platelets from normal, healthy humans, which appears to be based on a silent polymorphism in the $\alpha2$ gene influencing the expression rate of the integrin on platelets.[59] Thus, even individuals with very low $\alpha2\beta1$ levels have no obvious hemostatic defect, supporting the assumption that this integrin plays a minor role in human hemostasis. However, the situation may be different in pathological conditions, such as arterial thrombosis following rupture of an atherosclerotic plaque. In the process of atherogenesis, enhanced collagen synthesis by intimal smooth muscle cells and fibroblasts has been shown to significantly contribute to luminal narrowing.[60,61] Thus, the matrix exposed upon plaque rupture is enriched in collagens, indicating that platelet collagen receptors may be of particular importance in arterial thrombus formation. A large number of studies have examined a possible association of polymorphisms in the $\alpha2$-locus and cardiovascular disease, and the majority of them suggests such a correlation, although no overall consensus has yet been reached.[62-65] Therefore, in vivo studies using integrin $\alpha2$-deficient mice in models of cardiovascular disease, such as arterial thrombosis or atherogenesis, are warranted to provide a more detailed understanding of the role of this receptor in pathological conditions.

Concluding Remarks

Integrin-mediated cell-extracellular matrix interactions regulate cellular responses, many of which are de-regulated in pathological conditions. As a long-term goal, the detailed analysis of integrin functions at the molecular level should allow to specifically interfere with disease progression, e.g., by blocking ligand binding or by altering signaling cascades.

References

1. Hynes R. Integrins. Bidirectional, allosteric signaling machines. Cell 2002; 110:673-687.
2. Gullberg DE, Lundgren-Akerlund E. Collagen-binding I domain integrins—what do they do? Prog Histochem Cytochem 2002; 37:3-54.
3. Gardner H, Kreidberg J, Koteliansky V et al. Deletion of integrin alpha 1 by homologous recombination permits normal murine development but gives rise to a specific deficit in cell adhesion. Dev Biol 1996; 175:301-313.
4. Holtkotter O, Nieswandt B, Smyth N et al. Integrin alpha 2-deficient mice develop normally, are fertile, but display partially defective platelet interaction with collagen. J Biol Chem 2002; 277:10789-10794.
5. Chen J, Diacovo TG, Grenache DG et al. The alpha(2) integrin subunit-deficient mouse: a multifaceted phenotype including defects of branching morphogenesis and hemostasis. Am J Pathol 2002; 161:337-344.
6. Varner JA, Cheresh DA. Integrins and cancer. Curr Opin Cell Biol 1996; 8:724-730.
7. Sun H, Santoro SA, Zutter MM. Downstream events in mammary gland morphogenesis mediated by reexpression of the alpha2beta1 integrin: the role of the alpha6 and beta4 integrin subunits. Cancer Res 1998; 58:2224-2233.
8. Zutter MM, Santoro SA, Staatz WD et al. Reexpression of the alpha 2 beta 1 integrin abrogates the malignant phenotype of breast carcinoma cells. Proc Natl Acad Sci USA 1995; 92:7411-7415.
9. Klein CE, Dressel D, Steinmayer T et al. Integrin alpha 2 beta 1 is upregulated in fibroblasts and highly aggressive melanoma cells in three-dimensional collagen lattices and mediates the reorganization of collagen I fibrils. J Cell Biol 1991; 115:1427-1436.

10. Klein CE, Steinmayer T, Kaufmann D et al. Identification of a melanoma progression antigen as integrin VLA-2. J Invest Dermatol 1991; 96:281-284.
11. Friedl P, Maaser K, Klein CE et al. Migration of highly aggressive MV3 melanoma cells in 3-dimensional collagen lattices results in local matrix reorganization and shedding of alpha2 and beta1 integrins and CD44. Cancer Res 1997; 57:2061-2070.
12. Chan BM, Matsuura N, Takada Y et al. In vitro and in vivo consequences of VLA-2 expression on rhabdomyosarcoma cells. Science 1991; 251:1600-1602.
13. Langholz O, Rockel D, Mauch C et al. Collagen and collagenase gene expression in three-dimensional collagen lattices are differentially regulated by alpha 1 beta 1 and alpha 2 beta 1 integrins. J Cell Biol 1995; 131:1903-1915.
14. Saarialho-Kere UK, Kovacs SO, Pentland AP et al. Cell-matrix interactions modulate interstitial collagenase expression by human keratinocytes actively involved in wound healing. J Clin Invest 1993; 92:2858-2866.
15. Pilcher BK, Dumin JA, Sudbeck BD et al. The activity of collagenase-1 is required for keratinocyte migration on a type I collagen matrix. J Cell Biol 1997; 137:1445-1457.
16. Dumin JA, Dickeson SK, Stricker TP et al. Pro-collagenase-1 (matrix metalloproteinase-1) binds the alpha(2)beta(1) integrin upon release from keratinocytes migrating on type I collagen. J Biol Chem 2001; 276:29368-29374.
17. Zigrino P, Drescher C, Mauch C. Collagen-induced proMMP-2 activation by MT1-MMP in human dermal fibroblasts and the possible role of alpha2beta1 integrins. Eur J Cell Biol 2001; 80:68-77.
18. Strongin AY, Marmer BL, Grant GA et al. Goldberg GI. Plasma membrane-dependent activation of the 72-kDa type IV collagenase is prevented by complex formation with TIMP-2. J Biol Chem 1993; 268:14033-14039.
19. Tomasek JJ, Halliday NL, Updike DL et al. Gelatinase A activation is regulated by the organization of the polymerized actin cytoskeleton. J Biol Chem 1997; 272:7482-7487.
20. Zucker S, Drews M, Conner C et al. Tissue inhibitor of metalloproteinase-2 (TIMP-2) binds to the catalytic domain of the cell surface receptor, membrane type 1-matrix metalloproteinase 1 (MT1-MMP). J Biol Chem 1998; 273:1216-1222.
21. Lehti K, Lohi J, Valtanen H et al. Proteolytic processing of membrane-type-1 matrix metalloproteinase is associated with gelatinase A activation at the cell surface. Biochem J 1998; 334:345-353.
22. Kurschat P, Zigrino P, Nischt R et al. Tissue inhibitor of matrix metalloproteinase-2 regulates matrix metalloproteinase-2 activation by modulation of membrane-type 1 matrix metalloproteinase activity in high and low invasive melanoma cell lines. J Biol Chem 1999; 274:21056-21062.
23. Kurschat P, Wickenhauser C, Groth W et al. Identification of activated matrix metalloproteinase-2 (MMP-2) as the main gelatinolytic enzyme in malignant melanoma by in situ zymography. J Pathol 2002; 197:179-187.
24. Ravanti L, Heino J, Lopez-Otin C et al. Induction of collagenase-3 (MMP-13) expression in human skin fibroblasts by three-dimensional collagen is mediated by p38 mitogen- activated protein kinase. J Biol Chem 1999; 274:2446-2455.
25. Brooks PC, Stromblad S, Sanders LC et al. Localization of matrix metalloproteinase MMP-2 to the surface of invasive cells by interaction with integrin alpha v beta 3. Cell 1996; 85:683-693.
26. Yu Q, Stamenkovic I. Cell surface-localized matrix metalloproteinase-9 proteolytically activates TGF-beta and promotes tumor invasion and angiogenesis. Genes Dev 2000; 14:163-176.
27. Yu WH, Woessner JF Jr. Heparan sulfate proteoglycans as extracellular docking molecules for matrilysin (matrix metalloproteinase 7). J Biol Chem 2000; 275:4183-4191.
28. Senger DR, Claffey KP, Benes JE et al. Angiogenesis promoted by vascular endothelial growth factor: regulation through alpha1beta1 and alpha2beta1 integrins. Proc Natl Acad Sci USA 1997; 94:13612-13617.
29. Senger DR, Perruzzi CA, Streit M et al. The alpha(1)beta(1) and alpha(2)beta(1) integrins provide critical support for vascular endothelial growth factor signaling, endothelial cell migration, and tumor angiogenesis. Am J Pathol 2002; 160:195-204.
30. Pozzi A, Moberg PE, Miles LA et al. Elevated matrix metalloprotease and angiostatin levels in integrin alpha 1 knockout mice cause reduced tumor vascularization. Proc Natl Acad Sci USA 2000; 97:2202-2207.

31. Krieg T, Meurer M. Systemic scleroderma. Clinical and pathophysiologic aspects. J Am Acad Dermatol 1988; 18:457-481.
32. Eckes B, Kessler D, Aumailley M et al. Interactions of fibroblasts with the extracellular matrix: implications for the understanding of fibrosis. Springer Semin Immunopathol 1999; 21:415-429.
33. Mauch C, Kozlowska E, Eckes B et al. Altered regulation of collagen metabolism in scleroderma fibroblasts grown within three-dimensional collagen gels. Exp Dermatol 1992; 1:185-190.
34. Kulozik M, Hogg A, Lankat-Buttgereit B et al. Colocalization of transforming growth factor beta 2 with alpha 1(I) procollagen mRNA in tissue sections of patients with systemic sclerosis. J Clin Invest 1990; 86:917-922.
35. Peltonen J, Kahari L, Jaakkola S et al. Evaluation of transforming growth factor beta and type I procollagen gene expression in fibrotic skin diseases by in situ hybridization. J Invest Dermatol 1990; 94:365-371.
36. Mauch C, Hatamochi A, Scharffetter K et al. Regulation of collagen synthesis in fibroblasts within a three- dimensional collagen gel. Exp Cell Res 1988; 178:493-503.
37. Eckes B, Mauch C, Huppe G et al. Differential regulation of transcription and transcript stability of pro-alpha 1(I) collagen and fibronectin in activated fibroblasts derived from patients with systemic scleroderma. Biochem J 1996; 315:549-554.
38. Riikonen T, Westermarck J, Koivisto L et al. Integrin alpha 2 beta 1 is a positive regulator of collagenase (MMP-1) and collagen alpha 1(I) gene expression. J Biol Chem 1995; 270:13548-13552.
39. Ivarsson M, McWhirter A, Black CM et al. Impaired regulation of collagen pro-alpha 1(I) mRNA and change in pattern of collagen-binding integrins on scleroderma fibroblasts. J Invest Dermatol 1993; 101:216-221.
40. Herzhoff K, Sollberg S, Huerkamp C et al. Fibroblast expression of collagen integrin receptors alpha1beta1 and alpha2beta1 is not changed in systemic scleroderma. Br J Dermatol 1999; 141:218-223.
41. Gardner H, Broberg A, Pozzi A et al. Absence of integrin alpha1beta1 in the mouse causes loss of feedback regulation of collagen synthesis in normal and wounded dermis. J Cell Sci 1999; 112:263-272.
42. Pozzi A, Wary KK, Giancotti FG et al. Integrin alpha1beta1 mediates a unique collagen-dependent proliferation pathway in vivo. J Cell Biol 1998; 142:587-594.
43. Mochizuki T, Lemmink HH, Mariyama M et al. Identification of mutations in the alpha 3(IV) and alpha 4(IV) collagen genes in autosomal recessive Alport syndrome. Nat Genet 1994; 8:77-81.
44. Cosgrove D, Rodgers K, Meehan D et al. Integrin alpha1beta1 and transforming growth factor-beta1 play distinct roles in alport glomerular pathogenesis and serve as dual targets for metabolic therapy. Am J Pathol 2000; 157:1649-1659.
45. Sixma JJ, van Zanten GH, Huizinga EG et al. Platelet adhesion to collagen: an update. Thromb Haemost 1997; 78:434-438.
46. Pareti FI, Niiya K, McPherson JM et al. Isolation and characterization of two domains of human von Willebrand factor that interact with fibrillar collagen types I and III. J Biol Chem 1987; 262:13835-13841.
47. Rand JH, NPatel ND, Schwartz E et al. 150-kD von Willebrand factor binding protein extracted from human vascular subendothelium is type VI collagen. J Clin Invest 1991; 88:253-259.
48. Moroi M, Jung SM, Nomura S et al. Analysis of the involvement of the von Willebrand factor-glycoprotein Ib interaction in platelet adhesion to a collagen-coated surface under flow conditions. Blood 1997; 90:4413-4424.
49. Savage B, Almus-Jacobs F, Ruggeri ZM. Specific synergy of multiple substrate-receptor interactions in platelet thrombus formation under flow. Cell 1998; 94:657-666.
50. Santoro SA. Identification of a 160,000 dalton platelet membrane protein that mediates the initial divalent cation-dependent adhesion of platelets to collagen. Cell 1986; 46:913-920.
51. Moroi M, Jung SM, Okuma M et al. A patient with platelets deficient in glycoprotein VI that lack both collagen-induced aggregation and adhesion. J Clin Invest 1989; 84:1440-1445.
52. Nieuwenhuis HK, Akkerman JW, Houdijk WP et al. Human blood platelets showing no response to collagen fail to express surface glycoprotein Ia Nature 1985; 318:470-472.

53. Kehrel B, Balleisen L, Kokott R et al. Deficiency of intact thrombospondin and membrane glycoprotein Ia in platelets with defective collagen-induced aggregation and spontaneous loss of disorder. Blood 1988; 71:1074-1078.
54. Handa M, Watanabe K, Kawai Y et al. Platelet unresponsiveness to collagen: involvement of glycoprotein la- lla (alpha 2 beta 1 integrin) deficiency associated with a myeloproliferative disorder. Thromb Haemost 1995; 73:521-528.
55. Nieswandt B, Brakebusch C, Bergmeier W et al. Glycoprotein VI but not alpha2beta1 integrin is essential for platelet interaction with collagen. Embo J 2001; 20:2120-2130.
56. Santoro SA, Walsh JJ, Staatz WD Distinct determinants on collagen support alpha 2 beta 1 integrin-mediated platelet adhesion and platelet activation. Cell Regul 1991; 2:905-913.
57. Moroi M, Onitsuka I, Imaizumi T et al. Involvement of activated integrin alpha2beta1 in the firm adhesion of platelets onto a surface of immobilized collagen under flow conditions. Thromb Haemost 2000; 83:769-776.
58. Jung SM, Moroi M. Platelets interact with soluble and insoluble collagens through characteristically different reactions. J Biol Chem 1998; 273:14827-14837.
59. Kunicki TJ, Kritzik M, Annis DS, Nugent DJ. Hereditary variation in platelet integrin alpha 2 beta 1 density is associated with two silent polymorphisms in the alpha 2 gene coding sequence. Blood 1997; 89:1939-1943.
60. Rekhter MD. Collagen synthesis in atherosclerosis: too much and not enough. Cardiovasc Res 1999; 41:376-384.
61. Opsahl WP, DeLuca DJ, Ehrhart LA. Accelerated rates of collagen synthesis in atherosclerotic arteries quantified in vivo. Arteriosclerosis 1987; 7:470-476.
62. Moshfegh K, Wuillemin WA, Redondo M et al. Association of two silent polymorphisms of platelet glycoprotein Ia/IIa receptor with risk of myocardial infarction: a case-control study. Lancet 1999; 353:351-354.
63. Osterborg A, Bjorkholm M, Bjoreman M et al. Natural interferon-alpha in combination with melphalan/prednisone versus melphalan/prednisone in the treatment of multiple myeloma stages II and III: a randomized study from the Myeloma Group of Central Sweden. Blood 1993; 81:1428-1434.
64. Reiner AP, Kumar PN, Schwartz SM et al. Genetic variants of platelet glycoprotein receptors and risk of stroke in young women. Stroke. 2000; 31:1628-1633..
65. Roest M, Banga JD, Grobbee DE et al. Homozygosity for 807 T polymorphisms in alpha(2) subunit of platelet alpha(2)beta(1) is associated with increased risk of cardiovascular mortality in high-risk women. Circulation 2000; 102:1645-1650.

Chapter 12

Integrins in Immune and Inflammatory Disease

Antonin R. de Fougerolles

The adhesive interactions of cells with other cells and with extracellular matrix have a central role in regulating immune and inflammatory responses. Integrins are capable of acting at different points along the inflammatory cascade. They are important in initiation of immune responses, in promoting leukocyte-endothelial cell attachment and transmigration, and in migration and retention of cells within tissues. The strong mechanistic rationale coupled with solid in vivo animal model data has made integrin antagonists attractive candidates for clinical development. This chapter details recent advances around the biology of α chain I-domain-containing integrins with an emphasis on in vivo data supporting the importance of integrin-mediated interactions in the regulation of auto-immune and inflammatory disease. Recent clinical progress with integrin antagonists is also discussed.

Adhesive interactions of cells with other cells and with extracellular matrix play an important role in regulating immune and inflammatory responses. These adhesive interactions are mediated in part by integrins, a family of cell surface-expressed noncovalently linked α/β heterodimers. There are 8 β and 18 α subunits, divided into subfamilies that share β subunits, giving rise to 24 distinct heterodimers (Fig. 1). The integrin family can be further subdivided structurally based on the presence of an extra "inserted" or I domain within certain α subunits. Half of the α subunits contain an I domain, thereby resulting in 9 distinct I-domain containing integrins (α1β1, α2β1, α10β1, α11β1, αLβ2, αMβ2, αXβ2, αDβ2, and αEβ7) (Fig. 1). In these integrins, ligand binding is mediated by the 200 or so amino acid "inserted" domain. While I domains from different integrins show a high degree of sequence and structural similarity, it has clearly been shown that each of these I domains are sufficient in themselves to confer the distinct ligand specificity found for each integrin. The structural characteristics and functional importance of the I domain itself has been extensively reviewed elsewhere.[1,2] While α chain I-domain containing integrins represent only about a third of all known integrin heterodimers, they are disproportionately expressed on immune cell types and include, for instance, all of the β2 or 'leukocyte' integrins (Fig. 1). In this chapter, I will review recent advances around the biology of the α chain I-domain-containing integrins, with special emphasis on their role in autoimmune and inflammatory disease, as evidenced in animal models and where possible human clinical trials. Although integrins with I-domain containing α subunits play a large role in immune and inflammatory responses, other integrins that lack I domains in their α subunit, such as α4β1, also play important roles. As these fall outside of the scope of this chapter, they will only be mentioned peripherally. Several recent reviews nicely outline the immunological function and importance of these non-I domain containing integrins.[3,4]

I Domains in Integrins, edited by Donald Gullberg.

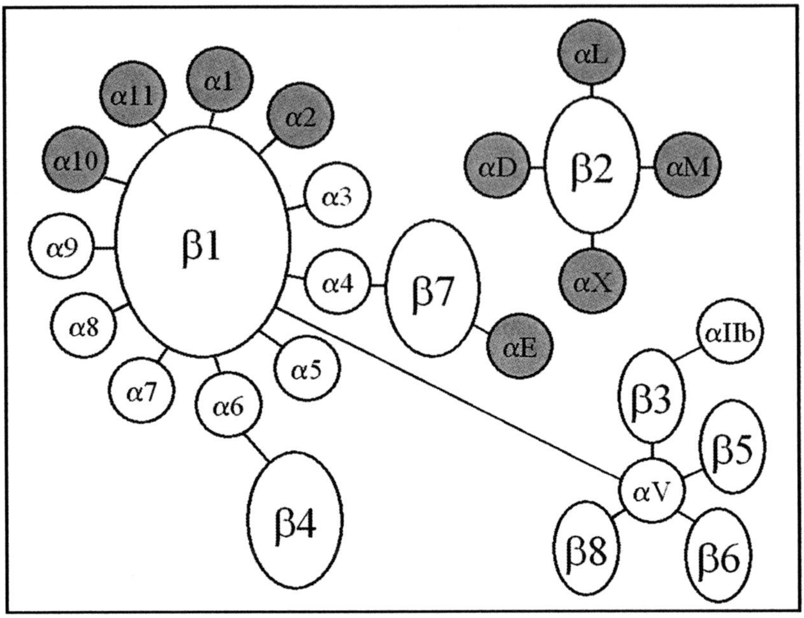

Figure 1. Integrin αβ heterodimer pairings. Integrin α chains containing an inserted or "I" domain are shown in grey.

αLβ2

Expression of αLβ2 (LFA-1, CD11a/CD18) is restricted to leukocytes, being well expressed on the surface of neutrophils, monocytes, macrophages, NK cells, and lymphocytes[3,5] (see also chapter by Gahmberg and Fagerholm in this book for detailed discussion of αLβ2 distribution and function). αLβ2 has several ligands including ICAM-1 (intercellular adhesion molecule 1; CD54), ICAM-2 (CD102), ICAM-3 (CD50), ICAM-4, ICAM-5 (telencephalin), as well as the recently described junctional adhesion molecule 1 (JAM-1).[5,6] The ICAMs are closely related members of the immunoglobulin (Ig) superfamily that are expressed on a number of different cell types including antigen-presenting cells (APC) and endothelial cells.[5] Engagement of αLβ2 by its ligands (primarily ICAM-1) has been shown to be important in stabilizing the interaction between T cells and APCs, in providing costimulatory and proliferative signals, and in recruiting leukocytes to sites of inflammation.[5,7] The αLβ2 interaction with JAM-1, expressed at tight junctions on endothelial cells, is involved in transendothelial migration of leukocytes.[6] Lastly, binding of αLβ2 to its ligands is greatly increased in response to ligation of cell surface receptors, such as T cell and chemokine receptors, which through activation of signaling pathways result in an increased affinity of αLβ2 for its ligands; this phenomenon being referred to as inside-out signaling.[2,5]

The in vivo importance of the αLβ2/ICAM1 pathway in inflammatory and auto-immune disease has been extensively reviewed.[3,8] Briefly, increased expression of αLβ2 has been detected in a wide variety of auto-immune and inflammatory conditions. The functional importance of the β2 integrins, and in part αLβ2, is exemplified by the clinical condition known as leukocyte adhesion deficiency type I (LAD-I), in which a mutation in the common β2 chain results in no or very low expression of β2 integrins. LAD-I is characterized by a number of leukocyte defects, the foremost of which is an almost complete lack of neutrophil emigration into inflammatory sites, leaving patients at severe risk from recurrent life-threatening bacterial infections.[5] By gene targeting and other techniques, it is now possible using animal models, to more precisely study the functions of individual integrins. While the β2-deficient mice have a phenotype similar to that seen in the LAD-I patients, the phenotype of the αL-deficient mice is somewhat different.[9] While in an air pouch inflammation model, neutrophil migration was equally reduced in β2- and αL-deficient mice, αL-deficient mice unlike β2-deficient mice are not susceptible to infections.[9,10] Studies with the αL-deficient mice confirm the significance of this molecule in T cell activation. Consistent with an important role for αLβ2 in generation of an immune response and cytotoxic T cell killing, alloantigen triggered T cell proliferation and cytotoxicity are severely impaired, leading to defective host-versus-graft reaction and abrogated tumor rejection.[11-13] Lastly, αL-deficient mice also show defective homing of lymphocytes to lymph nodes, particularly peripheral nodes.[14]

The importance of αLβ2 blockade in inflammatory and auto-immune disease has been extensively investigated and reviewed.[3,8,15] The generation of ICAM-1-deficient mice and the use of blocking anti-ICAM-1 mAbs have also confirmed the importance of the β2 integrins in numerous inflammatory disease settings.[3,8,15-17] While the phenotype seen in these mice are generally interpreted as primarily being the result of blockade of αLβ2-ICAM1 interactions, this cannot be firmly established given the fact that ICAM-1 is a ligand for many β2 integrin family members, including αMβ2. Thus, I will largely concentrate here on data derived from testing αLβ2 antagonists as these directly implicate αLβ2 in disease pathogenesis.

Auto-immune diseases, such as rheumatoid arthritis, psoriasis, and colitis, are the result of an on-going systemic inflammatory process, which is in part driven both by the continued generation of auto-immune responses within the lymphoid compartments and by the emigration of effector immune cells into the inflamed tissue. Given the role that αLβ2 interaction with ICAM-1 plays in both these processes, it is not surprising that antagonists of αL have been tried in many disease settings. Anti-αL mAbs have been shown to attenuate neutrophil and monocyte trafficking to inflamed joints in a model of arthritis,[18] and in a collagen-induced model of arthritis, treatment with either anti-αL or anti-ICAM-1 mAb was very effective at preventing disease development.[19] The importance of αLβ2 in rheumatoid arthritis is further supported by studies showing that ICAM-1-deficient mice are partially resistant to development of the disease.[16] Similarly, in a model of contact hypersensitivity, treatment with mAbs to either αL or ICAM-1 reduced inflammatory responses and recruitment of cells to inflammatory sites.[20] While anti-αL mAb inhibit effector T cell responses in contact hypersensitivity models, the inhibitory effect of anti-αL mAb treatment is most profound when given at the sensitization, or T cell priming, phase[21-23] (and A.D.F., unpublished results). These studies with αLβ2 in contact hypersensitivity have now been expanded to include work with anti-human αL mAb in a human psoriatic skin-SCID mouse transplant model.[24] The anti-αL mAb successfully reduced the epidermal thickness of the transplanted human psoriatic skin but not that of transplanted normal skin. These observations provide evidence for activity of anti-αL mAbs in an animal model of human psoriasis, and correspond nicely to the demonstrated efficacy of anti-αL mAbs in the clinical treatment of human psoriasis (see section on clinical trials). Lastly, while the importance of ICAM-1 in inflammatory bowel disease has been extensively demonstrated,[25] evaluation of αL antagonists in models of inflammatory bowel disease has been relatively limited. Prophylactic treatment with anti-β2 mAb in a trinitrobenzene sulfonic acid-induced colitis model resulted in decreased neutrophil influx into the inflamed colon and reduced disease severity.[26] The efficacy seen with ICAM-1 antagonists is likely not due solely to its interaction with αLβ2, as antagonists of αMβ2 are also effective in inflammatory bowel disease models.[27]

In contrast to the inflammatory disease models discussed above, inhibition of the αLβ2/ICAM pathway has had mixed results in experimental autoimmune encephalomyelitis (EAE), a widely used "model" for multiple sclerosis. In two sets of studies using the rat model of EAE, administration of anti-αL and anti-ICAM-1 mAbs alone had no effect, and only a marginal inhibition was seen when the mAbs were used in combination.[28,29] In sharp contrast to the rat model, administration of anti-αL mAb, anti-ICAM-1 mAb or the combination of mAbs in the murine model of EAE was generally found to augment disease severity.[30-32] In addition, disease severity was significantly increased in ICAM-1-deficient mice.[33] The reason for the discordant results using αLβ2/ICAM pathway antagonists in the rat and mouse EAE models is unclear and may be model-related. Overall where seen, blockade of the αLβ2/ICAM pathway was more successful in actively induced EAE models than in the passive transfer of encephalogenic T cell clones. These results are suggestive of the relatively greater importance of the αLβ2/ICAM pathway in antigen-presentation and lymphocyte activation as compared to leukocyte-endothelial attachment and transmigration through the blood-brain barrier, the latter of which can compensated for through other adhesive interactions such as α4β1/VCAM.

Similarly, the role of the αLβ2 integrin in asthma is unclear. Conflicting results have been found in different asthma models using anti-β2, anti-αL, and anti-ICAM-1 mAbs, with decreased leukocyte influx and attenuated airway hypersensitivity seen in some studies[34-36] but not others.[37,38] While the αLβ2/ICAM pathway does not seem to play a prominent role in antigen-induced asthma, other adhesive interactions such as α4β1/VCAM do.[39]

Anti-αL mAbs have also been tested in a range of other immune and inflammatory-based pathologic models, such as glomerulonephritis, lupus nephritis, ischemia-reperfusion injury, and transplantation. Autoimmune glomerulonephritis induced by transfer of serum containing anti-glomerular basement membrane (GBM) antibodies into rats was very significantly reduced in animals receiving either anti-αL or anti-ICAM-1 mAbs.[40,41] Similarly, kidney pathology in a model of lupus nephritis was reduced in mice receiving anti-αL mAb, with profound effects seen on autoantibody production.[42] In a graft-versus-host disease (GVHD)-based model of lupus nephritis, the combined administration of anti-αL and anti-ICAM-1 mAbs also profoundly attenuated development of renal disease, even when mAb treatment was started after the first detectable signs of lupus nephritis were evident.[43]

Monoclonal antibodies directed at αL have also been found to reduce local ischemia-reperfusion injury in the heart and brain; similar findings have also been found using anti-β2 and anti-αM mAbs.[15] In a thermal injury model anti-αL mAbs, like anti-β2 mAbs and anti-αM mAbs, were effective in reducing lung vascular edema and injury if given prior to thermal injury; however in contrast to anti-αM and anti-β2 mAbs, no efficacy was seen with anti-αL mAbs in a skin thermal injury model.[15] These results suggest that the relative contribution of αL in relation to other β2 integrins to leukocyte influx and disease pathology may differ amongst tissues.

While extensive work has been done showing a protective effect of anti-ICAM-1 mAb treatment on transplantation rejection,[15] treatment with anti-αL mAbs alone has generally failed to show much benefit.[44-48] With a few exceptions,[49-51] it is only when used in combination with cyclosporine[45,48] or anti-ICAM-1 Mabs[44,46,47,52] that anti-αL mAb demonstrate any significant increase in allograft survival. Clearly, while αL blockade has shown some efficacy alone in transplantation models, additional immunosuppressive regimens will need to be used in conjunction.

In summary, the beneficial effects of αL blockade are two-fold. It is thought to be due both to its ability to affect T cell costimulation and proliferation, thereby affecting immune responses, and to its ability to regulate leukocyte influx into inflamed tissues. While αLβ2 plays an important role in both processes, efficacy of αL antagonists varies among different inflammatory and auto-immune disease models. Understanding the reason for these differences may help us to learn more about the disease processes themselves and to be able to dissect the relative importance of αLβ2 function in regulating immune responses and leukocyte trafficking.

αMβ2

The αMβ2 (CD11b/CD18; Mac-1; CR3) integrin is expressed mainly on polymorphonuclear cells, monocytes, macrophages, and NK cells[53] (see also chapter by Gahmberg and Fagerholm in this book). αMβ2 is also found on peritoneal B-1 cells, peritoneal mast cells, and a subset of recently activated CD8+ T cells.[53-55] Unlike αLβ2, αMβ2 has many structurally diverse ligands that include complement fragment iC3b, ICAM-1(CD54), fibrinogen, factor X, CD23, bacterial lipopolysaccharide, and mostly recently, the platelet-expressed JAM-3.[53,56] There is a significant degree of overlap in ligand specificity amongst the different β2 integrins. This is true not only for ICAM-1 which is a ligand αLβ2, αMβ2, and αXβ2, but is also especially apparent by the fact that most αMβ2 ligands are also capable of interacting with αXβ2.

Engagement of αMβ2 has also been shown to promote phagocytosis, spreading, chemotaxis, and transduction of intracellular signals that lead to changes in calcium flux, gene expression, and the oxidative burst.[53] αMβ2 can also associate with other membrane proteins and this may account for the many signaling functions of αMβ2.[57]

The difficulty in assessing the function of αMβ2 lies in the fact that many of its ligands are shared among other β2 integrin family members. While this problem can be overcome in vitro by using purified proteins, it is particularly acute in the in vivo context. While both β2-deficient mice and LAD patients have striking phenotypes, the specific contribution that loss of αMβ2 has towards these phenotypes is hard to quantitate. LAD and β2-deficient mice are useful only in assessing the general importance of the β2 family of integrins. Dissection of in vivo αMβ2 function has been greatly helped by the generation of αM-deficient mice.[58] These mice showed reduced chemoattractant-induced leukocyte adhesion to microvascular endothelium, and while neutrophil emigration was largely normal, the extravasated neutrophils had nearly absent phagocytic function and reduced oxygen free radical generation. Consistent with a role for αMβ2 in facilitating neutrophil adhesion, phagocytosis, and respiratory burst, is the finding that αM-deficient mice are also less susceptible to cerebral ischemia/reperfusion injury[59] and have decreased neutrophil Fc receptor-mediated cytotoxicity.[60]

Granulocytes are not the only cell-type in which αMβ2 function is important. Lack of αMβ2 expression also has severe consequences for mast cell development and function. Studies using the αM-deficient mice showed that expression of αMβ2 is important in establishing resident mast cell populations and is required for maintaining adequate mast cell-dependent host defense against bacterial infection.[54] Also, studies using αM-deficient mice showed that expression of αMβ2 by liver macrophages (Kupffer cells) plays a critical role in clearance of IgM-opsonized erythocytes and soluble beta-glucans.[61] It is quite possible that other defects exist in αM-deficient mice that have yet to be identified, especially given expression of αMβ2 on other important cell types such as NK cells, peritoneal B cells, and subets of activated T cells.

Blockade of αMβ2 using mAbs to αM have been found to be successful in a variety of inflammatory and auto-immune models. Firstly, anti-αM mAbs are capable of significantly delaying myelomonocytic accumulation at inflammatory foci in a variety of trafficking models.[62,63] With regards to pathological disease, treatment with anti-αM mAbs inhibited recruitment of macrophages to autoimmune lesions in the islets of Langerhans in nonobese diabetic mice and prevented development of diabetes.[64] This mAb also prevents the immuno-inflammatory pathologic changes seen in viable moth-eaten bone marrow chimeric mice.[65] Disease severity was also significantly reduced in two different models of arthritis by treatment with blocking anti-αM mAbs.[66,67] Likewise, anti-αM mAbs were found to be effective in models of experimental autoimmune encephalomyelitis[68] and colitis;[27] in both these cases blockade of αMβ2 severely impacted the recruitment of neutrophils and monocytes to the inflammatory sites.

Blockade of αMβ2 may potentially also be useful in glomerulonephritis, as pretreatment with anti-αM mAbs in a model of nephrotoxic serum nephritis was found to reduce proteinuria, even though no effect on glomerular infiltration of neutrophils was seen.[69] Anti-αM mAb therapy has also been found to be protective in a number of acute lung injury models, including thermal and oxidant-mediated injury models.[15] Lastly, ischemia-reperfusion injury has been extensively studied, with mAbs to αM reducing injury in the heart, liver, lung, skeletal muscle, and brain.[15] In all cases, ischemia was temporary with mAb given prior to or shortly after induction of ischemia.

A concern with most anti-inflammatory drugs are potential complications arising from a higher incidence of infections. The effect of blocking β2 integrins is well illustrated by the fact that LAD patients develop recurrent life-threatening bacterial and fungal infections. The safety concerns are particularly acute for αMβ2 as this integrin plays a pivotal role on granulocytes, monocytes and mast cells, cell types which are key players in combating infectious microorganisms.[8,15,54,70] It was found that pretreating mice with mAb to αM prior to inoculation of *Listeria monocytogenes* into the liver and spleen significantly increased mortality.[71] Data with anti-β2 mAbs showed similar increases in the severity of bacterial infections.[15] In the end, the potential benefits of anti-adhesion therapy, such as αM blockade, need to be weighed on a case-by-case basis against potential increased risk of infectious complication.

αXβ2

The αXβ2 (CD11c/CD18; p150,95) integrin is mainly expressed on myeloid cells, including a subset of dendritic cells, with highest level of expression found on tissue macrophages. It is however also expressed on granulocytes, NK cells, subsets of activated T and B lymphocytes, lymphoid cell lines including hairy cell leukemias.[72] αXβ2 has several known ligands, including the complement fragment iC3b, ICAM-1, fibrinogen and bacterial lipopolysaccharide.[72] Other recently discovered, less well-defined, ligands for αXβ2 include collagen[73] and the platelet-expressed JAM-3 protein.[56]

Based on in vitro studies, αXβ2 is implicated in a number of different inflammatory processes.[8,72] The αXβ2 integrin is thought to mediate the adhesion of monocyte/granulocytes to the endothelium and their subsequent extravasation, the chemotaxis of myeloid cells, the binding of iC3b-opsonized particles, and cytotoxic T lymphocyte-mediated killing. Recently, αXβ2 integrin, along with αMβ2, has also been implicated in antibody-mediated complement-independent phagocytosis of the yeast *C. neoformans*.[70] The in vivo functional role of this integrin remains undefined. Mice deficient exclusively in the αX gene have not yet been derived, nor have in vivo studies using blocking anti-αX mAbs been performed.

αDβ2

The newest and last member of the β2 integrin family is αDβ2. The αD (CD11d) chain was cloned in 1995 and found to be structurally similar to the other β2-associated α chains, being most closely related to αM and αX.[74] Despite this similarity, the function and expression pattern of αDβ2 appears to be unique from the other β2 integrins. For instance, αDβ2 binds preferentially to ICAM-3 (with no binding to ICAM-1)[74] and has subsequently been shown to also function as an alternative ligand for VCAM-1.[75,76] The pattern of expression of αDβ2 also differs from the other β2 integrins, being primarily expressed on monocytes, tissue macrophages, NK cells, a subpopulation of T and B cells, basophils, eosinophils and neutrophils.[74,75,77,78] Like other β2 integrins though, expression of αDβ2 can be further increased in vitro by a variety of stimuli[75] and in vivo during inflammatory disease.[79,80] In general, expression of αDβ2 is strongest on tissue-compartmentalized cells, such as macrophage foam cells within atherosclerotic plaques,[74] alveolar macrophages within inflamed lungs[80] and synovial macrophages and macrophage-like lining cells from both normal and arthritic subjects.[79]

The role of αDβ2 in inflammatory disease remains largely unstudied. The preferential expression of this integrin among tissue-resident cells, combined with the strong expression of one of its counter-receptors, ICAM-3, on infiltrating leukocytes suggests a potential role for this adhesion pathway in cellular interactions within an inflammatory site. Likewise, through its interaction with VCAM-1, αDβ2 may contribute to the homing and retention of leukocytes in certain tissues. While αD deficient mice have not yet been described, the in vivo role of αD has been examined in a couple of inflammatory models using function-blocking anti-αD antibodies. In a model of immune complex-induced lung injury, blockade of αD was found to be protective and appeared to result in decreased activation of lung macrophages.[80] In a spinal cord injury model, treatment with anti-αD mAb caused a significant reduction in the number of macrophages and neutrophils recruited to the injury site.[81] Data from both of these models are consistent with the hypotheses that αDβ2 plays an important role in recruitment and activation of cells in inflammatory sites.

α1β1/α2β1

Adhesive interactions play a key role in inflammation by promoting leukocyte attachment and extravasation from the vasculature into the peripheral tissues.[82] As detailed above, members of the β2 integrin family play a key role in this process. A further critical step in the inflammatory cascade, and one that until recently has not been extensively explored, occurs within the peripheral tissues where infiltrating, as well as resident cells, need to migrate towards the site of infection, recognize foreign antigen, and undergo cellular activation in order to perform their effector functions.

The interstitial environment and basement membranes through which cells must migrate, and in which they are activated, is an environment rich in extracellular matrix (ECM) molecules. The most abundant of these are collagens, which in skin was found to account for ª75% of its dry weight.[83] Two major cell surface receptors for collagens are the α1β1 and α2β1 integrins, with α1β1 showing a preference for type IV collagen and α2β1 showing a preference for type I collagen.[84,85] α1β1 and α2β1 are also able to interact weakly with laminins.[86]

The distribution of the α1β1 and α2β1 integrins has been studied in detail in various species[86,87] (see also chapters by Gardner and, Zutter and Santoro in this book). The expression of α1β1 in adult is largely confined to mesenchymal cells, notably smooth muscle cells, fibroblasts, stellate cells, hepatocytes, and microvascular endothelium.[88-91] α2β1 is predominantly epithelial in distribution, though it is also well expressed on platelets, endothelial cells and fibroblasts.[92]

Analysis of α1β1 and α2β1 expression on immune cells initially revealed little detectable α1β1 or α2β1 on resting peripheral blood lymphocytes.[86] Consistent with the initial description of α1β1 and α2β1 as Very Late Antigens (VLA), long-term activated lymphoid cells can express these receptors: IL-2-activated T cell clones express both α1β1 and α2β1 after several weeks of in vitro culture,[86] and infiltrating T cells express α1β1 in a variety of chronic inflammatory settings, including in the rheumatoid synovium of arthritis patients,[86,93,94] in the lungs of sarcoidosis patients,[86] and in atherosclerotic plaques.[95] Besides activated T lymphocytes, α1β1 and α2β1 can also be expressed by other cells of the immune system. Monocytes and platelets both express moderate amounts of α2β1,[86] while α1β1 is expressed on monocytes only upon activation.[96] Both α1β1 and α2β1 can also be expressed by activated NK cells;[86] α2β1 is also well expressed on resting NK cells, and is in fact identical to DX5, a well-known NK cell marker[97] (and A.D.F., unpublished results). Neutrophils do not express α1β1 (A.D.F. unpublished results), but under certain activation conditions can express α2β1.[98]

Due to the restricted expression of α1β1 and α2β1 on immune cells, most of the information on the role of α1β1 and α2β1 relates to the function of these integrins on nonimmune cells[87] (see also chapters in this book). The α1β1 integrin, but not α2β1, is linked via the Shc adaptor protein to the MAP kinase pathway, and is responsible for regulating cell survival and cell cycle progression.[99] Both α1β1 and α2β1 in vitro play a role in cell migration,[100,101] and in reorganization and contraction of collagen matrices, important in wound healing.[90,102,103]

Recently, some data has begun to be generated concerning the function of α1β1 and α2β1 on immune cells. While collagen-binding integrins have been hypothesized based on in vitro data to play an important role in migration of activated leukocytes into peripheral tissues, this has not been directly demonstrated.[104] Using viral models, we have recently investigated the expression of α1β1 and α2β1 on activated and memory T cells and assessed their functional importance (A. de Fougerolles, D. Topham, and J. Christensen, unpublished results). Using either an influenza or lymphocytic choriomeningitis virus system, and the relevant MHC tetramers to define antigen-specific T cells, we have demonstrated for the first time that contrary to being "very late antigens", expression of α1β1 and α2β1 can be rapidly induced on acutely activated T cells, that expression of α1β1 remains elevated on memory T cells, and that expression of α1β1 parallels that of viral-specific effector CD8+ T cells (defined by tetramer and IFN-γ staining). MAb-mediated blockade of these integrins on activated effector and memory T cells inhibited antigen-specific delayed-type hypersensitivity (DTH) responses; similar decreased responses were also seen in α1 knock-out mice. These results demonstrate that expression of α1β1 and α2β1 integrins on activated T cells is directly functionally important for generation of inflammatory responses within tissues. The inhibitory effect of α1β1 blockade on the DTH response could be bypassed by direct injection of antigen-specific T cells to inflammatory sites, demonstrating for the first time in vivo that collagen-binding integrins are involved in leukocyte migration into tissues. Lastly, using the flu model, we have also shown that the localization and retention of peripheral tissue-resident activated and memory T cells (the vast majority of which are α1β1+) is regulated in part by this integrin, as mAb blockade or genetic deficiency in α1β1 resulted in reduced numbers of activated/memory T cells in peripheral tissues.

While engagement of collagen-binding integrins is important in allowing migration of leukocytes to inflammatory sites, it is also capable of promoting cellular activation.[104] For instance, integrin-mediated binding to collagen provides a costimulatory signal for in vitro T cell activation, resulting in increased proliferation and secretion of pro-inflammatory cytokines such as TNF-α and IFN-γ.[104,105] While attachment of effector T cells to collagen alone did not result in T cell proliferation, TCR-mediated proliferation and cytokine secretion was synergistically enhanced by α1β1 and α2β1-dependent attachment to collagen. A recent report has extended these results to activated human T cells, showing that coimmobilization of TCR with type I collagen resulted in synergistic T cell activation and proliferation that was dependent on α1β1 and α2β1 integrins.[106] Besides a proliferative role, engagement of the α2β1 integrin, but not α1β1, has also been found to reduce activation-induced cell death in T cells by inhibiting Fas ligand expression.[107] This in vitro-generated hypothesis that engagement of collagen-binding integrins can also profoundly affect leukocyte activation and proliferation, in addition to affecting migration, is in fact supported by data from several in vivo animal models (see details below).

Targeted null mutations for α1[100] and α2[108,109] have been generated (see also chapters on α1β1 and α2β1 in this book for more detailed discussion concerning these mice). While the α1-deficient mice showed no gross developmental effects, fibroblasts from these mice showed a specific deficiency in attachment to collagen type IV, as well as increased collagen expression due to loss of feedback regulation.[100,110] Like the α1-deficient mice, the α2-deficient mice surprisingly showed no obvious anatomical or histological differences under normal conditions.[108,109] Given the role that $\alpha2\beta1$ plays in hemostasis by mediating platelet adhesion and activation on subendothelial collagen, it was not surprising to find that $\alpha2$-deficient platelets had reduced capacity to adhere and aggregate in response to collagen.[108,109] This defect in platelet adhesion and aggregation does not however translate into any change in bleeding time in the $\alpha2$-deficient mice.[108,109] Despite in vitro evidence to the contrary, wound healing in $\alpha2$-deficient mice is normal.[108] The $\alpha2$-deficient mice do exhibit defects in mammary gland branching morphogenesis, presumably related to its expression on epithelial cells.[108]

While both $\alpha1$ and $\alpha2$-deficient mice are relatively normal under resting conditions, the situation is very different under challenged situations. In these instances, the $\alpha1$-deficient mice have a striking anti-inflammatory and anti-fibrotic phenotype; this will be discussed in greater detail below. The $\alpha2$-deficient mice have yet to be studied under challenged situations, though based on data using blocking anti-$\alpha2$ mAbs,[67] they are expected to also show reduced inflammatory responses.

The $\alpha1\beta1$ and $\alpha2\beta1$ integrins, by promoting attachment to collagen, may allow activated T cells and monocytes to be retained within tissues and be activated in a permissive environment. The importance of collagen-integrin interactions in modulating in vivo inflammatory processes has recently been described. Treatment of wild-type mice with mAb to either $\alpha1\beta1$ or $\alpha2\beta1$ was found to significantly inhibit effector phase inflammatory responses in models of both contact and delayed-type hypersensitivity.[67] Consistent with expression and function of these integrins on activated effector cells, blockade of neither $\alpha1\beta1$ nor $\alpha2\beta1$ had any effect on T cell priming or sensitization. Anti-$\alpha1$ mAb treatment also suppressed development of arthritis in two different animal models, having effects on both activated T cells and monocytes.[67,111] Likewise, $\alpha1$-deficient mice also showed decreased inflammatory responses in models of contact hypersensitivity and arthritis.[67]

Recently, $\alpha1\beta1$ has been shown to play an important role in two different mouse models of inflammatory bowel disease.[112,113] $\alpha1\beta1$ can be expressed on activated T cells and monocytes during disease. $\alpha1$-deficient mice were found to be resistant to development of disease, and blockade of $\alpha1\beta1$ with mAb was effective in both preventative and therapeutic modes. In vivo blockade of $\alpha1\beta1$ not only improved macroscopic and histologic scores, but also decreased inflammatory cytokine production, and profoundly affected the ability of lamina propria mononuclear cells to proliferate and produce inflammatory cytokines. In both models, development and $\alpha1$-mediated inhibition of disease could occur independently of lymphocytes, thus pointing to the activated monocyte as a key $\alpha1$-expressing cell type involved in the development of inflammatory bowel disease. In both models, monocyte accumulation and activation within the intestinal lamina propria were significantly reduced. The importance of monocytes and ECM in the inflammatory cascade is underscored by recently published gene expression data demonstrating that integrin-mediated attachment of monocytes to ECM resulted in increased expression of numerous inflammatory and immune response genes important in promoting cell recruitment and activation.[114,115] Lastly, further supporting a role for $\alpha1\beta1$ in leukocyte trafficking to the intestinal lamina propria, treatment with anti-$\alpha1$ mAb in an intestinal graft-versus-host disease model not only partially prevented mucosal pathology but also showed reduced donor T lymphocytic infiltration into the intestinal mucosa.[116]

In summary, these findings demonstrate that interaction of leukocyte-associated collagen-binding integrins, especially $\alpha1\beta1$, with the surrounding matrix plays a pivotal role in mediating local inflammatory responses in part via promotion of T cell and monocyte movement, retention, and activation within the inflamed tissue. It also supports an emerging paradigm shift with regards to the role of leukocyte adhesion molecules in inflammation. It extends the role of integrins in inflammation beyond leukocyte attachment and extravasation at the vascular endothelial interface, and reveals the ECM environment of peripheral tissues as a new point of intervention for adhesion-based therapies. Therapeutic strategies designed to disrupt leukocyte/matrix interactions may prove beneficial in the treatment of inflammatory disorders.

Collagen-binding integrins have also been shown to play an important role in other disease pathologies, notably fibrosis and angiogenesis. Blockade or loss of α1β1 has been shown to be protective in three different models of kidney fibrosis. When α1-deficient mice were crossed onto the fibrosis-prone COL4A3-deficient mice, a mouse model for human Alport's kidney disease, the double knock-out mice lived twice as long (8 weeks to 16 weeks) and showed delayed loss of renal function.[117] More detailed analysis of the mice revealed that loss of α1β1 resulted in decreased monocyte and myofibroblast accumulation within the inflamed kidney.[118] Similarly, recent studies using blocking anti-α1 mAbs demonstrated remarkable efficacy in two different models of glomerulonephritis. In an anti-glomerular basement membrane model of nephritis, α1β1 was found to be upregulated in glomeruli and around tubules during disease and mAb blockade of α1β1 on established disease reduced tubulointerstitial scarring, preserved renal function, and prolonged survival.[119] In a rat anti-thy1 model of glomerulonephritis, VLA1 was expressed on mesangial cells during disease, and anti-α1 mAb given after disease induction, reduced mesangial cell proliferation and matrix accumulation.[120] The key cell types involved in mediating fibrosis in these models are presumably the activated myofibroblast/mesangial cell and the monocyte. Antagonism of α1β1 could be important at several levels: at the level of migration of cells into the tissues, their proliferation and activation in situ, and lastly, the ability of α1β1 engagement to directly modulate collagen synthesis.

While α1β1 and α2β1 expressed on activated T cells and monocytes are thought to play an important role in inflammatory and immune-based diseases, and in fibrotic settings it is the expression of α1β1 integrin on activated myofibroblasts and monocytes that is thought to be critical, both these integrins are also very well expressed on capillary endothelial cells. Given the importance of these integrins for cell migration and activation, it is not therefore surprising that they play an important role in angiogenesis and tumorigenesis. Both these integrins provide critical support for vascular endothelial growth factor signaling and endothelial cell migration in vitro, and treatment of mice with mAbs to α1β1 and α2β1 not only inhibited vascular endothelial angiogenesis but also partially retarded tumor cell growth.[121,122] Besides, these direct effects on endothelial cells, blockade of the α1β1 pathway can also decrease tumor vascularization indirectly via elevation of matrix metalloprotease and angiostatin levels.[123]

Overall, antagonism of the α1β1 pathway has dramatic effects on attenuating inflammatory, fibrotic and oncologic disease. This is likely due to the fact that in each instance multiple key cell types and mechanisms are being inhibited. Clinically, antagonists of α1β1, both mAb- and small molecule-based are being actively pursued. The case for α2β1 is slightly less compelling, though this is largely a reflection of the fact that less in vivo data exists for this integrin.

α10β1/α11β1

Two other I-domain containing integrin receptors for matrix proteins have recently been described: α10β1 and α11β1.[87,124,125] While both these integrins mediate adherence to collagen, little is known about their role in the immune system. The α10β1 integrin was originally identified on chondrocytes, and northern blot and immunohistochemical analysis revealed that its expression is very restricted, being found only in cartilage, lung, heart, aorta, and spinal cord. The α11β1 integrin is highly expressed in mesenchymal cells in vitro and in vivo on fibroblasts in areas of highly organized interstitial collagen networks. With regards to the expression on immune cell types, no detectable expression of either integrin was found in any lymphoid organs, including spleen, lymph node or thymus. At present, no immunological function for these integrins has been described.

αEβ7

The αEβ7 (CD103/unclustered; HML-1 antigen) integrin is mainly expressed on intraepithelial lymphocytes, on 1-2% of peripheral blood lymphocytes, as well as on mucosal mast cells and mucosal dendritic antigen presenting cells[126,127] (see also chapter on αEβ7 by Kilshaw and Higgins in this book). This restricted pattern of expression has recently been extended to include subsets of CD4+ T regulatory cells.[128] Expression of αEβ7 can be upregulated by lymphocyte mitogens, and is transcriptionally regulated by TGF-β. Expression of the αE mRNA, which pairs only with the β7 subunit, is found in the thymus and spleen as well as tissues such as skin, lung, small intestine, and colon, which contain large populations of intraepithelial lymphocytes.

The ligand for αEβ7 is E-cadherin,[129-131] though recent evidence suggests that an E-cadherin-independent ligand may exist on intestinal endothelial cells.[132] Given the high level of expression of αEβ7 on intraepithelial lymphocytes and E-cadherin on epithelial cells, this interaction is postulated to play an important role in the homing and retention of αEβ7-expressing lymphocytes in the intestinal epithelium.[127,133]

Targeted disruption of the αE gene leads to a reduction of T cells in the lamina propria and gut epithelium but not in the Peyer's patches.[134] Consistent with this phenotype are the β7-deficient mice, which also show reduced numbers of lamina propria and intraepithelial T lymphocytes.[135] The additional phenotype of defective migration of T cells to Peyer's patches that is seen in β7-deficient mice is consistent with the fact that β7 integrin can also associate with the α4 integrin subunit. Migration of lymphocytes to the Peyer's patches is mediated by α4β7 binding to the endothelial MadCAM receptors on the high endothelial venules in the gut. In contrast to these results, studies performed with OVA-specific CD8 T cells failed to demonstrate any role for αEβ7 in retention of CD8 T cells in the intestinal epithelium.[136] Interpretation of this data is complicated by many factors, such as whether antigen was expressed within the intraepithelial lymphocyte compartment. Thus while much remains to be determined, the overall function of αEβ7 integrin in the gut is clearly more towards favoring retention or micro-localization of T cells in the vicinity of epithelia rather than in mucosal-specific homing.

Mucosal sites are not the only tissue, where αEβ7 integrin is thought to play an important role. In addition to epithelial cells, E-cadherin can also be expressed by other cell types including professional antigen-presenting cells such as langerhans cells in the skin. CD8 T cells preferentially reside in the skin epidermis (versus the dermis) and it is postulated that expression of αEβ7 and interaction with keratinocytes and/or langerhans cells might contribute to this skewed distribution. In support of this, epidermal, but not dermal, T cells, express significant amounts of αEβ7 in psoriatic lesions;[137] additionally, the numbers of epidermal T cells are diminished in the skin of αE-deficient mice.[138] Lastly, αE-deficient mice have been reported to develop exacerbated inflammatory cutaneous skin lesions, though this is dependent on genetic background and animal housing conditions.[139] Clearly, αEβ7 expression is involved in localization of CD8 T cells within the epidermis and may also be functionally important by promoting T cell interaction with other cells types in the epidermis.

Evidence is now accumulating that αEβ7 integrin interaction with E-cadherin is also implicated in localization of CD8 T cells in a variety of other tissues, besides gut and skin. This is particularly true in human inflammatory settings where infiltrating lymphocytes are in close proximity to epithelial cells. For instance, αEβ7-expressing CD8 T cells analogous to intraepithelial lymphocytes are found intercalating between ductal epithelial cells in chronic pancreatitis[140] and similarly around acinar epithelial cells in Sjogren's syndrome.[141] The interaction of αEβ7-expressing CD8 T cells with epithelial cells not only plays an adhesive role, but in some cases may also participate in inducing epithelial cell apoptosis, and thereby contribute to disease pathology.[141]

Table 1. Summary of in vivo animal models where blockade of α subunit I domain-containing integrins was effective in attenuating disease*

Integrin	Disease/Model	Agent	Reference
αLβ2	Inflammatory cell trafficking	mAb/KO	Reviewed in (9)
	Arthritis	mAb	(19)
	Contact Hypersensitivity	mAb	(20-23)
	Human psoriasis	mAb	(24, 145)
	Colitis	mAb	(26)
	EAE**	mAb	(28, 29)
	Kidney Fibrosis	mAb	(40, 41)
	Lupus nephritis	mAb	(42, 43)
	Ischemia/reperfusion	mAb	Reviewed in (15)
	Transplantation**	mAb	(49-51)
αMβ2	Inflammatory cell trafficking	mAb/KO	(59, 62, 63)
	Ischemia/reperfusion	mAb/KO	(59), reviewed in (15)
	Diabetes	mAb	(64)
	Inflammation-prone moth-eaten mouse	mAb	(65)
	Arthritis	mAb	(66, 67)
	EAE	mAb	(68)
	Colitis	mAb	(27)
	Kidney fibrosis	mAb	(69)
	Lung injury models	mAb	Reviewed in (15)
αDβ2	Immune complex-induced lung injury	mAb	(80)
	Spinal cord injury	mAb	(81)
α1β1	Delayed-type hypersensitivity	mAb/KO	(67, unpublished results)
	Contact hypersensitivity	mAb/KO	(67)
	Arthritis	mAb/KO	(67, 111)
	Colitis	mAb/KO	(112, 113)
	Intestinal GVHD	mAb	(116)
	Kidney Fibrosis	mAb/KO	(117-120)
	Angiogenesis/Tumorigenesis	mAb/KO	(121-123)
α2β1	Contact & Delayed-type hypersensitivity	mAb	(67)
αEβ7	Pancreatic islet allograft	KO	(142)

*Only disease models where knock-out (KO) mice or treatment with integrin-specific antagonists was successful at attenuating disease is shown. Data using KO or antagonists of multiple integrins, or where no effect or augmented disease was found, are not shown. Complete details on all data are found in text. **Contradictory results exist. See text for details.

A concrete link demonstrating a role for αEβ7/E-cadherin interactions in mediating epithelial cell lysis was recently demonstrated in a pancreatic islet allograft model.[142] In this model, wild-type hosts uniformly rejected islet allografts (with αEβ7-expressing effector CD8 T cells present concomitantly at the graft site), while the majority of islet allografts transplanted into αE-deficient hosts survived indefinitely. Using adoptive transfer of either wild-type or αE-deficient CD8 T cells, the authors further demonstrated that αEβ7-expressed on the CD8 T cells is responsible for mediating allograft rejection. Nicely dovetailing with the mouse studies is the finding in humans that αEβ7+ cytotoxic T lymphocytes accumulate within the graft epithelium during clinical renal allograft rejection.[143] Additional work by Feng et al showed that in contrast to the acceptance of islet allografts by αE-deficient mice, skin allografts were rejected normally in these mice.[142] Given that skin allograft rejection occurred due to T cell infiltration into the dermis and vascular bed, but not into the epithelium, this result is not surprising, but does serve to illustrate the very localized importance of αEβ7 interaction with E-cadherin-expressing epithelium.

Based on all of these results, development of antagonists to αEβ7 may prove useful in a variety of certain specific inflammatory and auto-immune settings where epithelial involvement is critical. Results using antagonists of αEβ7 in these settings will be eagerly anticipated.

Concluding Remarks and Clinical Perspective

Integrins with I-domain-containing α subunits are capable of acting at different points along the inflammatory cascade. They are important in initiation of immune responses (αLβ2), in promoting leukocyte-endothelial cell attachment and transmigration (αLβ2, αMβ2), and in migration and retention of cells within tissues (α1β1, α2β1, αEβ7). The strong mechanistic rationale coupled with solid in vivo animal disease model data (Table. 1) has made integrin antagonists attractive candidates for clinical development. It is therefore not surprising that almost all of these α chain I-domain-containing integrins are the subject of intensive drug development research. Many approaches are being used to antagonize integrin interaction with their ligands, including both mAb- and small molecule-based antagonists.[144] Of the I-domain containing integrins, αLβ2 is the furthest advanced, with recent positive phase III data reported for treatment of psoriasis with a humanized anti-αL mAb.[145] In addition, several companies are also pursuing orally active small molecule αL antagonists,[146,147] with one already in phase IIa trials for psoriasis.[148] Lastly, the statin family of compounds, commonly used for the treatment of hypercholesterolemia, show promise as leads on which to design selective inhibitors of αLβ2.[149,150]

Antagonists of I-domain integrins have also met with failure, the most notable being anti-β2 mAbs. Despite evidence supporting a role for β2 antagonists in the treatment of multiple sclerosis and ischemia-reperfusion disorders, several phase II and III trials with humanized anti-β2 mAbs failed to show efficacy.[144] The reason for failure of the β2 trials is unclear, though the experimental animal data for involvement of β2 integrin in these particular diseases is mixed. Nevertheless, these results serve to highlight the unpredictability inherent in development of drugs, and the imperfect nature of animal models in mimicking human disease. The challenge for the future will be to fully understand the in vivo mechanism of action of integrins and to use this information to identify the most relevant human disease settings in which to test potential drugs.

References

1. Arnaout MA. Integrin structure: new twists and turns in dynamic cell adhesion. Immunological Rev 2002; 186:125-140.
2. Takagi J, Springer TA. Integrin activation and structural rearrangement. Immunological Rev 2002; 186:141-163.
3. Yusuf-Makagiansar H, Anderson ME, Yakoleva TV et al. Inhibition of LFA-1/ICAM-1 and VLA-4/VCAM-1 as a therapeutic approach to inflammation and autoimmune diseases. Med Res Rev 2002; 22:146-167.
4. Rose DM, Han J, Ginsberg MH. α4 integrins and the immune response. Immunological Rev 2002; 186:118-124.
5. Gahmberg CG, Tolvanen M, Kotovuori P. Leukocyte adhesion. Structure and function of human leukocyte β2-integrins and their cellular ligands. Eur J Biochem 1997; 245:215-232.
6. Ostermann G, Weber KS, Zernecke A et al. JAM-1 is a ligand of the β (2) integrin LFA-1 involved in transendothelial migration of leukocytes. Nature Immunol 2002; 3:151-158.
7. Sims TN, Dustin ML. The immunological synapse: integrins take the stage. Immunological Rev 2002; 186:100-117.

8. Mazzone A, Ricevuti G. Leukocyte CD11/CD18 integrins: Biological and clinical relevance. Haematologica 1995; 80:161-175.
9. Bouvard D, Brakebusch C, Gustafsson E et al. Functional consequences of integrin gene mutations in mice. Circ Res 2001; 89:211-223.
10. Scharffetter-Kochanek K, Lu H, Norman K et al. Spontaneous skin ulceration and defective T cell function in CD18 null mice. J Exp Med 1998; 188:119-131.
11. Schmits R, Kundig TM, Baker DM et al. LFA-1-deficient mice show normal CTL responses to virus but fail to reject immunogenic tumor. J Exp Med 1996; 183:1415-1426.
12. Shier P, Otulakowski G, Ngo K et al. Impaired immune responses toward alloantigens and tumor cells but normal thymic selection in mice deficient in the beta2 integrin leukocyte function-associated antigen-1. J Immunol 1996; 157:5375-5386.
13. Shier P, Ngo K, Fung-Leung WP. Defective CD8+ T cell activation and cytolytic function in the absence of LFA-1 cannot be restored by increased TCR signaling. J Immunol 1999; 163:4826-4832.
14. Berlin-Rufenach C, Otto F, Mathies M et al. Lymphocyte migration in lymphocyte function-associated antigen (LFA)-1-deficient mice. J Exp Med 1999; 189:1467-1478.
15. Cornejo CJ, Winn RK, Harlan JM. Anti-adhesion therapy. Adv Pharmacol 1997; 39:99-142.
16. Bullard DC, Hurley LA, Lorenzo I et al. Reduced susceptibility to collagen-induced arthritis in mice deficient in intercellular adhesion molecule-1. J Immunol 1996; 157:3153-3158.
17. Wolyniec WW, De Sanctis GT, Nabozny G et al. Reduction of antigen-induced airway hyperreactivity and eosinophilia in ICAM-1-deficient mice. Am J Respir Cell Mol Biol 1998; 18:777-785.
18. Birner U, Issekutz TB, Walter U et al. The role of alpha(4) and LFA-1 integrins in selectin-independent monocyte and neutrophil migration to joints of rats with adjuvant arthritis. Int Immunol 2000; 12:141-150.
19. Kakimoto K, Nakamura T, Ishii K et al. The effect of anti-adhesion molecule antibody on the development of collagen-induced arthritis. Cell Immunol 1992; 142:326-337.
20. Scheynius A, Camp RL, Pure E. Reduced Contact Sensitivity Reactions in Mice treated with Monoclonal Antibodies to Leukocyte Function-Associated Molecule-1 and Intercellular Adhesion Molecule-1. J Immunol 1993; 150:655-663.
21. Ma J, Wang JH, Guo YJ E et al. In vivo treatment with anti-ICAM-1 and anti-LFA-1 antibodies inhibits contact sensitization-induced migration of epidermal Langerhans cells to regional lymph nodes. Cell Immunol 1994; 158:389-399.
22. Scheynius A, Camp RL, Pure E. Unresponsiveness to 2,4-dinitro-1-fluoro-benzene after treatment with monoclonal antibodies to leukocyte function-associated molecule-1 and intercellular adhesion molecule-1 during sensitization. J Immunol 1996; 156:1804-1809.
23. Murayama M, Yasuda H, Nishimura Y et al. Suppression of mouse contact hypersensitivity after treatment with antibodies to leukocyte function-associated antigen-1 and intracellular adhesion molecules-1. Arch Dermatol Res 1997; 289:98-103.
24. Zeigler M, Chi Y, Tumas DB et al. Anti-CD11a ameliorates disease in the human psoriatic skin-SCID mouse transplant model: comparison of antibody to CD11a with Cyclosporin A and clobetasol propionate. Lab Invest 2001; 81:1253-1261.
25. van Assche G, Rutgeerts P. Antiadhesion molecule therapy in inflammatory bowel disease. Inflamm Bowel Dis 2002; 8:291-300.
26. Wallace JL, Higa A, McKnight GW et al. Prevention and reversal of experimental colitis by a monoclonal antibody which inhibits leukocyte adherence. Inflammation 1992; 16:343-354.
27. Palmen MJ, Dijkstra CD, van der Ende MBAS et al. Anti-CD11b/CD18 antibodies reduce inflammation in acute colitis in rats. Clin Exp Immunol 1995; 101:351-356.
28. Kobayashi Y, Kawai K, Honda H et al. Antibodies against leukocyte function-associated antigen-1 and against intercellular adhesion molecule-1 together suppress the progression of experimental allergic encephalomyelitis. Cell Immunol 1995; 164:295-305.
29. Kawai K, Kobayashi Y, Shiratori M et al. Intrathecal administration of antibodies against LFA-1 and against ICAM-1 suppresses experimental allergic encephalomyelitis in rats. Cell Immunol 1996; 171:262-268.
30. Cannella B, Cross AH, Raine CS. Anti-adhesion molecule therapy in experimental autoimmune encephalomyelitis. J Neuroimmunol 1993; 46:43-55.

31. Welsh CT, Rose JW, Hill KE et al. Augmentation of adoptively transferred experimental allergic encephalomyelitis by administration of a monoclonal antibody specific for LFA-1 alpha. J Neuroimmunol 1993; 43:161-167.
32. Rose JW, Welsh CT, Hill KE et al. Contrasting effects of anti-adhesion molecule therapy in experimental allergic encephalomyelitis and Theiler's murine encephalomyelitis. J Neuroimmunol 1999; 97:110-118.
33. Samoilova EB, Horton JL, Chen Y. Experimental autoimmune encephalomyelitis in intercellular adhesion molecule-1-deficient mice. Cell Immunol 1998; 190:83-89.
34. Noonan TC, Gundel RH, Desai SN et al. The effects of anti-CD18 antibody (R15.7) in antigen-induced airway hyperresponsiveness (DH) and cell influx in guinea pigs. Agents Actions 1991; 34:211-213.
35. Milne AA, Piper PJ. The effects of two anti-CD18 antibodies on antigen-induced airway hyperresponsiveness and leukocyte accumulation in the guinea pig. Am J Respir Cell Mol Biol 1994; 11:337-343.
36. Wegner CD, Gundel RH, Reilly P et al. Intercellular adhesion molecule-1 (ICAM-1) in the pathogenesis of asthma. Science 1990; 247:456-459.
37. Gundel RH, Wegner CD, Torcellini CA et al. Endothelial leukocyte adhesion molecule-1 mediates antigen-induced acute airway inflammation and late-phase airway obstruction in monkeys. J Clin Invest 1991; 88:1407-1411.
38. Nakajima H, Sano H, Nishimura T et al. Role of vascular cell adhesion molecule 1/very late activation antigen 4 and intercellular adhesion molecule 1/lymphocyte function-associated antigen 1 interactions in antigen-induced eosinophil and T cell recruitment into the tissue. J Exp Med 1994; 179:1145-1154.
39. Lobb RR, Abraham WM, Burkly LC et al. Pathophysiologic role of alpha 4 integrins in the lung. Ann N Y Acad Sci 1996; 796:113-123.
40. Kawasaki K, Yaoita E, Yamamoto T et al. Antibodies against intercellular adhesion molecule-1 and lymphocyte function-associated antigen-1 prevent glomerular injury in rat experimental crescentic glomerulonephritis. J Immunol 1993; 150:1074-1083.
41. Nishikawa K, Guo YJ, Miyasaka M et al. Antibodies to intercellular adhesion molecule 1/lymphocyte function-associated antigen 1 prevent crescent formation in rat autoimmune glomerulonephritis. J Exp Med 1993; 177:667-677.
42. Connolly MK, Kitchens EA, Chan B et al. Treatment of murine lupus with monoclonal antibodies to lymphocyte function-associated antigen-1: dose-dependent inhibition of autoantibody production and blockade of the immune response to therapy. Clin Immunol Immunopathol 1994; 72:198-203.
43. Kootstra CJ, Van Der Giezen DM, Van Krieken JH et al. Effective treatment of experimental lupus nephritis by combined administration of anti-CD11a and anti-CD54 antibodies. Clin Exp Immunol 1997; 108:324-332.
44. Isobe M, Yagita H, Okumura K et al. Specific acceptance of cardiac allograft after treatment with antibodies to ICAM-1 and LFA-1. Science 1992; 255:1125-1127.
45. Komori A, Nagata M, Ochiai T et al. Role of ICAM-1 and LFA-1 in cardiac allograft rejection of the rat. Transplant Proc 1993; 25:831-832.
46. Kameoka H, Ishibashi M, Tamatani T et al. Comparative immunosuppressive effect of anti-CD18 and anti-CD11a monoclonal antibodies on rat heart allotransplantation. Transplant Proc 1993; 25:833-836.
47. Kato Y, Yamataka A, Yagita H et al. Specific acceptance of fetal bowel allograft in mice after combined treatment with anti-intercellular adhesion molecule-1 and leukocyte function-associated antigen-1 antibodies. Ann Surg 1996; 223:94-100.
48. Sarnacki S, Auber F, Cretolle C et al. Blockade of the integrin alphaLbeta2 but not of integrins alpha4 and/or beta7 significantly prolongs intestinal allograft survival in mice. Gut 2000; 47:97-104.
49. Nakakura EK, McCabe SM, Zheng B et al. Potent and effective prolongation by anti-LFA-1 monoclonal antibody monotherapy of nonprimarily vascularized heart allograft survival in mice without T cell depletion. Transplantation 1993; 55:412-417.

50. Nakakura EK, Shorthouse RA, Zheng B et al. Long-term survival of solid organ allografts by brief anti-lymphocyte function-associated antigen-1 monoclonal antibody monotherapy. Transplantation 1996; 62:547-552.
51. Poston RS, Robbins RC, Chan B et al. Effects of humanized monoclonal antibody to rhesus CD11a in rhesus monkey cardiac allograft recipients. Transplantation 2000; 69:2005-2013.
52. Jendrisak M, Jendrisak G, Gamero J et al. Prolongation in murine cardiac allograft survival with monoclonal antibodies to LFA-1, ICAM-1, and CD4. Transplant Proc 1993; 25:825-827.
53. Ross GD. Regulation of the adhesion versus cytotoxic functions of the Mac-1/CR3/αMβ2-integrin glycoprotein. Critical Rev Immunol 2000; 20:197-222.
54. Rosenkranz AR, Coxon A, Maurer M et al. Impaired mast cell development and innate immunity in Mac-1 (CD11b/CD18, CR3)-deficient mice. J Immunol 1998; 161:6463-6467.
55. Christensen JE, Andreasen SO, Christensen JP et al. CD11b expression as a marker to distinguish between recently activated effector CD8+ T cells and memory cells. Int Immunol 2001; 13:593-600.
56. Santoso S, Sachs UJH, Kroll H et al. The junctional adhesion molecule 3 (JAM-3) on human platelets is a counter-receptor for the leukocyte integrin Mac-1. J Exp Med 2002; 196:679-691.
57. Petty HR, Worth RG, Todd RF 3rd. Interactions of integrins with their partner proteins in leukocyte membranes. Immunol Res 2002; 25:75-95.
58. Coxon A, Rieu P, Barkalow FJ et al. A novel role for the b2 integrin CD11b/CD18 in neutrophil apoptosis: A homeostatic mechanism in inflammation. Immunity. 1996; 5:653-666.
59. Soriano SG, Coxon A, Yanming FW et al. Mice deficient in Mac-1 (CD11b/CD18) are less susceptible to cerebral ischemia/reperfusion injury. Stroke 1999; 30:134-139.
60. Van Spriel AB, Leusen JHW, van Egmond M et al. Mac-1 (CD11b/CD18) is essential for Fc receptor-mediated neutrophil cytotoxicity and immunologic synapse formation. Blood 2001; 97:2478-2486.
61. Yan J, Vetvicka V, Xia Y et al. Critical role of Kupffer cell CR3 (CD11b/CD18) in the clearance of IgM-opsonized erythrocytes or soluble beta-glucan. Immunopharmacology 2000; 46:39-54.
62. Rosen H, Gordon S. Monoclonal antibody to the murine type 3 complement receptor inhibits adhesion of myelomonocytic cells in vitro and inflammatory cell recruitment in vivo. J Exp Med 1987; 166:1685-701.
63. Rosen H, Milon G, Gordon S. Antibody to the murine type 3 complement receptor inhibits T lymphocyte-dependent recruitment of myelomonocytic cells in vivo. J Exp Med 1989; 169:535-548.
64. Hutchings P, Rosen H, O'Reilly L et al. Transfer of diabetes in mice prevented by blockade of adhesion-promoting receptor on macrophages. Nature 1990; 348:639-642.
65. Koo GC, Rosen H, Sirotina A et al. Anti-CD11b antibody prevents immunopathologic changes in viable moth-eaten bone marrow chimeric mice. J Immunol 1993; 151:6733-6741.
66. Taylor PC, Chu CQ, Plater-Zyberk C et al. Transfer of type II collagen-induced arthritis from DBA/1 to severe combined immunodeficiency mice can be prevented by blockade of Mac-1. Immunology 1996; 88:315-321.
67. de Fougerolles AR, Sprague AG, Nickerson-Nutter CL et al. Regulation of inflammation by collagen-binding integrins alpha1beta1 and alpha2beta1 in models of hypersensitivity and arthritis. J Clin Invest 2000; 105:721-729.
68. Huitinga I, Damoiseaux JG, Dopp EA et al. Treatment with anti-CR3 antibodies ED7 and ED8 suppresses experimental allergic encephalomyelitis in Lewis rats. Eur J Immunol 1993; 23:709-715.
69. Wu X, Pippin J, Lefkowith JB. Attenuation of immune-mediated glomerulonephritis with an anti-CD11b monoclonal antibody. Am J Physiol 1993; 264:F715-F721.
70. Taborda CP, Casadevall A. CR3(CD11b/CD18) and CR4 (CD11c/CD18) are involved in complement-independent antibody-mediated phagocytosis of Cryptococcus neoformans. Immunity 2002; 16:791-802.
71. Rosen H, Gordon S, North RJ. Exacerbation of murine listeriosis by a monoclonal antibody specific for the type 3 complement receptor of myelomonocytic cells. Absence of monocytes at infective foci allows Listeria to multiply in nonphagocytic cells. J Exp Med 1989; 170:27-37.
72. Cabanas C, Sanchez-Madrid F. CD11c (leukocyte integrin CR4 alpha subunit). J Biol Regul Homeost Agents 1999; 13:134-136.
73. Garnotel R, Rittie L, Poitevin S et al. Human blood monocytes interact with type I collagen through αXβ2 (CD11c-CD18, gp150-95). J Immunol 2000; 164:5928-5934.

74. Van der Vieren M, Trong HL, Wood CL et al. A novel leukointegrin, αDβ2, binds preferentially to ICAM-3. Immunity 1995; 3:683-690.
75. Grayson MH, Van der Vieren M, Sterbinsky SA et al. αDβ2 integrin is expressed on human eosinophils and functions as an alternative ligand for vascular cell adhesion molecule 1 (VCAM-1). J Exp Med 1998; 188:2187-2191.
76. Van der Vieren M, Crowe DT, Hoekstra D et al. The leukocyte integrin αDβ2 binds VCAM-1: Evidence for a binding interface between I domain and VCAM-1. J Immunol 1999; 163:1984-1990.
77. Danilenko DM, Rossitto PV, Van der Vieren M et al. A novel canine leukointegrin, aDb2, is expressed by specific macrophage subpopulations in tissue and a minor CD8+ lymphocyte subpopulation in peripheral blood. J Immunol 1995; 155:35-44.
78. Grayson MH, Van der Vieren M, Gallatin WM et al. Expression of a novel β2 integrin (αDβ2) on human leukocytes and mast cells. J Allergy Clin Immunol 1997; 99:5386.
79. El-Gabalawy H, Canvin J, Ma GM et al. Synovial distribution of alphad/CD18, a novel leukointegrin. Comparison with other integrins and their ligands. Arthritis Rheum 1996; 39:1913-1921.
80. Shanley TP, Warner RL, Crouch LD et al. Requirements for αd in IgG immune complex-induced rat lung injury. J Immunol 1998; 160:1014-1020.
81. Mabon PJ, Weaver LC, Dekaban GA. Inhibition of monocyte/macrophage migration to a spinal cord injury site by an antibody to the integrin alphaD: a potential new anti-inflammatory treatment. Exp Neurol 2000; 166:52-64.
82. Springer TA. Traffic signals for lymphocyte recirculation and leukocyte emigration: the multistep paradigm. Cell 1992; 76:301-314.
83. Fitzpatrick TB, Eisen AZ, Wolff K et al. In: J.D. Jeffers, E. Scott, J. White, eds. The Structure and Development of Skin. New York: McGraw-Hill,1987.
84. Gullberg D, Gehlsen KR, Turner DC et al. Analysis of alpha1 beta 1, alpha 2 beta 1 and alpha 3 beta 1 integrins in cell-collagen interactions: identification of conformation dependent alpha 1 beta 1 binding sites in collagen type I. EMBO J 1992; 11:3865-3873.
85. Kern A, Eble J, Golbik R et al. Interaction of type IV collagen with the isolated integrins α1β1 and α2β1. Eur J Biochem 1993; 215:151-159.
86. Hemler ME. VLA proteins in the integrin family: Structures, functions, and their role on leukocytes. Ann Rev Immunol 1990; 8:365-400.
87. Gullberg DE, Lundgren-Akerlund E. Collagen-binding I domain integrins—what do they do? Progr Histochem Cytochem 2002; 37:3-54.
88. Belkin VM, Belkin AM, Koteliansky VE. Human smooth muscle VLA-1 integrin: purification, substrate specificity, localization in aorta, and expression during development. J Cell Biol 1990; 111:2159-2170.
89. Duband JL, Belkin AM, Syfrig J et al. Expression of α1 integrin, a laminin-collagen receptor, during myogenesis and neurogenesis in the avian embryo. Development 1992; 116:585-600.
90. Racine-Samson L, Rockey DC, Bissell DM. The role of α1β1 integrin in wound contraction: A quantitative analysis of liver myofibroblasts in vivo and in primary culture. J Biol Chem 1997; 272:30911-30917.
91. Kocken JM, de Heer E, Rinkes IB et al. Blocking of α1β1 integrin strongly improves survival of hepatocytes in allogeneic transplantation. Lab Invest 1997; 77:19-28.
92. Wu JE. Santoro SA. Complex patterns of expression suggest extensive roles for the alpha 2 beta 1 integrin in murine development. Dev Dynam 1994; 199:292-314.
93. Bank I, Roth D, Book M et al. Expression and functions of Very Late Antigen 1 in inflammatory joint diseases. J Clin Immunol 1991; 11:29-38.
94. Takahashi H, Soderstrom K, Nilsson E et al. Integrins and other adhesion molecules on lymphocytes from synovial fluid and peripheral blood rheumatoid arthritis patients. Eur J Immunol 1992; 22:2879-2885.
95. Stemme S, Holm J, Hansson GK. T lymphocytes in human atherosclerotic plaques are memory cells expressing CD45RO and the integrin VLA-1. Arteriosclerosis and Thrombosis 1992; 12:206-211.
96. Rubio MA, Sotillos M, Jochems G et al. Monocyte activation: rapid induction of α1β1 (VLA-1) integrin expression by lipopolysaccharide and inteferon-γ. Eur J Immunol 1995; 25:2701-2705.

97. Arase H, Saito T, Phillips JH et al. Cutting edge: the mouse NK cell-associated antigen recognized by DX5 monoclonal antibody is CD49b (alpha2 integrin, very late antigen-2) J Immunol 2001; 167:1141-1144.
98. Werr J, Johansson J, Eriksson EE et al. Integrin alpha(2)beta(1) (VLA-2) is a principal receptor used by neutrophils for locomotion in extravascular tissue. Blood 2000; 95:1804-1809.
99. Wary KK, Mainiero F, Isakoff SJ et al. The adaptor protein Shc couples a class of integrins to the control of cell cycle progression. Cell 1996; 87:733-743.
100. Gardner H, Kreidberg J, Koteliansky V et al. Deletion of integrin alpha 1 by homologous recombination permits normal murine development but gives rise to a specific deficit in cell adhesion. Dev Biol 1996; 175:301-313.
101. Keely PJ, Fong AM, Zutter MM et al. Alteration of collagen-dependent adhesion, motility, and morphogenesis by the expression of antisense alpha 2 integrin mRNA in mammary cells. J Cell Sci 1995; 108:595-607.
102. Schiro JA, Chan BMC, Roswit WT et al. Integrin alpha 2 beta 1 (VLA-2) mediates reorganization and contraction of collagen matrices by human cells. Cell 1991; 67:403-410.
103. Gotwals PJ, Chi-Rosso G, Lindner V et al. The $\alpha 1\beta 1$ integrin is expressed during neointima formation in rat arteries and mediates collagen matrix reorganization. J Clin Invest 1996; 97:2469-2477.
104. Dustin ML, de Fougerolles AR. Reprogramming T cells: the role of extracellular matrix in coordination of T cell activation and migration. Curr Opin Immunol 2001; 13:286-290.
105. Miyake S, Sakurai T, Okumura K et al. Identification of collagen and laminin receptor integrins on murine T lymphocytes. Eur J Immunol 1994; 24:2000-2005.
106. Rao WH, Hales JM, Camp RD. Potent costimulation of effector T lymphocytes by human collagen type I. J Immunol 2000; 165:4935-4940.
107. Aoudjit F, Vuori, K. Engagement of the alpha2beta1 integrin inhibits Fas ligand expression and activation-induced cell death in T cells in a focal adhesion kinase-dependent manner. Blood 2000; 95:2044-2051.
108. Chen J, Diacovo TG, Grenache DG et al. The alpha(2) integrin subunit-deficient mouse: a multifaceted phenotype including defects of branching morphogenesis and hemostasis. Am J Pathol 2002; 161:337-344.
109. Holtkotter O, Nieswandt B, Smyth N et al. Integrin alpha 2-deficient mice develop normally, are fertile, but display partially defective platelet interaction with collagen. J Biol Chem 2002; 277:10789-10794.
110. Gardner H, Broberg A, Pozzi A et al. Absence of $\alpha 1\beta 1$ in the mouse causes loss of feedback regulation of collagen synthesis in normal and wounded dermis. J Cell Sci 1999; 112:263-272.
111. Ianaro A, Cicala C, Calignano A et al. Anti-very late antigen-1 monoclonal antibody modulates the development of secondary lesion and T-cell response in experimental arthritis. Lab Invest 2000; 80:73-80.
112. Krieglstein CF, Cerwinka WH, Sprague AG et al. Collagen-binding integrin $\alpha 1\beta 1$ regulates intestinal inflammation in experimental colitis.J Clin Invest 2002; 110:1773-1782.
113. Fiorucci S, Mencarelli A, Palazzetti B et al. Importance of innate immunity and collagen-binding integrin $\alpha 1\beta 1$ in TNBS-induced colitis. Immunity 2002; 17:769-780.
114. De Fougerolles AR, Chi-Rosso G, Bajardi A et al. Global expression analysis of extracellular matrix-integrin interactions in monocytes. Immunity 2000; 13:749-758.
115. De Fougerolles AR, Koteliansky VE. Regulation of monocyte gene expression by the extracellular matrix and its functional implications. Immunol Rev 2002; 186:208-220.
116. Tanaka T, Ohtsuka Y, Yagita H et al. Involvement of alpha 1 and alpha 4 integrins in gut mucosal injury of graft-versus-host disease. Int Immunol 1995; 7:1183-1189.
117. Cosgrove D, Rodgers K, Meehan D et al. Integrin alpha1beta1 and transforming growth factor-beta1 play distinct roles in alport glomerular pathogenesis and serve as dual targets for metabolic therapy. Am J Pathol 2000; 157:1649-1659.
118. Sampson NS, Ryan ST, Enke DA et al. Global gene expression analysis reveals a role for the alpha 1 integrin in renal pathogenesis. J Biol Chem 2001; 276:34182-34188.

119. Cook HT, Khan SB, Allen A et al. Treatment with an antibody to VLA-1 integrin reduces glomerular and tubulointerstitial scarring in a rat model of crescentic glomerulonephritis. Am J Pathol 2002; 161:1265-1272.
120. Kagami S, Urushihara M, Kondo S et al. Effects of anti-alpha1 integrin subunit antibody on anti-Thy-1 glomerulonephritis. Lab Invest 2002; 82:1219-1227.
121. Senger DR, Claffey KP, Benes JE et al. Angiogenesis promoted by vascular endothelial growth factor: Regulation through $\alpha 1\beta 1$ and $\alpha 2\beta 1$ integrins. Proc Natl Acad Sci USA 1997; 94:13612-13617.
122. Senger DR, Perruzzi CA, Streit M et al. The alpha(1)beta(1) and alpha(2)beta(1) integrins provide critical support for vascular endothelial growth factor signaling, endothelial cell migration, and tumor angiogenesis. Am J Pathol 2002; 160:195-204.
123. Pozzi A, Moberg PE, Miles LA et al. Elevated matrix metalloprotease and angiostatin levels in integrin alpha 1 knockout mice cause reduced tumor vascularization. Proc Natl Acad Sci USA 2000; 97:2202-2207.
124. Camper L, Hellman U, Lundgren-Akerlund E. Isolation, cloning, and sequence analysis of the integrin subunit alpha 10, a beta 1-associated collagen binding integrin expressed on chondrocytes. J Biol Chem 1998; 273:20383-20389.
125. Velling T, Kusche-Gullberg M, Sejersen T et al. cDNA cloning and chromosomal localization of human alpha(11) integrin. A collagen-binding, I domain-containing, beta(1)-associated integrin alpha-chain present in muscle tissues. J Biol Chem 1999; 274:25735-25742.
126. Cerf-Bensussan N, Jarry A, Brousse N et al. A monoclonal antibody (HML-1) defining a novel membrane molecule present on human intestinal lymphocytes. Eur J Immunol 1987; 17:1279-1285.
127. Shaw SK, Breener MB. The beta7 integrins in mucosal homing and retention. Semin Immunol 1995; 7:335-342.
128. Lehmann J, Huehn M, de la Rosa M et al. Expression of the integrin $\alpha E\beta 7$ identifies unique subsets of CD25+ as well as CD25- regulatory T cells. Proc Natl Acad Sci USA 2002; 99:13031-13036.
129. Cepek KL, Parker CM, Madara JL et al. Integrin $\alpha E\beta 7$ mediates adhesion of T lymphocytes to epithelial cells. J Immunol 1993; 150:3459-3470.
130. Karecla PI, Bowden SJ, Green SJ et al. Recognition of E-cadherin on epithelial cells by the mucosal T cell integrin $\alpha M290\beta 7$ ($\alpha E\beta 7$). Eur J Immunol 1995; 25:852-856.
131. Higgins JMG, Mandlebrot DA, Shaw SK et al. Direct and regulated interaction of integrin $\alpha E\beta 7$ with E-cadherin. J Cell Biol 1998; 140:197-210.
132. Strauch UG, Mueller RC, Li XY et al. Integrin αE(CD103)$\beta 7$ mediates adhesion to intestinal microvascular endothelial cell lines via an E-cadherin-independent interaction. J Immunol 2001; 166:3506-3514.
133. Cepek KL, Shaw SK, Parker CM et al. Adhesion between epithelial cells and T lymphocytes mediated by E-cadherin and the alpha E beta 7 integrin. Nature 1994; 372:190-193.
134. Schon MP, Arya A, Murphy EA et al. Mucosal T lymphocyte numbers are selectively reduced in integrin αE (CD103)-deficient mice. J Immunol 1999; 162:6641-6649.
135. Wagner N, Lohler J, Kunkel EJ et al. Critical role for $\beta 7$ integrins in the formation of the gut-associated lymphoid tissue. Nature 1996; 382:366-370.
136. Lefrancois L, Parker CM, Olson S et al. The role of $\beta 7$ integrins in CD8 T cell trafficking during antiviral immune response. J Exp Med 1999; 189:1631-1638.
137. Pauls K, Schon M, Kubitza RC et al. Role of integrin alphaE(CD103)beta7 for tissue-specific epidermal localization of CD8+ T lymphocytes. J Invest Dermatol 2001; 117:569-575.
138. Schon MP, Schon M, Parker CM et al. Dendritic epidermal T cells (DETC) are diminished in integrin alphaE(CD103)-deficient mice. J Invest Dermatol 2002; 119:190-193.
139. Schon MP, Schon M, Warren HB et al. Cutaneous inflammatory disorder in integrin alphaE (CD103)-deficient mice. J Immunol 2000; 165:6583-6589.
140. Ebert MP, Ademer K, Muller-Ostermeyer F et al. CD8+CD103+ T cells analogous to intestinal intraepithelial lymphocytes infiltrate the pancreas in chronic pancreatitis. Am J Gastroenterol 1998; 93:2141-2147.

141. Fujihara T, Fujita H, Tsubota K et al. Preferential localization of CD8+ αEβ7+ T cells around acinar epithelial cells with apoptosis in patients with Sjogren's syndrome. J Immunol 1999; 163:2226-2235.
142. Feng Y, Wang D, Yuan R et al. CD103 expression is required for destruction of pancreatic islet allografts by CD8+ T cells. J Exp Med 2002; 196:877-886.
143. Hadley GA, Charandee C, Weir MR et al. CD103+ CTL accumulate within the graft epithelium during clinical renal allograft rejection. Transplantation 2001; 72:1548-1555.
144. Harlan JM, Winn RK. Leukocyte-endothelial interactions: Clinical trials of anti-adhesion therapy. Crit Care Med 2002; 30:S214-S219.
145. Genentech: Press releases: Genentech And XOMA Announce Phase III Study With Raptiva(TM) Achieves Primary Efficacy Endpoint. Available at http://www.genentech.com/gene/news/press-releases/detail.jsp?detail=5447. Posted September 17, 2002.
146. Winquist RJ, Desai S, Fogal S et al. The role of leukocyte function-associated antigen-1 in animal models of inflammation. Eur J Pharmacol 2001; 429:297-302.
147. Gadek TR, Burdick DJ, McDowell RS et al. Generation of an LFA-1 antagonist by the transfer of the ICAM-1 immunoregulatory epitope to a small molecule. Science 2002; 295:1086-1089.
148. ICOS: Press releases: ICOS Corporation Reports Results for Third Quarter of 2002. Available at http://www.corporate-ir.net/irey/ir_site.zhtml?ticker=ICOS&script=410&layout=0&item_id=353375. Posted November 5, 2002.
149. Frenette PS. Locking a leukocyte integrin with statins. N Engl J Med 2001; 345:1419-1421.
150. Weitz-Schmidt G, Welzenbach K, Brinkmann V et al. Statins selectively inhibit leukocyte function antigen-1 by binding to a novel regulatory integrin site. Nat Med 2001; 7:687-692.

Index

L

M

N

O

P

R

S

T

V

W

X